Acute Neuronal Injury

Denson G. Fujikawa
Editor

Acute Neuronal Injury

The Role of Excitotoxic Programmed Cell Death Mechanisms

Second Edition

Editor
Denson G. Fujikawa
Department of Neurology
VA Greater Los Angeles Healthcare System
North Hills, CA, USA

Department of Neurology and Brain Research Institute
David Geffen School of Medicine
University of California at Los Angeles
Los Angeles, CA, USA

ISBN 978-3-319-77494-7 ISBN 978-3-319-77495-4 (eBook)
https://doi.org/10.1007/978-3-319-77495-4

Library of Congress Control Number: 2018942497

Cover Caption: AIF translocation in vivo following global ischemia is prevented by overexpression of calpastatin. Representative immunofluorescence of apoptosis-inducing factor (AIF, red) from non-ischemic CA1 (a) or 72 h after global ischemia (b– d). AAV–calpastatin, a calpain inhibitory protein (c, d), or the empty vector (b) was infused 14 d before ischemia, and brain sections were double-label immunostained for AIF (red) and calpastatin overexpression (green, d). Note that the majority of CA1 neurons lost normal localization of AIF after ischemia (b, arrows), but AIF translocation was rare in calpastatin-overexpressed CA1 (c, d, arrows). Scale bars, 50 μm. (From Cao et al. 2007)

Printed on acid-free paper

This Springer imprint is published by the registered company Springer International Publishing AG part of Springer Nature.
The registered company address is: Gewerbestrasse 11, 6330 Cham, Switzerland

This book is dedicated to the memory of John W. Olney, M.D. (1932–2015), the father of excitotoxicity.

Introduction

In the Introduction to the first edition of this book, the history behind the concept of excitotoxicity was described, from John Olney's initial descriptions of the phenomenon (Olney 1969, 1971; Olney et al. 1974) through subsequent studies identifying the mechanisms by which excessive glutamate release presynaptically and by reversal of astrocytic glutamate uptake results in postsynaptic neuronal death. A synaptic mechanism was identified in 1983 (Rothman 1983), followed by identification of calcium entry via the *N*-methyl-D-aspartate (NMDA) subtype of glutamate receptor (MacDermott et al. 1986) and excessive calcium entry through the NMDA receptor-operated cation channel as the mechanism by which neurons died (Choi 1987; Choi et al. 1987). Early electron-microscopic studies of neuronal death from experimental cerebral ischemia (McGee-Russell et al. 1970), hypoglycemia (Auer et al. 1985a, b; Kalimo et al. 1985) and status epilepticus (Griffiths et al. 1983; Ingvar et al. 1988) showed electron-dense, shrunken neurons with pyknotic nuclei containing irregular chromatin clumps and dilated mitochondria, which the authors called "dark-cell degeneration." These were morphologically identical to the neurons that Olney found following exposure to glutamate or its analogues. We now call these neurons "necrotic," to differentiate them from "apoptotic" neurons (Fujikawa 2000, 2002), both of which die from different programmed mechanisms.

As was emphasized in the First Edition, excitotoxicity underlies all acute neuronal injuries, from cerebral ischemia, traumatic CNS injury, status epilepticus and hypoglycemia (Fujikawa 2010). Excessive intracellular calcium activates the cytosolic calcium-dependent enzymes calpain I and neuronal nitric oxide synthase (nNOS). Among other actions, calpain I is responsible for mitochondrial release of cytochrome *c*, apoptosis-inducing factor (AIF) and endonuclease G (endoG), and lysosomal release of cathepsins B and D and DNase II, all of which translocate to the neuronal nucleus and participate in its destruction. nNOS uses L-arginine as a substrate to form nitric oxide (NO), which reacts with superoxide (O_2-) to form the toxic free radical peroxynitrite (ONOO−). Peroxynitrite, with other free radicals generated by mitochondria exposed to the high intracellular calcium concentration (Beal 1996), damages the plasma membrane, mitochondrial and lysosomal membranes and causes double-stranded nuclear DNA cleavage.

The nuclear DNA repair enzyme poly(ADP-ribose) polymerase-1 (PARP-1) produces poly(ADP-ribose) (PAR) polymers to repair DNA double-strand breaks, and excess PAR polymers exit nuclei and translocate to mitochondria membranes, where they, in addition to calpain I, trigger the exit of AIF from mitochondrial membranes to neuronal nuclei (Andrabi et al. 2006; Yu et al. 2006), where it recruits migration inhibitory factor (MIF), a PARP-1-dependent, AIF-associated nuclease (PAAN) to the nucleus, where MIF cleaves single-stranded DNA into large-scale DNA fragments (Wang et al. 2016). Ted and Valina Dawson in their chapter provide details of the PARP-1 pathway.

Two new areas are covered in the current edition: the role of extra-synaptic NMDA receptors in excitotoxic necrosis and a separate necrotic pathway uncovered by inhibition of caspase-8 *in vitro*: necroptosis. The first topic was first described 14 years ago (Hardingham and Bading 2002) but was not covered in the first edition of the book. Evidence was put forth that it is extra-synaptic NMDA receptors that are responsible for excitotoxicity by inhibiting cAMP-response element binding protein (CREB) activity and brain-derived neurotrophic factor (BDNF) gene expression, whereas synaptic NMDA receptors did the opposite and actually provides a neuroprotective effect that is overwhelmed by extra-synaptic NMDA-receptor activation (Hardingham and Bading 2002). Dr. Michel Baudry in his chapter reinforces this concept and gives evidence that calpain I (also known as μ-calpain) is activated by synaptic NMDA receptors, whereas calpain II (also known as m-calpain) is activated by extra-synaptic NMDA receptors. On the other hand, Dr. Jun Chen's group in their chapter of the First Edition provided evidence that calpain I activation activates AIF by cleaving it in the mitochondrial membrane, which results in its exit from the mitochondrial membrane and translocation to neuronal nuclei; they have updated their chapter for this edition.

In recent years another necrotic pathway has been described, which has been dubbed "necroptosis" (Degterev et al. 2005). In cell culture, after inhibition of caspases with a broad-spectrum caspase inhibitor (z-VAD.fmk), investigators have found that cells subjected to a lethal insult had a necrotic morphology, and that the pathway involved three key proteins: receptor-interacting protein 1 and 3 (RIP1 and RIP3; also known as RIP1 kinase and RIP3 kinase) and mixed lineage kinase domain-like protein (MLKL) (Degterev et al. 2008; Sun et al. 2012). This pathway has been shown to occur *in vivo* in cerebral ischemia (Yin et al. 2015; Miao et al. 2015; Xu et al. 2016; Vieira et al. 2014) and traumatic brain injury (Liu et al. 2016). Drs. Vieira and Carvalho in their chapter provide evidence that oxygen-glucose deprivation (OGD) of hippocampal neurons *in vitro* and transient global cerebral ischemia (TGCI) *in vivo* up-regulate RIP3 and induce necroptotic neuronal necrosis. Overexpression of RIP3 worsened and knock-down of RIP3 reduced necroptosis in OGD. Dr. Tao in his chapter shows that the necroptotic pathway is activated following traumatic brain injury and that Necrostatin-1 (Nec-1), a RIP1 inhibitor, is neuroprotective.

If both the excitotoxic and necroptotic pathways are activated following acute neuronal injury, do each contribute separately to neuronal necrosis, producing an additive effect, or are there interactions between the two, and if so, what are they and what is the outcome? The non-competitive NMDA-receptor antagonist MK-801 and the nNOS inhibitor 7-nitroindazole both reduced RIP3 nitrosylation and neuronal necrosis in the hippocampal CA1 region following TGCI (Miao et al. 2015). On the other hand, cathepsin B release from lysosomes, which occurs in excitotoxicity, was reduced by Nec-1 following TGCI (Yin et al. 2015). So there appears to be cross-talk between the two programmed necrotic pathways. Further interactions between the two pathways and their consequences will undoubtedly be elucidated in future research.

Department of Neurology
VA Greater Los Angeles Healthcare System
North Hills, CA, USA

Department of Neurology and Brain Research Institute
David Geffen School of Medicine,
University of California at Los Angeles
Los Angeles, CA, USA

Denson G. Fujikawa

References

Andrabi SA, Kim S-W, Wang H, Koh DW, Sasaki M, Klaus JA, Otsuka T, Zhang Z, Koehler RC, Hurn PD, Poirier GG, Dawson VL, Dawson TM (2006) Poly(ADP-ribose) (PAR) polymer is a death signal. Proc Natl Acad Sci U S A 103:18308–18313

Auer RN, Kalimo H, Olsson Y, Siesjo BK (1985a) The temporal evolution of hypoglycemic brain damage. II. Light- and electron-microscopic findings in the hippocampal gyrus and subiculum of the rat. Acta Neuropathol 67(1–2):25–36

Auer RN, Kalimo H, Olsson Y, Siesjo BK (1985b) The temporal evolution of hypoglycemic brain damage. I. Light- and electron-microscopic findings in the rat cerebral cortex. Acta Neuropathol 67(1–2):13–24

Beal MF (1996) Mitochondria free radicals, and neurodegeneration. Curr Opin Neurobiol 6(5):661–666

Choi DW (1987) Ionic dependence of glutamate neurotoxicity. J Neurosci 7(2):369–379

Choi DW, Maulucci-Gedde M, Kriegstein AR (1987) Glutamate neurotoxicity in cortical cell culture. J Neurosci 7(2):357–368

Degterev A, Huang Z, Boyce M, Li Y, Jagtap P, Mizushima N, Cuny GD, Mitchison TJ, Moskowitz MA, Yuan J (2005) Chemical inhibitor of nonapoptotic cell death with therapeutic potential for ischemic brain injury. Nat Chem Biol 1(2):112–119. https://doi.org/10.1038/nchembio711

Degterev A, Hitomi J, Germscheid M, Ch'en IL, Korkina O, Teng X, Abbott D, Cuny GD, Yuan C, Wagner G, Hedrick SM, Gerber SA, Lugovskoy A, Yuan J (2008) Identification of RIP1 kinase as a specific cellular target of necrostatins. Nat Chem Biol 4(5):313–321. https://doi.org/10.1038/nchembio.83

Fujikawa DG (2000) Confusion between neuronal apoptosis and activation of programmed cell death mechanisms in acute necrotic insults. Trends Neurosci 23:410–411

Fujikawa DG (2002) Apoptosis: ignoring morphology and focusing on biochemical mechanisms will not eliminate confusion. Trends Pharmacol Sci 23:309–310

Fujikawa DG (ed) (2010) Acute neuronal injury: the role of excitotoxic programmed cell death mechanisms. Springer, New York, p 306

Griffiths T, Evans M, Meldrum BS (1983) Intracellular calcium accumulation in rat hippocampus during seizures induced by bicuculline or L-allylglycine. Neuroscience 10:385–395

Hardingham GE, Bading H (2002) Coupling of extrasynaptic NMDA receptors to a CREB shut-off pathway is developmentally regulated. Biochim Biophys Acta 1600(1–2):148–153

Ingvar M, Morgan PF, Auer RN (1988) The nature and timing of excitotoxic neuronal necrosis in the cerebral cortex, hippocampus and thalamus due to flurothyl-induced status epilepticus. Acta Neuropathol 75:362–369

Kalimo H, Auer RN, Siesjo BK (1985) The temporal evolution of hypoglycemic brain damage. III. Light and electron microscopic findings in the rat caudoputamen. Acta Neuropathol 67(1–2):37–50

Liu T, Zhao DX, Cui H, Chen L, Bao YH, Wang Y, Jiang JY (2016) Therapeutic hypothermia attenuates tissue damage and cytokine expression after traumatic brain injury by inhibiting necroptosis in the rat. Sci Rep 6:24547. https://doi.org/10.1038/srep24547

MacDermott AB, Mayer ML, Westbrook GL, Smith SJ, Barker JL (1986) NMDA-receptor activation increases cytoplasmic calcium concentration in cultured spinal cord neurones. Nature 321(6069):519–522

McGee-Russell SM, Brown AW, Brierley JB (1970) A combined light and electron microscope study of early anoxic-ischaemic cell change in rat brain. Brain Res 20(2):193–200

Miao W, Qu Z, Shi K, Zhang D, Zong Y, Zhang G, Zhang G, Hu S (2015) RIP3 S-nitrosylation contributes to cerebral ischemic neuronal injury. Brain Res 1627:165–176. https://doi.org/10.1016/j.brainres.2015.08.020

Olney JW (1969) Brain lesions, obesity and other disturbances in mice treated with monosodium glutamate. Science 164:719–721

Olney JW (1971) Glutamate-induced neuronal necrosis in the infant mouse hypothalamus. An electron microscopic study. J Neuropathol Exp Neurol 30(1):75–90

Olney JW, Rhee V, Ho OL (1974) Kainic acid: a powerful neurotoxic analogue of glutamate. Brain Res 77(3):507–512

Rothman SM (1983) Synaptic activity mediates death of hypoxic neurons. Science 220:536–537

Sun L, Wang H, Wang Z, He S, Chen S, Liao D, Wang L, Yan J, Liu W, Lei X, Wang X (2012) Mixed lineage kinase domain-like protein mediates necrosis signaling downstream of RIP3 kinase. Cell 148(1–2):213–227. https://doi.org/10.1016/j.cell.2011.11.031

Vieira M, Fernandes J, Carreto L, Anuncibay-Soto B, Santos M, Han J, Fernandez-Lopez A, Duarte CB, Carvalho AL, Santos AE (2014) Ischemic insults induce necroptotic cell death in hippocampal neurons through the up-regulation of endogenous RIP3. Neurobiol Dis 68:26–36. https://doi.org/10.1016/j.nbd.2014.04.002

Wang Y, An R, Umanah GK, Park H, Nambiar K, Eacker SM, Kim BW, Bao L, Harraz MM, Chang C, Chen R, Wang JE, Kam T-I, Jeong JS, Xie Z, Neifert S, Qian J, Andrabi SA, Blackshaw S, Zhu H, Song H, Ming G-H, Dawson VL, Dawson TM (2016) A nuclease that mediates cell death induced by DNA damage and poly(ADP-ribose) polymerase-1. Science 354(6308):aad6872-1-13

Xu Y, Wang J, Song X, Qu L, Wei R, He F, Wang K, Luo B (2016) RIP3 induces ischemic neuronal DNA degradation and programmed necrosis in rat via AIF. Sci Rep 6:29362. https://doi.org/10.1038/srep29362

Yin B, Xu Y, Wei RL, He F, Luo BY, Wang JY (2015) Inhibition of receptor-interacting protein 3 upregulation and nuclear translocation involved in Necrostatin-1 protection against hippocampal neuronal programmed necrosis induced by ischemia/reperfusion injury. Brain Res 1609:63–71. https://doi.org/10.1016/j.brainres.2015.03.024

Yu S-W, Andrabi SA, Wang H, Kim NS, Poirier GG, Dawson TM, Dawson VL (2006) Apoptosis-inducing factor mediates poly(ADP-ribose) (PAR) polymer-induced cell death. Proc Natl Acad Sci U S A 103:18314–18319

Contents

Part I
General Considerations

Chapter 1
Excitotoxic Programmed Cell Death Involves Caspase-Independent Mechanisms

Ted M. Dawson and Valina L. Dawson

Abstract Excitotoxicity is a common pathological process in many neurologic and neurodegenerative disorders, and this process involves over-stimulation of glutamate receptors and an excessive influx of calcium into cells. Cell death in excitotoxicity is unique in that, for the most part, it does not involve caspase-dependent pathways. Overactivation of poly (ADP-ribose) polymerase-1 (PARP-1) is an early pathological event in excitotoxicity that leads to a unique form of cell death called parthanatos. Biochemical events in parthanatos include early accumulation of poly (ADP-ribose) (PAR) and nuclear translocation of apoptosis inducing factor (AIF) from the mitochondria followed by nuclear accumulation of macrophage migration inhibitory factor (MIF). MIF's nuclease activity serves as the final executioner in excitotoxicity by shredding genomic DNA. Interfering with PARP activation, PAR signaling or MIF nuclease activity offers therapeutic interventions that could protect against a variety of neuronal injury due to a variety of insults involving glutamate excitotoxicity.

Keywords Poly (ADP-ribose) polymerase-1 (PARP-1) · Parthanatos · Poly (ADP-ribose) (PAR) · Apoptosis inducing factor (AIF) · Macrophage migration inhibitory factor (MIF) · Glutamate

1.1 Introduction

In conditions like brain trauma, ischemia/reperfusion, stroke, neurodegenerative diseases, brain injury, and cellular stress, neurons, glia and the vasculature succumb to cell death processes. Cell death, in general, is a complex process involving multiple pathways (Galluzzi et al. 2012, 2015). Apoptosis and necrosis are the two traditional described pathways of cell death based on morphology. Apoptosis involves the formation of apoptotic bodies that are cleared through mechanisms that avoid

T. M. Dawson (✉) · V. L. Dawson
Institute for Cell Engineering, Johns Hopkins University School of Medicine, Baltimore, MD, USA
e-mail: vdawson@jhmi.edu; tdawson@jhmi.edu

D. G. Fujikawa (ed.), *Acute Neuronal Injury*,
https://doi.org/10.1007/978-3-319-77495-4_1

eliciting inflammatory responses. Necrosis, on the other hand, involves massive cell swelling, inflammation and rupture of cellular structures. The morphologic classification of cell death has now been replaced by biochemical definitions (Galluzzi et al. 2012) in which there are different forms of regulated and programmed cell death (Galluzzi et al. 2015). Accidental cell death has been coined as the form of cell death that involved rupture of cellular structures. During the development of living organisms or homeostatic control of cell number in mature organisms, cell death is a programmed physiological process that is executed by the activation of caspases; and has an apoptotic morphology. Regulated necrosis is defined by biochemical assessments that are quantifiable and these pathways share many common biochemical events as well as morphologic features of both necrosis and apoptosis in cell death execution (Dawson and Dawson 2017). This chapter reviews the underlying mechanisms accounting for glutamate excitotoxicity that primarily involves regulated necrosis and in particular, parthanatos (Fatokun et al. 2014).

1.2 Excitotoxicity

Excitotoxicity is a process of neuronal injury mediated by excitatory amino acids (Arundine and Tymianski 2004; Lai et al. 2014; Mehta et al. 2013; Olney and Sharpe 1969). Glutamate is the most abundant amino acid in the brain and is an essential neurotransmitter in the central nervous system. It plays a primary role in excitotoxicity (Arundine and Tymianski 2003). Overstimulation of glutamate receptors results in an excessive influx of calcium to mediate excitotoxic responses in nerve cells. Glutamate can act on many receptor types in the nervous system, including ionotropic and metabotropic receptors. Ionotropic NMDA-receptors are activated by the glutamate analogue N-methyl-D-aspartate (NMDA) and play a major role in excitotoxicity (Lai et al. 2014; Mehta et al. 2013; Meldrum 1992). α-Amino-3-hydroxy-5-methyl-4- isoxalone propionic acid (AMPA) and kainate receptors are activated by AMPA and by kainate, respectively (Nakanishi 1992). These non-NMDA-type receptors are also involved in excessive intracellular calcium influx and excitotoxicity (Arundine and Tymianski 2003, 2004; Dawson and Dawson 2017). Ionotropic receptors are ion channel-linked receptors and cause ion-influx when stimulated. Metabotropic glutamate receptors (mGluR) are G-protein coupled receptors. Glutamate receptors may take part in excitotoxicity by modulating the function of other receptors either directly or indirectly. The Ca^{2+} influx following glutamate receptor activation in excitotoxicity can induce cell death by activating Ca^{2+}-dependent enzyme systems, such as nitric oxide (NO) synthase (nNOS), calpains, phospholipases and other Ca^{2+}-dependent enzymes (Dawson and Dawson 2017). The activation of nNOS occurs through postsynaptic density-95 (PSD-95) dependent mechanisms since PSD-95 inhibitors prevent the downstream consequences of glutamate excitotoxicity (Aarts et al. 2002; Sattler et al. 1999;

Soriano et al. 2008; Sun et al. 2008). nNOS activation leads to the overproduction of NO through the conversion of L-arginine to L-citrulline. NO can exert many roles as a signaling molecule in neurons (Dawson et al. 1992). Generation of excess NO can be neurotoxic (Dawson et al. 1991, 1993; Samdani et al. 1997). NO can combine with $O_2^{\bullet-}$ in the mitochondria to generate more toxic peroxynitrite ($ONOO^-$), which can cause oxidative or nitrosative injury to cellular proteins, lipids and DNA (Beckman and Koppenol 1996; Szabo and Dawson 1998; Xia et al. 1996). Injury to DNA causes a massive activation of poly (ADP-ribose) polymerase-1, which ultimately triggers cell death via the process of parthanatos (Dawson and Dawson 2017; Fatokun et al. 2014) (Fig. 1.1). PARP-1 can also be activated through mechanisms that do not involve DNA damage, in that aminoacyl-tRNA synthetase complex interacting multifunctional protein-2 (AIMP2) directly activates PARP-1 initiating parthanatos (see Fig. 1.3) (Lee et al. 2013).

1.3 Role of PARP-1 and PAR Polymer in Excitotoxicity

Poly (ADP-ribose) polymerases (PARPs) are known to play key roles in DNA repair (de Murcia and Menissier de Murcia 1994; de Murcia et al. 1994). PARP-1 is the founding member of the PARP family, which includes 18 different isoforms based on protein sequence homology to the PARP-1 catalytic domain (Hottiger et al. 2010). PARP-1 accounts for more than 90% of PARP activity in living cells. In response to DNA damage, PARP-1 uses NAD+ as a substrate and attaches polymers of PAR on different acceptor proteins (hetero-modification) or on PARP-1 itself (auto-modification) (D'Amours et al. 1999; Virag and Szabo 2002). PARP-1 is considered a "genome guardian," because it takes part in DNA repair under physiological conditions (Chatterjee et al. 1999). Under mild genomic stress, PARP-1 is activated to induce DNA repair, whereas severe cell stress induces massive PARP-1 activation that ultimately leads to cell death (Beck et al. 2014) (Fig. 1.2). Both gene deletion and pharmacological inhibition studies have shown that PARP-1 activation plays a key role in cytotoxicity following ischemia/reperfusion, neurodegeneration, spinal cord injury, ischemic injury in heart, liver, and lungs, and in retinal degeneration, arthritis, diabetes and many other disorders (Fatokun et al. 2014; Virag and Szabo 2002). In the nervous system, PARP-1 activation is triggered by excitotoxic stimuli (Eliasson et al. 1997; Mandir et al. 2000; Zhang et al. 1994).

It was originally presumed that cell death in PARP-1 toxicity was induced by the intracellular energy depletion from PARP-1's use of NAD^+ (Fatokun et al. 2014; Virag and Szabo 2002). NAD^+ is an important cellular molecule for many physiological processes. Energy-generating processes, like glycolysis, the Krebs cycle and the pentose phosphate pathway, utilize NAD^+ as a cofactor (Belenky et al. 2007).

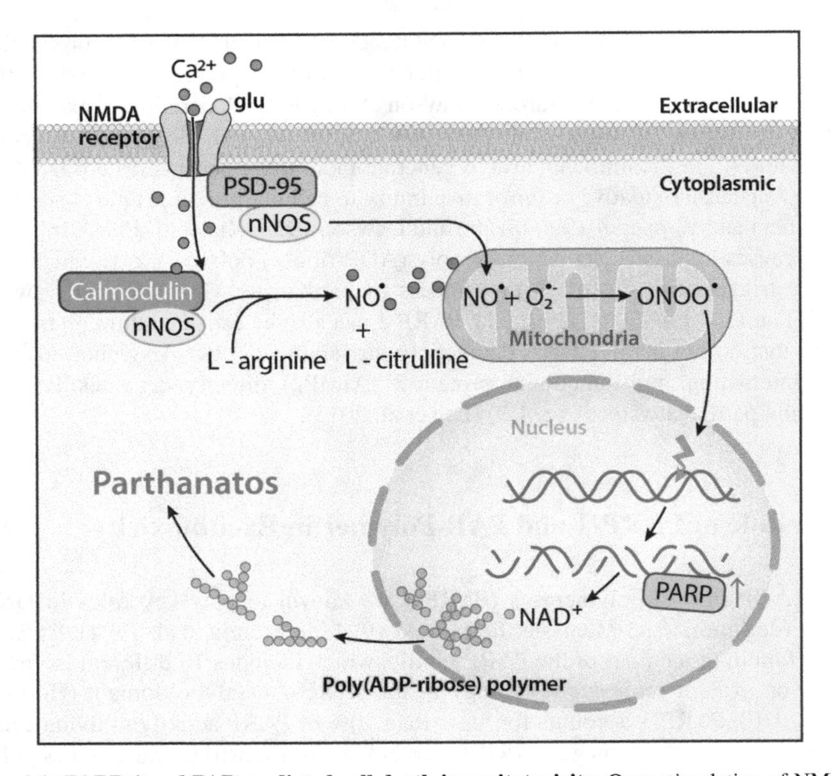

Fig. 1.1 PARP-1 and PAR mediated cell death in excitotoxicity. Over-stimulation of NMDA receptors by glutamate (yellow circle) results in the influx of Ca^{2+} (red circles), which binds calmodulin and activates neuronal nitric oxide (NO) synthase (nNOS), to convert L-arginine to NO and L-citrulline. nNOS is tethered to the NMDA receptor via postsynaptic density-95 (PSD-95) protein. Even though NO is an essential molecule in neuronal signal transduction, excess NO can be neurotoxic. Neuronal toxicity by excess NO is mediated by peroxynitrite, a reaction product from NO and superoxide anion ($O_2^{\bullet-}$). Peroxynitrite damages DNA, which results in over activation of PARP (PARP↑), depletion of NAD+, and generation of PAR polymer, leading to parthanatos

While PARP-1 activation leads to decreased cellular NAD+ and energy levels (Ha and Snyder 1999), it is difficult to obtain evidence that proves that PARP-1 activation depletes enough cellular energy to kill the cell (Fossati et al. 2007; Goto et al. 2002). Numerous studies show that cellular ATP and NAD^+ levels drop significantly following PARP-1 activation (Eliasson et al. 1997; Yu et al. 2002). The drop in cellular energy levels following PARP-1 activation may primarily be due to alterations in mitochondrial function and defective oxidative phosphorylation as opposed to PARP-1 mediated catabolism of NAD^+ (Dawson and Dawson 2017; Virag and Szabo 2002). Along these lines, it was shown by many studies that mitochondrial depolarization, loss of mitochondrial function and increased mitochondrial membrane permeability are required factors for PARP-1-dependent cell death (Alano et al. 2004). Conclusions that NAD^+ utilization by PARP-1 is a death inducer were

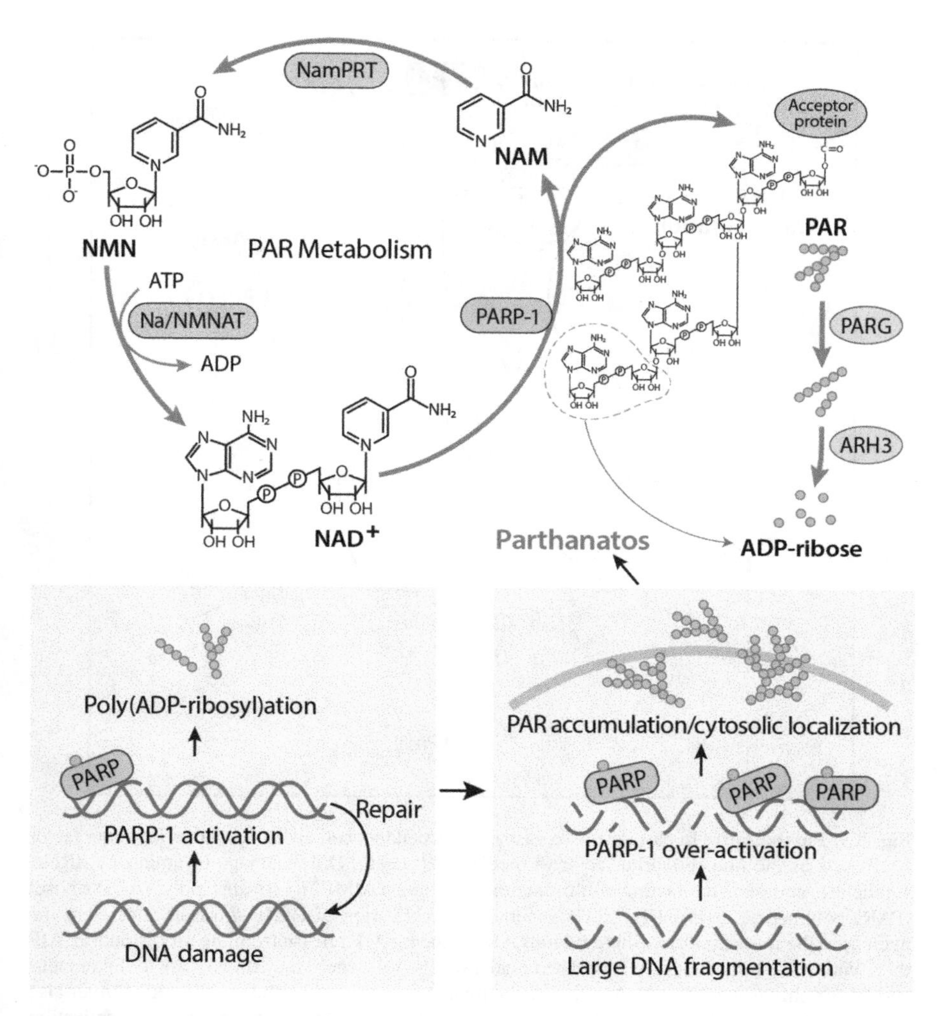

Fig. 1.2 **PAR metabolism**. PARP-1 utilizes NAD+ as a substrate for synthesis of PAR polymers. In the process of PAR formation, nicotinamide (NAM), a product of NAD+ hydrolysis, is first converted into nicotinamide mononucleotide (NMN) and then into NAD+ by nicotinamide phosphoribosyl transferase (NamPRT) and nicotinamide mononucleotide adenylyl transferases (Na/NMNAT-1, -2, and -3), respectively. Mild DNA damage or breaks activate the activation of PARP proteins, where they play a role in the DNA repair process. Under conditions of severe DNA damage, parthanatos is initiated through excessive PAR polymer formation. Poly(ADP-ribose) glycohydrolase (PARG) and ADP-ribose-(arginine) protein hydrolase-3 (ARH3) degrade PAR polymers

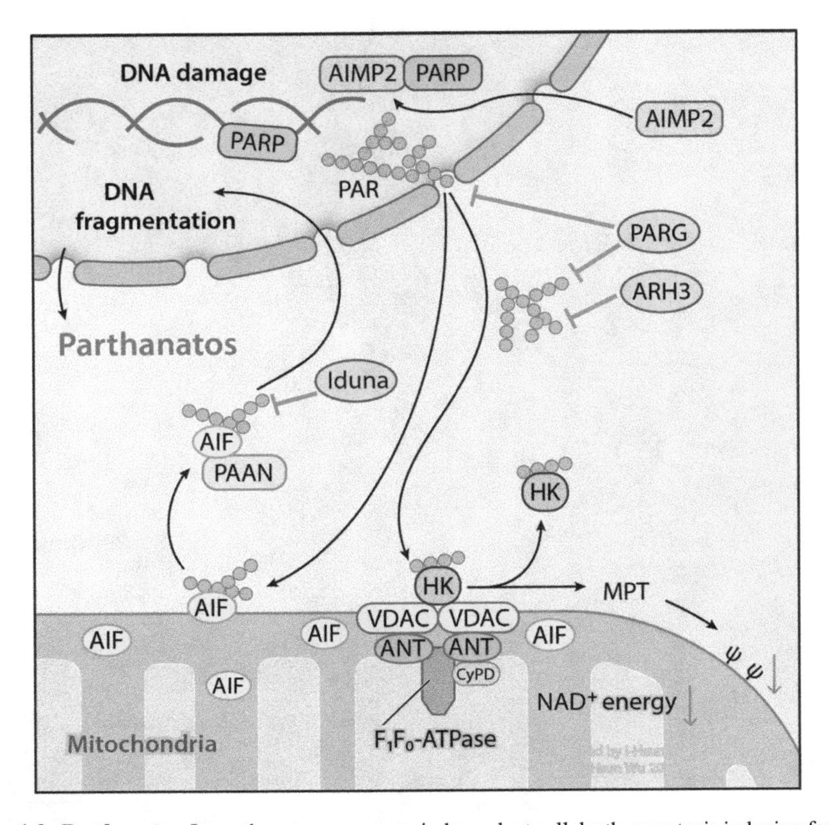

Fig. 1.3 Parthanatos. In parthanatos, a caspase-independent cell death, apoptosis inducing factor (AIF) acts as the mitochondrial factor to mediate cell death. DNA damage or aminoacyl tRNA synthetase complex interacting multifunctional protein (AIMP2) activates poly (ADP-ribose) (PAR) polymerase 1 (PARP-1). PAR generated by activation of PARP-1 translocates from the nucleus to the mitochondria, where it binds AIF, inducing AIF release from the mitochondria. AIF then binds the parthanatos AIF associated nuclease (PAAN) where they translocate to the nucleus and PAAN shreds genomic DNA acting as the final executioner in parthanatos. PAR polymer also binds and inactivates hexokinase (HK), which accounts for the energy depletion due to activation of PARP-1. PARG or ARH3 degrade PAR polymer, preventing parthanatos. Iduna, a PAR- dependent ubiquitin E3 ligase, is an inhibitor of parthanatos. Abbreviations: *ANT* adenine nucleotide translocase, *CyPD* cyclophilin D, *F_1F_0-ATPase* ATP synthase, *MPT* mitochondrial membrane permeability transition, *VDAC* voltage-dependent anion channel, *Ψ* mitochondrial membrane potential. Figure adapted from Dawson and Dawson (2017)

drawn from studies that used direct exogenous delivery of NAD^+ or energy substrates as cytoprotective agents. It is important to note that the off-target effects of these substrates may contribute to the observed effects. For example, consumption of NAD^+ by PARP-1 generates nicotinamide (NAM) as a by-product. NAM is a potent PARP-1 inhibitor, so the protective mechanism mediated by exogenous NAD^+ should be interpreted with caution.

Recent studies indicate that energy depletion following PARP-1 activation is not a critical factor for cell death. Following PARP-1 activation, we recently demonstrated that cells die due to a toxic accumulation of PAR. PAR, generated by PARP-1 in the nucleus, travels to the cytosol to induce cell death (Fig. 1.1). Neutralization of cytosolic PAR by PAR-specific antibodies protects against NMDA-induced cell death in mouse primary neurons (Andrabi et al. 2006). Conversely, exogenous delivery of purified PAR kills cells (Andrabi et al. 2006). The toxic potential of PAR increases with dose and polymer complexity. Highly complex and long chain polymers are more toxic than shorter and less complex polymers (Andrabi et al. 2006). Among the PARP family members, there are several different PARPs that are confirmed to synthesize PAR. The heterogeneity in the complexity and structure of PAR may vary depending upon the PARP involved. This may contribute to the possible different roles of individual PARP isoforms in cell survival or cell death.

PARP-1-dependent cell death, known as parthanatos, is distinct from classic necrosis or apoptosis in its biochemical and morphological features, although many of the morphologic features are similar to those previously described for neuronal excitotoxicity (Dawson and Dawson 2017; Fatokun et al. 2014). The biochemical features of parthanatos are distinct from classically defined pathways of cell death, and include rapid PARP-1 activation, early PAR accumulation, mitochondrial depolarization, early nuclear AIF translocation, loss of cellular NAD^+ and ATP, and late caspase activation (Yu et al. 2002). Caspase activation, which is a hallmark of apoptotic cell death, does not play a primary role in parthanatos, as broad-spectrum caspase inhibitors, knockout of Apoptotic protease activating factor-1 (APAF-1) or BAX are unable to protect cells (Cregan et al. 2004; Yu et al. 2002). Morphological features of parthanatos include shrunken and condensed nuclei, disintegrating membranes and cells becoming propidium iodide-positive within a few hours after the onset of parthanatos.

1.4 Role of PARG in Excitotoxicity and PAR-Mediated Cell Death

Poly(ADP-ribose) glycohydrolase (PARG) is an important cellular enzyme that together with PARPs plays an important role in balancing PAR levels in cells (Fig. 1.2). Many genes encoding different PARPs have been identified, whereas only a single gene encoding PARG has been identified so far. The full-length nuclear PARG in humans is 111 kDa with two cytosolic splice variants, 102 and 99 kDa (Meyer-Ficca et al. 2004). PARG catalyzes the hydrolysis of PAR to ADP-ribose units through its glycosidic activity (Davidovic et al. 2001). Evidence from recent data shows that PARG is critical for cell survival. Genetic deletion of PARG results in accumulation of PAR, which leads to early embryonic lethality in drosophila and mice (Hanai et al. 2004; Koh et al. 2004). Conversely, overexpression of PARG leads to protection against excitotoxicity and PARP-1 dependent cell death (Andrabi et al. 2006; Cozzi et al. 2006). Mouse trophoblasts from E3.5 PARG null mice survive only in the presence of the PARP inhibitor benzamide. Withdrawal of the PARP inhibitor results in cell death in the PARG trophoblasts via toxic accumulation of PAR (Andrabi et al. 2006). The inactivation of PAR by PARG predigestion, shows that PARG is important for cell survival and that PAR is a death signaling molecule.

Although only one gene for PARG has been discovered, recent data shows that a 39-kDa ADP-ribose-(arginine) protein hydrolase (ARH3) has PARG-like activity (Oka et al. 2006). ARH3 seems to play a neuroprotective role as well by reducing PAR levels in the cytoplasm that are generated by PARG, preventing the release of AIF by PAR polymer (Mashimo et al. 2013).

1.5 Mitochondria in PAR-Induced Cell Death: Role of AIF and Hexokinase

Mitochondria have important roles in cellular energy generating processes. However, in cellular stress, mitochondria participate in cell death signaling by releasing pro-death proteins such as cytochrome c (Cyt C), AIF, Smac/Diablo, and Omi/HtrA2 (Green and Kroemer 2004; Kroemer et al. 2007; Newmeyer and Ferguson-Miller 2003; Tait and Green 2010). Among these, Smac/Diablo and Omi/HtrA2 proteins act as inhibitors of cytosolic inhibitor apoptosis proteins (IAPs), which act by inhibiting caspase 9, 3, and 7 (Richter and Duckett 2000). Thus, the release of Smac/Diablo and Omi/HtrA2 into the cytoplasm ensures that the brake that IAPs provide on caspase activation is removed. Release of Cyt-C into the cytosol leads to APAF-1 binding and initiates cell death through assembly of the apoptosome complex (Li et al. 1997; Liu et al. 1996; Zou et al. 1997). Besides Cyt C and APAF-1, the apoptosome requires pro-caspase 9 as the initiator caspase and dATP. In this complex, caspase 9 is cleaved and activated, which in turn activates downstream caspases that include the effector caspase, caspase 3 (Green and Kroemer 2004; Kroemer et al. 2007; Newmeyer and Ferguson-Miller 2003; Tait and Green 2010). Active caspase 3 has many cellular substrates, including α-foldrin, PARP-1, plasma membrane Ca^{2+} pump (PMCA) and inhibitor of caspase- activated DNase (ICAD). Caspase-activated DNase (CAD) is normally sequestered to an inactive form in a complex with ICAD (CAD-ICAD complex). On ICAD degradation by caspases, CAD is activated to induce large scale DNA-fragmentation and cell death (Liu et al. 1997; Sakahira et al. 1998). Although there is evidence that caspase activation occurs during excitotoxic cell death and in a variety of injuries to the nervous system, it seems to play a secondary role in excitotoxic cell death (Dawson and Dawson 2017; Lipton 1999).

AIF, on the other hand, directly translocates to the nucleus to initiate large scale chromatin condensation and caspase-independent cell death during glutamate excitotoxicity (Fig. 1.3) (Cregan et al. 2004; Wang et al. 2004, 2011, 2016; Yu et al. 2002).

AIF is a mitochondrial flavoprotein with important functions in oxidative phosphorylation (Pospisilik et al. 2007). Originally, AIF was discovered as a death inducing factor (Susin et al. 1999). Numerous studies have clearly demonstrated that AIF induces cell death upon its translocation to the nucleus (Krantic et al. 2007; Modjtahedi et al. 2006). AIF as a cell death effector in PARP-1 toxicity became evident through studies using AIF-neutralizing antibodies or genetic knock down of AIF (Culmsee et al. 2005; Wang et al. 2004; Yu et al. 2002, 2006). Moreover, PARP-1 activation is required for AIF release from the mitochondria since knockout of PARP-1 completely prevents AIF translocation from the mitochondria to the nucleus (Wang et al. 2004; Yu et al. 2002). PAR generated from PARP-1 is necessary and sufficient to induce the release of AIF from the mitochondria where PAR directly binds to mitochondrial AIF (Andrabi et al. 2006; Wang et al. 2011; Yu et al. 2006). Mutating the PAR binding domain of AIF prevents the release of AIF from the mitochondria and is protective against glutamate excitotoxicity (Wang et al. 2011). In addition to PARG, Iduna (RNF146) prevents glutamate excitotoxicity by preventing AIF translocation from the mitochondria to the nucleus by interfering with PAR signaling (Andrabi et al. 2011). Iduna is a PAR-dependent ubiquitin E3 ligase that targets PARylated or PAR binding proteins for proteasomal degradation (Callow et al. 2011; Kang et al. 2011; Zhang et al. 2011). Future experiments are required to identify Iduna substrates that play a role in parthanatos. How PAR leaves the nucleus and translocates to the mitochondria also remains an unresolved issue, but there are likely PAR binding proteins or PARylated proteins that translocate from the nucleus to the mitochondria after PARP-1 activation (Dawson and Dawson 2017). Some of these may be Iduna substrates.

PAR also binds to hexokinase (HK) during parthanatos, where PAR inhibits HK activity (Andrabi et al. 2014; Fouquerel et al. 2014). This leads to inhibition of glycolysis, which accounts for the reduction in NAD^+ and ATP that has been observed following PARP-1 activation (Alano et al. 2010; Ha and Snyder 1999). Thus, the energy collapse following PARP-1 activation is not directly due the consumption of NAD^+ by PARP-1 activity, but instead is due to PAR- dependent inhibition of HK (Fig. 1.3). The mitochondrial substrates glutamine and pyruvate can rescue the PAR-HK dependent defects in glycolysis, supporting the notion that the reduction in NAD^+ is due to PAR inhibition of HK (Andrabi et al. 2014; Dawson and Dawson 2017; Fouquerel et al. 2014).

1.6 Parthanatos Associated AIF Nuclease (PAAN)

Once AIF enters the nucleus it causes the nucleus to undergo nuclear condensation and genomic DNA is cleaved into large (20–50 Kb) fragments. Although AIF binds to DNA, it does not have any nuclease activity, and it was assumed that AIF binds to and activates a DNA nuclease. The identity of this nuclease has remained a mystery for almost two decades. Recently our group definitively identified the first parthanatos- associated AIF nuclease (PAAN) as macrophage migration inhibitory factor (MIF) (Fig. 1.3) (Wang et al. 2016). MIF is a Mg^{2+}/Ca^{2+} dependent nuclease. Following PARP-1 activation AIF binds to MIF and carries it into the nuclease where MIF cleaves genomic DNA into large fragments. Neuronal culture containing MIF mutants that lack nuclease activity or fail to bind to AIF are resistant to glutamate excitotoxicity and stimuli that induce parthanatos. Moreover, mice with MIF mutants that lack nuclease activity or fail to bind to AIF are resistant to stroke (Wang et al. 2016). The development of MIF nuclease inhibitors as neuroprotective agents holds particular promise as they would not interfere with DNA repair, like PARP inhibitors (Dawson and Dawson 2017).

1.7 Conclusion

Glutamate excitotoxicity is largely a caspase-independent process. Depending on the length and strength of the insult, PARP-1 plays a primary role in the death process. Parthanatos is a unique form of cell death mediated by cytotoxic PAR polymer in cytosol due to overactivation of PARP-1. PAR polymer is synthesized primarily in the nucleus and translocates into the cytosol to induce cell death by regulating mitochondria function. Mitochondria act as the core organelle to release pro-cell death factors. In the case of parthanatos, cell death is initiated by nuclear translocation and mitochondrial release of AIF. PAR polymer induces the structural change of a number of cellular proteins by either the process of PARylation by PARP or through non-covalent interactions. Recent studies indicate that human neurons predominantly die via parthanatos in response to glutamate excitotoxicity indicating that this form of cell death is particularly important to human neurologic injury (Xu et al. 2016). Inhibition of PARP-1 and identification of PAR-binding proteins and their characterization may provide a novel opportunity to understand the PAR-signaling mechanisms and to identify novel therapeutics that interfere with PAR-dependent cell death.

Acknowledgements This work was supported by US National Institutes of Health grants NS38377, NS67525, and DA00266. The authors thank I-Hsun Wu for assistance with the illustrations. T. M. D. is the Leonard and Madlyn Abramson Professor in Neurodegenerative Diseases.

References

Aarts M, Liu Y, Liu L, Besshoh S, Arundine M, Gurd JW, Wang YT, Salter MW, Tymianski M (2002) Treatment of ischemic brain damage by perturbing NMDA receptor- PSD-95 protein interactions. Science 298(5594):846–850. https://doi.org/10.1126/science.1072873

Alano CC, Garnier P, Ying W, Higashi Y, Kauppinen TM, Swanson RA (2010) NAD+ depletion is necessary and sufficient for poly(ADP-ribose) polymerase-1-mediated neuronal death. J Neurosci 30(8):2967–2978. https://doi.org/10.1523/JNEUROSCI.5552-09.2010

Alano CC, Ying W, Swanson RA (2004) Poly(ADP-ribose) polymerase-1-mediated cell death in astrocytes requires NAD+ depletion and mitochondrial permeability transition. J Biol Chem 279(18):18895–18902. https://doi.org/10.1074/jbc.M313329200

Andrabi SA, Kang HC, Haince JF, Lee YI, Zhang J, Chi Z, West AB, Koehler RC, Poirier GG, Dawson TM, Dawson VL (2011) Iduna protects the brain from glutamate excitotoxicity and stroke by interfering with poly(ADP-ribose) polymer-induced cell death. Nat Med 17(6):692–699. https://doi.org/10.1038/nm.2387

Andrabi SA, Kim NS, Yu SW, Wang H, Koh DW, Sasaki M, Klaus JA, Otsuka T, Zhang Z, Koehler RC, Hurn PD, Poirier GG, Dawson VL, Dawson TM (2006) Poly(ADP-ribose) (PAR) polymer is a death signal. Proc Natl Acad Sci U S A 103(48):18308–18313. https://doi.org/10.1073/pnas.0606526103

Andrabi SA, Umanah GK, Chang C, Stevens DA, Karuppagounder SS, Gagne JP, Poirier GG, Dawson VL, Dawson TM (2014) Poly(ADP-ribose) polymerase-dependent energy depletion occurs through inhibition of glycolysis. Proc Natl Acad Sci U S A 111(28):10209–10214. https://doi.org/10.1073/pnas.1405158111

Arundine M, Tymianski M (2003) Molecular mechanisms of calcium-dependent neurodegeneration in excitotoxicity. Cell Calcium 34(4–5):325–337

Arundine M, Tymianski M (2004) Molecular mechanisms of glutamate-dependent neurodegeneration in ischemia and traumatic brain injury. Cell Mol Life Sci 61(6):657–668. https://doi.org/10.1007/s00018-003-3319-x

Beck C, Robert I, Reina-San-Martin B, Schreiber V, Dantzer F (2014) Poly(ADP-ribose) polymerases in double-strand break repair: focus on PARP1, PARP2 and PARP3. Exp Cell Res 329(1):18–25. https://doi.org/10.1016/j.yexcr.2014.07.003

Beckman JS, Koppenol WH (1996) Nitric oxide, superoxide, and peroxynitrite: the good, the bad, and ugly. Am J Phys 271(5 Pt 1):C1424–C1437

Belenky P, Bogan KL, Brenner C (2007) NAD+ metabolism in health and disease. Trends Biochem Sci 32(1):12–19. https://doi.org/10.1016/j.tibs.2006.11.006

Callow MG, Tran H, Phu L, Lau T, Lee J, Sandoval WN, Liu PS, Bheddah S, Tao J, Lill JR, Hongo JA, Davis D, Kirkpatrick DS, Polakis P, Costa M (2011) Ubiquitin ligase RNF146 regulates tankyrase and Axin to promote Wnt signaling. PLoS One 6(7):e22595. https://doi.org/10.1371/journal.pone.0022595

Chatterjee S, Berger SJ, Berger NA (1999) Poly(ADP-ribose) polymerase: a guardian of the genome that facilitates DNA repair by protecting against DNA recombination. Mol Cell Biochem 193(1–2):23–30

Cozzi A, Cipriani G, Fossati S, Faraco G, Formentini L, Min W, Cortes U, Wang ZQ, Moroni F, Chiarugi A (2006) Poly(ADP-ribose) accumulation and enhancement of postischemic brain damage in 110-kDa poly(ADP-ribose) glycohydrolase null mice. J Cereb Blood Flow Metab 26(5):684–695. https://doi.org/10.1038/sj.jcbfm.9600222

Cregan SP, Dawson VL, Slack RS (2004) Role of AIF in caspase-dependent and caspase-independent cell death. Oncogene 23(16):2785–2796. https://doi.org/10.1038/sj.onc.1207517

Culmsee C, Zhu C, Landshamer S, Becattini B, Wagner E, Pellecchia M, Blomgren K, Plesnila N (2005) Apoptosis-inducing factor triggered by poly(ADP-ribose) polymerase and Bid mediates neuronal cell death after oxygen-glucose deprivation and focal cerebral ischemia. J Neurosci 25(44):10262–10272. https://doi.org/10.1523/JNEUROSCI.2818-05.2005

D'Amours D, Desnoyers S, D'Silva I, Poirier GG (1999) Poly(ADP-ribosyl)ation reactions in the regulation of nuclear functions. Biochem J 342(Pt 2):249–268

Davidovic L, Vodenicharov M, Affar EB, Poirier GG (2001) Importance of poly(ADP-ribose) glycohydrolase in the control of poly(ADP-ribose) metabolism. Exp Cell Res 268(1):7–13. https://doi.org/10.1006/excr.2001.5263

Dawson TM, Dawson VL (2017) Mitochondrial mechanisms of neuronal cell death: potential therapeutics. Annu Rev Pharmacol Toxicol 57:437–454. https://doi.org/10.1146/annurev-pharmtox-010716-105001

Dawson TM, Dawson VL, Snyder SH (1992) A novel neuronal messenger molecule in brain: the free radical, nitric oxide. Ann Neurol 32(3):297–311. https://doi.org/10.1002/ana.410320302

Dawson VL, Dawson TM, Bartley DA, Uhl GR, Snyder SH (1993) Mechanisms of nitric oxide-mediated neurotoxicity in primary brain cultures. J Neurosci 13(6):2651–2661

Dawson VL, Dawson TM, London ED, Bredt DS, Snyder SH (1991) Nitric oxide mediates glutamate neurotoxicity in primary cortical cultures. Proc Natl Acad Sci U S A 88(14):6368–6371

de Murcia G, Menissier de Murcia J (1994) Poly(ADP-ribose) polymerase: a molecular nick-sensor. Trends Biochem Sci 19(4):172–176

de Murcia G, Schreiber V, Molinete M, Saulier B, Poch O, Masson M, Niedergang C, Menissier de Murcia J (1994) Structure and function of poly(ADP-ribose) polymerase. Mol Cell Biochem 138(1–2):15–24

Eliasson MJ, Sampei K, Mandir AS, Hurn PD, Traystman RJ, Bao J, Pieper A, Wang ZQ, Dawson TM, Snyder SH, Dawson VL (1997) Poly(ADP-ribose) polymerase gene disruption renders mice resistant to cerebral ischemia. Nat Med 3(10):1089–1095

Fatokun AA, Dawson VL, Dawson TM (2014) Parthanatos: mitochondrial-linked mechanisms and therapeutic opportunities. Br J Pharmacol 171(8):2000–2016. https://doi.org/10.1111/bph.12416

Fossati S, Cipriani G, Moroni F, Chiarugi A (2007) Neither energy collapse nor transcription underlie in vitro neurotoxicity of poly(ADP-ribose) polymerase hyper-activation. Neurochem Int 50(1):203–210. https://doi.org/10.1016/j.neuint.2006.08.009

Fouquerel E, Goellner EM, Yu Z, Gagne JP, Barbi de Moura M, Feinstein T, Wheeler D, Redpath P, Li J, Romero G, Migaud M, Van Houten B, Poirier GG, Sobol RW (2014) ARTD1/PARP1 negatively regulates glycolysis by inhibiting hexokinase 1 independent of NAD+ depletion. Cell Rep 8(6):1819–1831. https://doi.org/10.1016/j.celrep.2014.08.036

Galluzzi L, Bravo-San Pedro JM, Vitale I, Aaronson SA, Abrams JM, Adam D, Alnemri ES, Altucci L, Andrews D, Annicchiarico-Petruzzelli M, Baehrecke EH, Bazan NG, Bertrand MJ, Bianchi K, Blagosklonny MV, Blomgren K, Borner C, Bredesen DE, Brenner C, Campanella M, Candi E, Cecconi F, Chan FK, Chandel NS, Cheng EH, Chipuk JE, Cidlowski JA, Ciechanover A, Dawson TM, Dawson VL, De Laurenzi V, De Maria R, Debatin KM, Di Daniele N, Dixit VM, Dynlacht BD, El-Deiry WS, Fimia GM, Flavell RA, Fulda S, Garrido C, Gougeon ML, Green DR, Gronemeyer H, Hajnoczky G, Hardwick JM, Hengartner MO, Ichijo H, Joseph B, Jost PJ, Kaufmann T, Kepp O, Klionsky DJ, Knight RA, Kumar S, Lemasters JJ, Levine B, Linkermann A, Lipton SA, Lockshin RA, Lopez-Otin C, Lugli E, Madeo F, Malorni W, Marine JC, Martin SJ, Martinou JC, Medema JP, Meier P, Melino S, Mizushima N, Moll U, Munoz-Pinedo C, Nunez G, Oberst A, Panaretakis T, Penninger JM, Peter ME, Piacentini M, Pinton P, Prehn JH, Puthalakath H, Rabinovich GA, Ravichandran KS, Rizzuto R, Rodrigues CM, Rubinsztein DC, Rudel T, Shi Y, Simon HU, Stockwell BR, Szabadkai G, Tait SW, Tang HL, Tavernarakis N, Tsujimoto Y, Vanden Berghe T, Vandenabeele P, Villunger A, Wagner EF, Walczak H, White E, Wood WG, Yuan J, Zakeri Z, Zhivotovsky B, Melino G, Kroemer G (2015) Essential versus accessory aspects of cell death: recommendations of the NCCD 2015. Cell Death Differ 22(1):58–73. https://doi.org/10.1038/cdd.2014.137

Galluzzi L, Vitale I, Abrams JM, Alnemri ES, Baehrecke EH, Blagosklonny MV, Dawson TM, Dawson VL, El-Deiry WS, Fulda S, Gottlieb E, Green DR, Hengartner MO, Kepp O, Knight RA, Kumar S, Lipton SA, Lu X, Madeo F, Malorni W, Mehlen P, Nunez G, Peter ME, Piacentini M, Rubinsztein DC, Shi Y, Simon HU, Vandenabeele P, White E, Yuan J, Zhivotovsky B, Melino G, Kroemer G (2012) Molecular definitions of cell death subroutines: recommendations of the Nomenclature Committee on Cell Death 2012. Cell Death Differ 19(1):107–120. https://doi.org/10.1038/cdd.2011.96

Goto S, Xue R, Sugo N, Sawada M, Blizzard KK, Poitras MF, Johns DC, Dawson TM, Dawson VL, Crain BJ, Traystman RJ, Mori S, Hurn PD (2002) Poly(ADP-ribose) polymerase impairs early and long-term experimental stroke recovery. Stroke 33(4):1101–1106

Green DR, Kroemer G (2004) The pathophysiology of mitochondrial cell death. Science 305(5684):626–629. https://doi.org/10.1126/science.1099320

Ha HC, Snyder SH (1999) Poly(ADP-ribose) polymerase is a mediator of necrotic cell death by ATP depletion. Proc Natl Acad Sci U S A 96(24):13978–13982

Hanai S, Kanai M, Ohashi S, Okamoto K, Yamada M, Takahashi H, Miwa M (2004) Loss of poly(ADP-ribose) glycohydrolase causes progressive neurodegeneration in Drosophila melanogaster. Proc Natl Acad Sci U S A 101(1):82–86. https://doi.org/10.1073/pnas.2237114100

Hottiger MO, Hassa PO, Luscher B, Schuler H, Koch-Nolte F (2010) Toward a unified nomenclature for mammalian ADP-ribosyltransferases. Trends Biochem Sci 35(4):208–219. https://doi.org/10.1016/j.tibs.2009.12.003

Kang HC, Lee YI, Shin JH, Andrabi SA, Chi Z, Gagne JP, Lee Y, Ko HS, Lee BD, Poirier GG, Dawson VL, Dawson TM (2011) Iduna is a poly(ADP-ribose) (PAR)-dependent E3 ubiquitin ligase that regulates DNA damage. Proc Natl Acad Sci U S A 108(34):14103–14108. https://doi.org/10.1073/pnas.1108799108

Koh DW, Lawler AM, Poitras MF, Sasaki M, Wattler S, Nehls MC, Stoger T, Poirier GG, Dawson VL, Dawson TM (2004) Failure to degrade poly(ADP-ribose) causes increased sensitivity to cytotoxicity and early embryonic lethality. Proc Natl Acad Sci U S A 101(51):17699–17704. https://doi.org/10.1073/pnas.0406182101

Krantic S, Mechawar N, Reix S, Quirion R (2007) Apoptosis-inducing factor: a matter of neuron life and death. Prog Neurobiol 81(3):179–196. https://doi.org/10.1016/j.pneurobio.2006.12.002

Kroemer G, Galluzzi L, Brenner C (2007) Mitochondrial membrane permeabilization in cell death. Physiol Rev 87(1):99–163. https://doi.org/10.1152/physrev.00013.2006

Lai TW, Zhang S, Wang YT (2014) Excitotoxicity and stroke: identifying novel targets for neuroprotection. Prog Neurobiol 115:157–188. https://doi.org/10.1016/j.pneurobio.2013.11.006

Lee Y, Karuppagounder SS, Shin JH, Lee YI, Ko HS, Swing D, Jiang H, Kang SU, Lee BD, Kang HC, Kim D, Tessarollo L, Dawson VL, Dawson TM (2013) Parthanatos mediates AIMP2-activated age-dependent dopaminergic neuronal loss. Nat Neurosci 16(10):1392–1400. https://doi.org/10.1038/nn.3500

Li P, Nijhawan D, Budihardjo I, Srinivasula SM, Ahmad M, Alnemri ES, Wang X (1997) Cytochrome c and dATP-dependent formation of Apaf-1/caspase-9 complex initiates an apoptotic protease cascade. Cell 91(4):479–489

Lipton P (1999) Ischemic cell death in brain neurons. Physiol Rev 79(4):1431–1568

Liu X, Kim CN, Yang J, Jemmerson R, Wang X (1996) Induction of apoptotic program in cell-free extracts: requirement for dATP and cytochrome c. Cell 86(1):147–157

Liu X, Zou H, Slaughter C, Wang X (1997) DFF, a heterodimeric protein that functions downstream of caspase-3 to trigger DNA fragmentation during apoptosis. Cell 89(2):175–184

Mandir AS, Poitras MF, Berliner AR, Herring WJ, Guastella DB, Feldman A, Poirier GG, Wang ZQ, Dawson TM, Dawson VL (2000) NMDA but not non-NMDA excitotoxicity is mediated by poly(ADP-ribose) polymerase. J Neurosci 20(21):8005–8011

Mashimo M, Kato J, Moss J (2013) ADP-ribosyl-acceptor hydrolase 3 regulates poly (ADP-ribose) degradation and cell death during oxidative stress. Proc Natl Acad Sci U S A 110(47):18964–18969. https://doi.org/10.1073/pnas.1312783110

Mehta A, Prabhakar M, Kumar P, Deshmukh R, Sharma PL (2013) Excitotoxicity: bridge to various triggers in neurodegenerative disorders. Eur J Pharmacol 698(1–3):6–18. https://doi.org/10.1016/j.ejphar.2012.10.032

Meldrum BS (1992) Excitatory amino acid receptors and disease. Curr Opin Neurol Neurosurg 5(4):508–513

Meyer-Ficca ML, Meyer RG, Coyle DL, Jacobson EL, Jacobson MK (2004) Human poly(ADP-ribose) glycohydrolase is expressed in alternative splice variants yielding isoforms that localize to different cell compartments. Exp Cell Res 297(2):521–532. https://doi.org/10.1016/j.yexcr.2004.03.050

Modjtahedi N, Giordanetto F, Madeo F, Kroemer G (2006) Apoptosis-inducing factor: vital and lethal. Trends Cell Biol 16(5):264–272. https://doi.org/10.1016/j.tcb.2006.03.008

Nakanishi S (1992) Molecular diversity of glutamate receptors and implications for brain function. Science 258(5082):597–603

Newmeyer DD, Ferguson-Miller S (2003) Mitochondria: releasing power for life and unleashing the machineries of death. Cell 112(4):481–490

Oka S, Kato J, Moss J (2006) Identification and characterization of a mammalian 39-kDa poly(ADP-ribose) glycohydrolase. J Biol Chem 281(2):705–713. https://doi.org/10.1074/jbc.M510290200

Olney JW, Sharpe LG (1969) Brain lesions in an infant rhesus monkey treated with monsodium glutamate. Science 166(3903):386–388

Pospisilik JA, Knauf C, Joza N, Benit P, Orthofer M, Cani PD, Ebersberger I, Nakashima T, Sarao R, Neely G, Esterbauer H, Kozlov A, Kahn CR, Kroemer G, Rustin P, Burcelin R, Penninger JM (2007) Targeted deletion of AIF decreases mitochondrial oxidative phosphorylation and protects from obesity and diabetes. Cell 131(3):476–491. https://doi.org/10.1016/j.cell.2007.08.047

Richter BW, Duckett CS (2000) The IAP proteins: caspase inhibitors and beyond. Sci STKE 2000(44):pe1. https://doi.org/10.1126/stke.2000.44.pe1

Sakahira H, Enari M, Nagata S (1998) Cleavage of CAD inhibitor in CAD activation and DNA degradation during apoptosis. Nature 391(6662):96–99. https://doi.org/10.1038/34214

Samdani AF, Dawson TM, Dawson VL (1997) Nitric oxide synthase in models of focal ischemia. Stroke 28(6):1283–1288

Sattler R, Xiong Z, Lu WY, Hafner M, MacDonald JF, Tymianski M (1999) Specific coupling of NMDA receptor activation to nitric oxide neurotoxicity by PSD-95 protein. Science 284(5421):1845–1848

Soriano FX, Martel MA, Papadia S, Vaslin A, Baxter P, Rickman C, Forder J, Tymianski M, Duncan R, Aarts M, Clarke P, Wyllie DJ, Hardingham GE (2008) Specific targeting of pro-death NMDA receptor signals with differing reliance on the NR2B PDZ ligand. J Neurosci 28(42):10696–10710. https://doi.org/10.1523/JNEUROSCI.1207-08.2008

Sun HS, Doucette TA, Liu Y, Fang Y, Teves L, Aarts M, Ryan CL, Bernard PB, Lau A, Forder JP, Salter MW, Wang YT, Tasker RA, Tymianski M (2008) Effectiveness of PSD95 inhibitors in permanent and transient focal ischemia in the rat. Stroke 39(9):2544–2553. https://doi.org/10.1161/STROKEAHA.107.506048

Susin SA, Lorenzo HK, Zamzami N, Marzo I, Snow BE, Brothers GM, Mangion J, Jacotot E, Costantini P, Loeffler M, Larochette N, Goodlett DR, Aebersold R, Siderovski DP, Penninger JM, Kroemer G (1999) Molecular characterization of mitochondrial apoptosis-inducing factor. Nature 397(6718):441–446. https://doi.org/10.1038/17135

Szabo C, Dawson VL (1998) Role of poly(ADP-ribose) synthetase in inflammation and ischaemia-reperfusion. Trends Pharmacol Sci 19(7):287–298

Tait SW, Green DR (2010) Mitochondria and cell death: outer membrane permeabilization and beyond. Nat Rev Mol Cell Biol 11(9):621–632. https://doi.org/10.1038/nrm2952

Virag L, Szabo C (2002) The therapeutic potential of poly(ADP-ribose) polymerase inhibitors. Pharmacol Rev 54(3):375–429

Wang H, Yu SW, Koh DW, Lew J, Coombs C, Bowers W, Federoff HJ, Poirier GG, Dawson TM, Dawson VL (2004) Apoptosis-inducing factor substitutes for caspase executioners in NMDA-triggered excitotoxic neuronal death. J Neurosci 24(48):10963–10973. https://doi.org/10.1523/JNEUROSCI.3461-04.2004

Wang Y, An R, Umanah GK, Park H, Nambiar K, Eacker SM, Kim B, Bao L, Harraz MM, Chang C, Chen R, Wang JE, Kam TI, Jeong JS, Xie Z, Neifert S, Qian J, Andrabi SA, Blackshaw S, Zhu H, Song H, Ming GL, Dawson VL, Dawson TM (2016) A nuclease that mediates cell death induced by DNA damage and poly(ADP-ribose) polymerase-1. Science 354(6308):aad6872. https://doi.org/10.1126/science.aad6872

Wang Y, Kim NS, Haince JF, Kang HC, David KK, Andrabi SA, Poirier GG, Dawson VL, Dawson TM (2011) Poly(ADP-ribose) (PAR) binding to apoptosis-inducing factor is critical for PAR polymerase-1-dependent cell death (parthanatos). Sci Signal 4(167):ra20. https://doi.org/10.1126/scisignal.2000902

Xia Y, Dawson VL, Dawson TM, Snyder SH, Zweier JL (1996) Nitric oxide synthase generates superoxide and nitric oxide in arginine-depleted cells leading to peroxynitrite-mediated cellular injury. Proc Natl Acad Sci U S A 93(13):6770–6774

Xu JC, Fan J, Wang X, Eacker SM, Kam TI, Chen L, Yin X, Zhu J, Chi Z, Jiang H, Chen R, Dawson TM, Dawson VL (2016) Cultured networks of excitatory projection neurons and inhibitory interneurons for studying human cortical neurotoxicity. Sci Transl Med 8(333):333ra348. https://doi.org/10.1126/scitranslmed.aad0623

Yu SW, Andrabi SA, Wang H, Kim NS, Poirier GG, Dawson TM, Dawson VL (2006) Apoptosis-inducing factor mediates poly(ADP-ribose) (PAR) polymer-induced cell death. Proc Natl Acad Sci U S A 103(48):18314–18319. https://doi.org/10.1073/pnas.0606528103

Yu SW, Wang H, Poitras MF, Coombs C, Bowers WJ, Federoff HJ, Poirier GG, Dawson TM, Dawson VL (2002) Mediation of poly(ADP-ribose) polymerase-1-dependent cell death by apoptosis-inducing factor. Science 297(5579):259–263. https://doi.org/10.1126/science.1072221

Zhang J, Dawson VL, Dawson TM, Snyder SH (1994) Nitric oxide activation of poly(ADP-ribose) synthetase in neurotoxicity. Science 263(5147):687–689

Zhang Y, Liu S, Mickanin C, Feng Y, Charlat O, Michaud GA, Schirle M, Shi X, Hild M, Bauer A, Myer VE, Finan PM, Porter JA, Huang SM, Cong F (2011) RNF146 is a poly(ADP-ribose)-directed E3 ligase that regulates axin degradation and Wnt signalling. Nat Cell Biol 13(5):623–629. https://doi.org/10.1038/ncb2222

Zou H, Henzel WJ, Liu X, Lutschg A, Wang X (1997) Apaf-1, a human protein homologous to C. Elegans CED-4, participates in cytochrome c-dependent activation of caspase-3. Cell 90(3):405–413

Chapter 2
To Survive or to Die: How Neurons Deal with it

Yubin Wang, Xiaoning Bi, and Michel Baudry

Abstract Unlike the majority of cells in the organism, neurons have only two options during their entire existence, to survive or to die. As a result, they have evolved elaborate mechanisms to determine which path they will follow in response to a multitude of internal and external signals, and to the wear-and-tear associated with the aging process. Until recently, activation of the calcium-dependent protease, calpain, had been traditionally associated with neurodegeneration. This chapter will review recent findings that indicate that two of the major calpain isoforms present in the brain, calpain-1 and calpain-2, play opposite functions in neuronal survival/death. Thus, calpain-1 activation, downstream of synaptic NMDA receptors, is part of a neuronal survival pathway through the truncation of PHLPP1 and the stimulation of the Akt pathway. In contrast, calpain-2 activation is downstream of extrasynaptic NMDA receptors and is neurodegenerative through the truncation of the phosphatase, STEP, and the activation of the p38 protein kinase. These findings have major significance for our understanding of neurological conditions associated with neurodegeneration and for the development of new therapeutic approaches to prevent neuronal death in these disorders.

Keywords Calpain-1 · Calpain-2 · Neuronal death · Neuronal survival · NMDA receptors · Akt · STEP

Y. Wang · M. Baudry (✉)
Graduate College of Biomedical Sciences, Western University of Health Sciences, Pomona, CA, USA
e-mail: mbaudry@westernu.edu

X. Bi
College of Osteopathic Medicine of the Pacific, Western University of Health Sciences, Pomona, CA, USA

D. G. Fujikawa (ed.), *Acute Neuronal Injury*,
https://doi.org/10.1007/978-3-319-77495-4_2

2.1 Introduction

Neurons have to perform several basic functions, including growing (from the time of differentiation), migrating, responding and adapting to external and internal stimuli, and surviving or dying, as a result of continuous challenges and the deleterious effects of the aging process. Numerous reviews have discussed the role of calpain in neurodegeneration in general (Vosler et al. 2008; Yildiz-Unal et al. 2015), and in stroke (Anagli et al. 2009; Koumura et al. 2008) and in traumatic brain injury (TBI) (Kobeissy et al. 2015; Liu et al. 2014). Likewise, numerous studies have attempted to use calpain inhibitors to reduce neurodegeneration in both stroke and TBI (Anagli et al. 2009; Bartus et al. 1994a, b; Cagmat et al. 2015; Hong et al. 1994; Li et al. 1998; Markgraf et al. 1998; Siklos et al. 2015; Tsubokawa et al. 2006). While some studies have reported some positive effects of calpain inhibitors in TBI (Thompson et al. 2010), other studies have not confirmed these results. In particular, overexpression of the endogenous calpain inhibitor, calpastatin, was reported to reduce the formation of the Spectrin Breakdown Product (SBDP), resulting from calpain-mediated truncation of spectrin, a widely used biomarker of calpain activation and potentially neurodegeneration (Yan and Jeromin 2012), but had no effect on neurodegeneration (Schoch et al. 2012). Another recent study concluded that even a blood-brain barrier- and cell-permeable calpain inhibitor, SNJ-1945, did not have a sufficient efficacy and a practical therapeutic window in a model of controlled cortical impact (Bains et al. 2013).

Several reasons could account for the failure to develop clinical applications of such inhibitors, including their lack of specificity/potency/selectivity (Donkor 2011), and the incomplete knowledge regarding the functions of the major calpain isoforms in the brain, calpain-1 and calpain-2 (aka μ- and m-calpain). Work from our laboratory over the last 5 years has revealed new features of these two enzymes, which significantly changed our understanding of their functions in the brain. Specifically, we found that calpain-1 and calpain-2 play opposite functions in both synaptic plasticity and neuroprotection/neurodegeneration (Baudry and Bi 2016). Thus, calpain-1 activation is required for theta burst stimulation-induced long-term potentiation (LTP) and for certain types of learning and memory, and is neuroprotective (Wang et al. 2013, 2014). Calpain-1 is neuroprotective due to the degradation of the PH domain and Leucine rich repeat Protein Phosphatase 1 (PHLPP1β) and the resulting activation of the Akt survival pathway. On the other hand, calpain-2 activation limits the magnitude of LTP and restricts learning, and is neurodegenerative due to the cleavage of STEP and the stimulation of death pathways (Wang et al. 2013, 2014). In addition, we found that ischemia-induced damage to retinal ganglion cells was exacerbated in calpain-1 knock-out mice, indicating that calpain-1 inhibition is likely to counteract the potential beneficial effects of calpain-2 inhibition if non-selective calpain inhibitors are used (Wang et al. 2016b). These findings could account for the failure of the previous studies to convincingly demonstrate the role of calpain in neurodegeneration, and for the lack of clear efficacy of the previously tested calpain inhibitors, which did not discriminate between calpain-1 and calpain-2. It is also important to stress that calpain activation has also been implicated in diffuse axonal injury (Wang et al. 2012a), which has been proposed to represent an important component of the pathophysiology of TBI (Xiong et al. 2013), although at this point, there is no information regarding which calpain isoform is involved.

In this chapter, we will first discuss how calpain-1 and calpain-2 activation appear to be closely related to the stimulation of synaptic and extra-synaptic NMDA receptors, respectively. We will then review the mechanisms underlying the neuroprotective effects of calpain-1 activation, which will be followed by a discussion of the mechanisms involved in calpain-2-mediated neurodegeneration. These two aspects will be illustrated by studies using intra-ocular NMDA injection to produce acute neurodegeneration of retinal ganglion cells. Finally, we will discuss the potential clinical implications of these findings and our current efforts to develop selective calpain-2 inhibitors as a new approach for neuroprotection in conditions associated with acute neurodegeneration.

2.2 Calpains and NMDA Receptors

NMDARs play critical roles in both physiological and pathological conditions, and several studies have shown that NMDA receptor localization is responsible for opposite consequences of NMDA receptor stimulation for neuronal survival or death; thus, synaptic NMDAR activation provides neuroprotection, while extrasynaptic NMDARs are linked to pro-death pathways (Hardingham and Bading 2010). The Akt and MAP kinase/extracellular signal-regulated kinase (ERK1/2) pathways are two key pro-survival pathways downstream of synaptic NMDARs (Hardingham et al. 2001; Papadia et al. 2005; Wang et al. 2012b). Akt phosphorylates and inhibits various pro-apoptotic substrates, such as glycogen synthase kinase-3 (GSK3), forkhead box O (FOXO) (Soriano et al. 2006), apoptosis signal-regulating kinase 1 (ASK1) (Kim et al. 2001), p53 (Yamaguchi et al. 2001), and Bcl2-associated death promoter (BAD) (Downward 1999). On the other hand, ERK1/2 activates the survival nuclear transcription factor, cyclic-AMP response element binding protein (CREB) (Hardingham et al. 2001). Although some protein kinases linking NMDARs to Akt and ERK have been found (Krapivinsky et al. 2003; Perkinton et al. 2002), how Akt and ERK1/2 were activated by synaptic but not extrasynaptic NMDARs was not clearly understood until recently.

PH domain and Leucine rich repeat Protein Phosphatase 1 (PHLPP1) exhibits two splice variants, PHLPP1α and PHLPP1β, which share amino acid sequence similarity but have different sizes (140 kDa and 190 kDa, respectively). PHLPP1α dephosphorylates Akt at Ser473 in cancer cells (Gao et al. 2005) and neurons (Jackson et al. 2010), and its down-regulation is related to cell survival in CNS (Chen et al. 2013; Liu et al. 2009; Saavedra et al. 2010). PHLPP1β inhibits ERK1/2

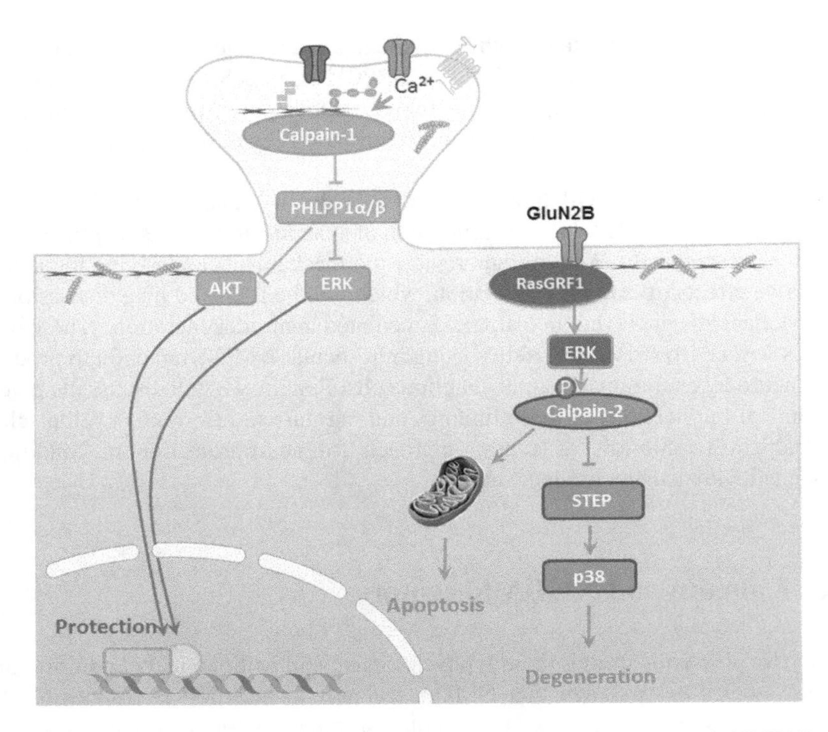

Fig. 2.1 Schematic representation of the links between synaptic and extrasynaptic NMDARs and calpain-1 and calpain-2. Calpain-1 is rapidly stimulated by the calcium influx generated by synaptic NMDA receptor activation, resulting in PHLPP1α/β degradation. This produces the activation of Akt and ERK, which triggers the stimulation of neuroprotective cascades. On the other hand, extrasynaptic NMDA receptors containing NR2B subunits trigger ERK activation, calpain-2 phosphorylation/activation and the activation of STEP and p38, leading to neurodegeneration. Moreover, calpain-2 activation has been linked to apoptosis through the truncation of anti-apoptotic factors

by binding and trapping its activator Ras in the inactive form (Shimizu et al. 2003). PHLPP1β was previously shown to be degraded by calpain in hippocampus, and its degradation contributes to novel object recognition memory (Shimizu et al. 2007). Thus, PHLPP1 was a good candidate to link NMDA receptor stimulation to Akt and ERK regulation.

Using primary neuronal cultures, we showed that calpain-1 and calpain-2 are activated by different NMDAR populations (synaptic vs. extrasynaptic NMDARs) and regulate different substrates (PHLPP1 and STEP) to produce opposite effects on neuronal fate (neuroprotection and neurodegeneration) (Fig. 2.1). Interestingly, calpain-induced cleavage of PHLPP1β and the resulting ERK activation were previously shown to regulate synaptic plasticity (Shimizu et al. 2007). We showed that calpain-1-mediated PHLPP1β degradation was specifically triggered by synaptic but not extra-synaptic NMDAR activation and contributed to the neuroprotective effects of synaptic NMDAR activation. In addition, PHLPP1α, which dephosphorylates and inhibits Akt, was also cleaved by calpain-1 following synaptic NMDAR activation. Calpain cleavage of PHLPP1 1α and β was necessary and sufficient for synaptic NMDAR-induced activation of the Akt and ERK pathways, since calpain inhibition blocked, while PHLPP1 knockdown mimicked, the effects of synaptic NMDAR activation on Akt and ERK pathways. PHLPP1 suppressed Akt and ERK pathways under basal conditions; following synaptic NMDAR activation, calpain cleaves PHLPP1α and β, thus releasing the inhibition of these two major pro-survival signaling cascades in neurons. Consistently, calpain-1-mediated cleavage of PHLPP1 was required for the neuroprotective effects of synaptic NMDARs, as calpain inhibition blocked the neuroprotection elicited by synaptic NMDAR activation. We further confirmed these results using PHLPP1 knockdown, as down-regulation of PHLPP1 not only suppressed the blockade of neuroprotection caused by calpain inhibition but also induced neuroprotection without synaptic NMDAR activation. Consistent with our results, a recent study reported that PHLPP1 knockout mice are more resistant to ischemic brain injury (Chen et al. 2013). Thus, PHLPP1 should be considered as a novel potential target for the treatment of neurodegenerative diseases.

As previously reported (Xu et al. 2009), we found that calpain activated by extrasynaptic NMDAR stimulation cleaved STEP and caused neuronal death (Wang et al. 2013). It had previously been proposed that prolonged or excessive activation of calpain was responsible for calpain-mediated neurotoxicity, whereas brief and limited calpain activation could be involved in the regulation of synaptic plasticity. However, prolonged activation of synaptic NMDARs (by Bic and 4-AP treatment) for as long as 3 days did not result in STEP cleavage, nor in neuronal damage, but produced neuroprotection against starvation and oxidative stress. On the other hand, activation of extrasynaptic NMDARs did not affect PHLPP1 or its downstream pathways, strongly suggesting that there are two separate pools of calpain downstream of synaptic and extrasynaptic NMDARs, which regulate different substrates and therefore exert separate functions.

The possibility that calpain-1 and calpain-2 could exert different roles in CNS had not been extensively discussed. However, the discovery that calpain-2 could be activated by phosphorylation (Zadran et al. 2010), coupled with the identification of PTEN as a specific calpain-2 substrate (Briz et al. 2013), raised the possibility that calpain-1 and calpain-2 could play distinct functions. Interestingly, synaptic NMDAR activation did not result in the degradation of PTEN, a specific calpain-2 substrate, further supporting the idea that synaptic NMDAR activation does not activate calpain-2. The use of calpain-1 and calpain-2 specific inhibitors also confirmed this idea, as a calpain-2 specific inhibitor did not affect synaptic NMDAR-dependent PHLPP1 cleavage and neuroprotection but blocked extrasynaptic NMDAR-dependent STEP cleavage and neurotoxicity. In contrast, a calpain-1 specific inhibitor blocked synaptic NMDAR-mediated effects but not extrasynaptic NMDAR-mediated neurotoxicity. Down-regulation of calpain-1 and calpain-2 by specific siRNAs in cultured neurons also indicated that only calpain-1 knockdown blocked synaptic NMDAR-mediated neuroprotective pathways. In addition, knockdown of calpain-2 but not calpain-1, by AAV-shRNA transfection increased survival of primary hippocampal neurons following NMDA treatment (Bevers et al. 2009). Results obtained in cultured neurons were further confirmed using a model of NMDA-induced neurotoxicity in acute hippocampal slices from young mice, which had previously indicated that NMDA treatment of acute hippocampal slices caused neurotoxicity in young but not adult rats (Zhou and Baudry 2006), probably because young rats have more NR2B-containing NMDARs, which are preferentially localized extrasynaptically (Tovar and Westbrook 1999). In hippocampal slices prepared from young calpain-1 knock-out mice, NMDA induced the degradation of STEP but not PHLPP1, and exacerbated neurotoxicity, as compared to slices prepared from wild-type mice. On the other hand, calpain-2 specific inhibition by applying either a selective calpain-2 inhibitor in slices from wild-type mice or a non-selective calpain inhibitor in slices from calpain-1 knock-out mice blocked NMDA-induced degradation of STEP and suppressed neurotoxicity (Wang et al. 2013).

Together, these results demonstrate that calpain-1 is preferentially activated by synaptic NMDAR stimulation, whereas calpain-2 is preferentially activated by extrasynaptic NMDAR stimulation. Calpain-1 was shown to be localized in synaptic compartments (Perlmutter et al. 1988), where it could regulate synaptic function through its action on synaptic elements such as cytoskeletal and scaffolding proteins, as well as glutamate receptors (Liu et al. 2008). Little is known regarding the ultrastructural localization of calpain-2 in neurons. One of the newly discovered physiological roles of calpain-2 is to regulate activity-dependent local protein synthesis (Briz et al. 2013; Wang and Huang 2012), which takes place not in synapses but in nearby extrasynaptic areas (Frey and Morris 1998; Steward and Wallace 1995). In addition, calpain-2 has been reported to control synaptogenesis in dendritic shafts through constitutive proteolysis of the cytoskeletal protein, cortactin (Mingorance-Le Meur and O'Connor 2009). These findings would suggest that calpain-2 is localized, at least in part, in extrasynaptic domains (Fig. 2.1).

The existence of separate signaling pathways for calpain-1 and calpain-2 suggested that these two calpain isoforms belong to different protein scaffolds, which could segregate them in different neuronal compartments. PHLPP1 could be cleaved by both purified calpain-1 and calpain-2 in membrane fractions, yet it was cleaved only by calpain-1 following synaptic NMDAR activation in hippocampal slices, suggesting that substrate specificity for calpains depends not only on amino acid sequences within substrates, but also on localization and scaffolding of both substrates and calpains in neurons. Co-immunoprecipitation experiments confirmed that NR2A-containing NMDARs, PSD95, calpain-1 and PHLPP1, form a complex in neurons. Furthermore, synaptic NMDAR activity recruited calpain-1 to this NMDAR multi-protein complex; such recruitment could facilitate the proteolysis of PHLPP1 and possibly other calpain-1 substrates in the complex. In contrast, calpain-2 was not present in this complex under basal conditions nor was it recruited by activity, consistent with the absence of calpain-2 activation following synaptic NMDAR activation. It is likely that a calpain-2-containing multi-protein complex is associated with extrasynaptic NMDARs. How could activation of extrasynaptic NMDARs results in calpain-2 activation? It has been repeatedly shown that NR2B subunits are enriched in extrasynaptic NMDARs (Papouin and Oliet 2014), and that their activation is critical for excitotoxicity (Chazot 2004). Interestingly, NR2B directly binds RasGRF1, which provides a link between NMDAR activation and ERK activation (Krapivinsky et al. 2003). As we have shown that ERK activation directly phosphorylates and activates calpain-2 (Zadran et al. 2010), this pathway is likely responsible for the prolonged activation of calpain-2 following stimulation of extrasynaptic NMDA receptors (Fig. 2.1). In addition, we discussed elsewhere the existence of different PDZ binding domains in the C-terminal of calpain-1 and calpain-2, which could account for their differential subcellular distribution (Baudry and Bi 2016).

2.3 Calpain-1 Activation and Neuroprotection

As discussed above, calpain-1 is downstream of synaptic NMDARs and as such, we postulated that it has a neuroprotective function. This notion was supported by results obtained in cultured neurons, where we demonstrated that calpain-1 activation following stimulation of synaptic NMDARs was neuroprotective against starvation- and oxidative stress-mediated neurotoxicity (Wang et al. 2013). Previous studies have shown that normal stimulation of synaptic NMDA receptors is required to limit the extent of apoptotic neuronal death during the postnatal period, as blockade of these receptors during this period increases the extent of apoptotic neuronal death (Monti and Contestabile 2000). Calpain activity is higher in cerebellum than in cortex or hippocampus across different mammalian species (Baudry et al. 1986). An immunohistochemical study revealed that the major calpain isoform expressed in cerebellar neurons is calpain-1 (Hamakubo et al. 1986). Calpain-1 activity in cerebellum during prenatal and early postnatal period is high, as compared to that in adulthood (Simonson et al. 1985), suggesting a potential role for calpain-1 in cerebellar development. Interestingly, a *CAPN1* missense mutation in the Parson Russell Terrier dog breed has been associated with spinocerebellar ataxia (Forman et al. 2013).

Loss of cerebellar granule cells (CGCs) induced by different mechanisms results in ataxia (Hashimoto et al. 1999; Kim et al. 2009; Pennacchio et al. 1998; Shmerling et al. 1998). NMDAR activity is essential for CGC survival during the critical stage of cerebellar development (Monti and Contestabile 2000; Balazs et al. 1988; Monti et al. 2002; Moran and Patel 1989), although the underlying mechanism has remained elusive. NMDAR-induced activation of the nuclear factor CREB is required (Monti et al. 2002), and CREB is a target of the pro-survival kinase Akt (Du and Montminy 1998).

As discussed above, synaptic NMDAR-mediated calpain-1 activation results in the degradation of PHLPP1. PHLPP1 dephosphorylates and inhibits Akt, and is involved in tumorigenesis (Chen et al. 2011), circadian clock (Masubuchi et al. 2010), learning and memory process (Wang et al. 2014; Shimizu et al. 2007), and autophagy (Arias et al. 2015). Calpain-1-mediated degradation of PHLPP1 activates Akt and promotes neuronal survival (Wang et al. 2013), and we postulated that calpain-1 mediated regulation of PHLPP1 and Akt could be involved in NMDAR-dependent CGC survival during postnatal development.

We analyzed apoptosis in the brain during the postnatal period in wild-type and calpain-1 KO mice (Wang et al. 2016a). Calpain-1 KO mice exhibited abnormal cerebellar development, including enhanced apoptosis of CGCs during the early postnatal period, and reduced granule cell density and impaired synaptic transmission from parallel fiber to Purkinje cells in adulthood, resulting in an ataxia phenotype. All these defects are due to deficits in the calpain-1/PHLPP1/Akt pro-survival pathway in developing granule cells, since treatment with an Akt activator during the postnatal period or crossing calpain-1 KO mice with PHLPP1 KO mice restored most of the observed alterations in cerebellar structure and function in calpain-1 KO mice (Wang et al. 2016a). To reverse reduced pAkt levels in cerebellum of calpain-1 KO mice during the early postnatal period, we treated them from PND1 to PND7 with a PTEN inhibitor, bisperoxovanadium (bpV) (0.5 mg/kg, i.p., twice daily), which has been shown to activate Akt (Boda et al. 2014; Li et al. 2009; Mao et al. 2013). BpV injection significantly increased pAkt levels in cerebellum of developing KO mice, and completely prevented the enhanced apoptosis in cerebellum and cerebrum of calpain-1 KO mice at PND7 (Fig. 2.2).

Thus the NMDAR/calpain-1/PHLLP1/Akt pro-survival pathway is active in developing CGCs, where it limits the extent of CGC apoptosis. Increased PHLPP1 and decreased pAkt levels were found in cerebellar homogenates of calpain-1 KO mice, indicating that calpain-1 activity normally reduces PHLPP1 levels and maintains Akt activated during the postnatal period in cerebellum. The density of pAkt-positive puncta was reduced in cerebellar granular layer but not in Purkinje or molecular layer of calpain-1 KO mice, suggesting that calpain-1-dependent regulation of Akt only takes place in CGCs but not in other cerebellar cell types. Down-regulation of PHLPP1 restored normal levels of pAkt in developing cerebellum of calpain-1 KO mice, indicating that PHLPP1 is downstream of calpain-1 and that its level is important for Akt regulation. Finally, reduced Akt activity was associated with enhanced CGC apoptosis in calpain-1 KO mice, while increased Akt activity was associated with reduced CGC apoptosis in bpV-injected WT and in mice lacking both calpain-1 and PHLPP1.

NMDAR- and calpain-1-mediated neuronal survival during brain development was not limited to CGCs, as enhanced apoptosis was present in other brain regions such as cortex, striatum and hippocampus in developing calpain-1 KO mice (Fig. 2.2). Importantly, calpain-1-mediated neuroprotection is also present in human brain, as calpain-1 mutations resulting in lack of function are associated with cerebellar ataxia (Wang et al. 2016a; Gan-Or et al. 2016). Furthermore, the important roles of calpain-1 in hippocampal neuronal survival during development and in synaptic plasticity in the adult (Wang et al. 2014; Zhu et al. 2015) may contribute to the cognitive decline found in ataxia patients with *CAPN1* mutations.

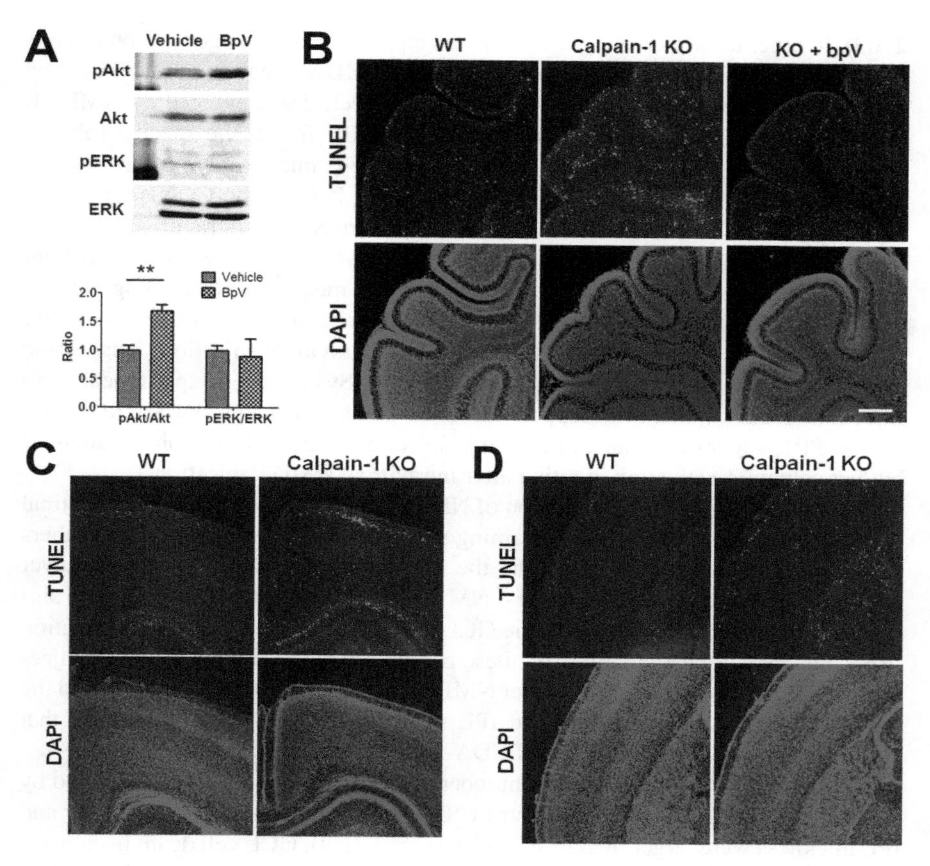

Fig. 2.2 Effects of bpV on apoptosis and Akt in telencephalon of calpain-1 (CAPN1) KO mice during the postnatal period. (**a–c**) TUNEL and DAPI staining of coronal sections at various anterior-posterior levels of PND7 calpain-1 KO mice injected from PND1 to PND7 with vehicle or a PTEN inhibitor, bisperoxovanadium (bpV) (0.5 mg/kg, i.p., twice daily). Note the clear decrease in TUNEL staining in bpv-injected calapin-1 KO mice. (**d, e**) Levels of Akt and p-Akt and ERK and p-ERK in cortex of PND7 calpain-1 KO mice injected from PND1 to PND7 with vehicle or bpv (0.5 mg/kg, i.p., twice daily). Results are expressed as means ± SEM of four experiments. $^{**}p < 0.05$, Student's t-test

2.4 Calpain-2 and Neurodegeneration

As mentioned above, there is abundant literature linking calpain activation with neurodegeneration. However, very few studies have explored the specific contributions of calpain-1 and calpain-2 in neurodegeneration. Our *in vitro* studies clearly indicated that calpain-2 activation, but not calpain-1 activation was responsible for NMDA-induced excitotoxicity through the activation of STEP. A similar study indicated that down-regulation of calpain-2 but not calpain-1 also increased neuronal survival following NMDA treatment of cultured hippocampal neurons (Bevers et al.

2009). In order to further analyze the role of calpain-2 in neurodegeneration in vivo, we used a model consisting of direct intraocular NMDA injection in mice. Calpain activation had been previously involved in retinal cell death induced by NMDAR activation (Chiu et al. 2005; Shimazawa et al. 2010). To test the specific roles of calpain-1 and calpain-2 in this process, wild-type (WT) mice were injected systemically with a calpain-2 selective inhibitor (C2I), Z-Leu-Abu-CONH-CH_2-C_6H_3 (3, 5-$(OMe)_2$) (Wang et al. 2013, 2014), 30 min before NMDA intravitreal injection. Levels of SBDP and of PHLPP1, were determined in retinal extracts 6 h after NMDA injection (Fig. 2.3a–c). Akt levels were also measured as a loading control. Levels of SBDP were significantly increased and those of PHLPP1 decreased after NMDA injection, as compared to control (PBS intravitreal injection), suggesting that calpain was activated after NMDA injection. Systemic (intraperitoneal; i.p.) injection of C2I significantly suppressed NMDA-induced changes in SBDP but not in PHLPP1, suggesting that C2I systemic injection selectively inhibited calpain-2 but not calpain-1 activation in retina after intravitreal NMDA injection.

Six days after intravitreal injection of NMDA or PBS to WT mice, frozen retinal sections were prepared and H&E staining was performed to evaluate cell numbers in the ganglion cell layer (GCL) and the thickness of the Inner Plexiform Layer (IPL), which contains RGC dendrites. NMDA injection (NMDA plus Vehicle) significantly reduced cell numbers in the GCL and IPL thickness, while PBS injection (PBS plus Vehicle) had no effect on these parameters (Fig. 2.3d–f). Systemic injection of C2I 30 min before and 6 h after NMDA injection significantly suppressed the reduction in GCL cell numbers and IPL thickness (Fig. 2.3d–f), suggesting that calpain-2 activation contributes to NMDA-induced cell death in GCL.

In calpain-1 KO mice, GCL cell number and IPL thickness were not affected by vehicle injection. However, the effects of NMDA injection on GCL cell number and IPL thickness were larger than in WT mice (Fig. 2.3g–i). GCL cell death in calpain-1 KO mice after NMDA injection was significantly more severe than that in WT mice (Fig. 2.3j), suggesting that calpain-1 supports cell survival in GCL after NMDA injection. Systemic injection of C2I to calpain-1 KO mice partially but significantly reversed NMDA-induced decrease in GCL cell number and IPL thickness

Fig. 2.3 (continued) (2 μl of 2.5 mM). Mice were injected i.p. with vehicle (10% DMSO) or C2I (0.3 mg/kg) 30 min before intravitreal injection. Quantification of the ratios of SBDP/Akt (**b**) and PHLPP1/Akt (**c**). n = 4. *$p < 0.05$, ***$p < 0.001$. One-way ANOVA followed by Bonferroni test. (**d**) H&E staining of naive, PBS- (control) or NMDA- (2 μl of 2.5 mM) treated retina from WT mice injected i.p. with vehicle (10% DMSO) or C2I (0.3 mg/kg) 30 min before and 6 h after NMDA injection. H&E staining was performed 7 days after injection. Scale bar = 30 μm. Quantification of cell numbers in GCL (**e**) and thickness of IPL (**f**). Six sections in each eye were analyzed. n = 4–8 (eyes). *$p < 0.05$, **$p < 0.01$, One-way ANOVA followed by Bonferroni test. (**g**) H&E staining of PBS- (control) and NMDA- (2 μl of 2.5 mM) treated retina from calpain-1 KO mice injected i.p. with vehicle or C2I (0.3 mg/kg) 30 min before and 6 h after NMDA injection. H&E stain was done 7 days after injection. Scale bar = 30 μm. Quantification of cell number in GCL (**h**) and thickness of IPL (**i**). n = 6. *$p < 0.05$, **$p < 0.01$, ***$p < 0.001$, One-way ANOVA followed by Bonferroni test. (**j**) GCL cell numbers in NMDA-treated WT and KO mice without and with C2I treatment. n = 6. **$p < 0.01$. Two-tailed t-test

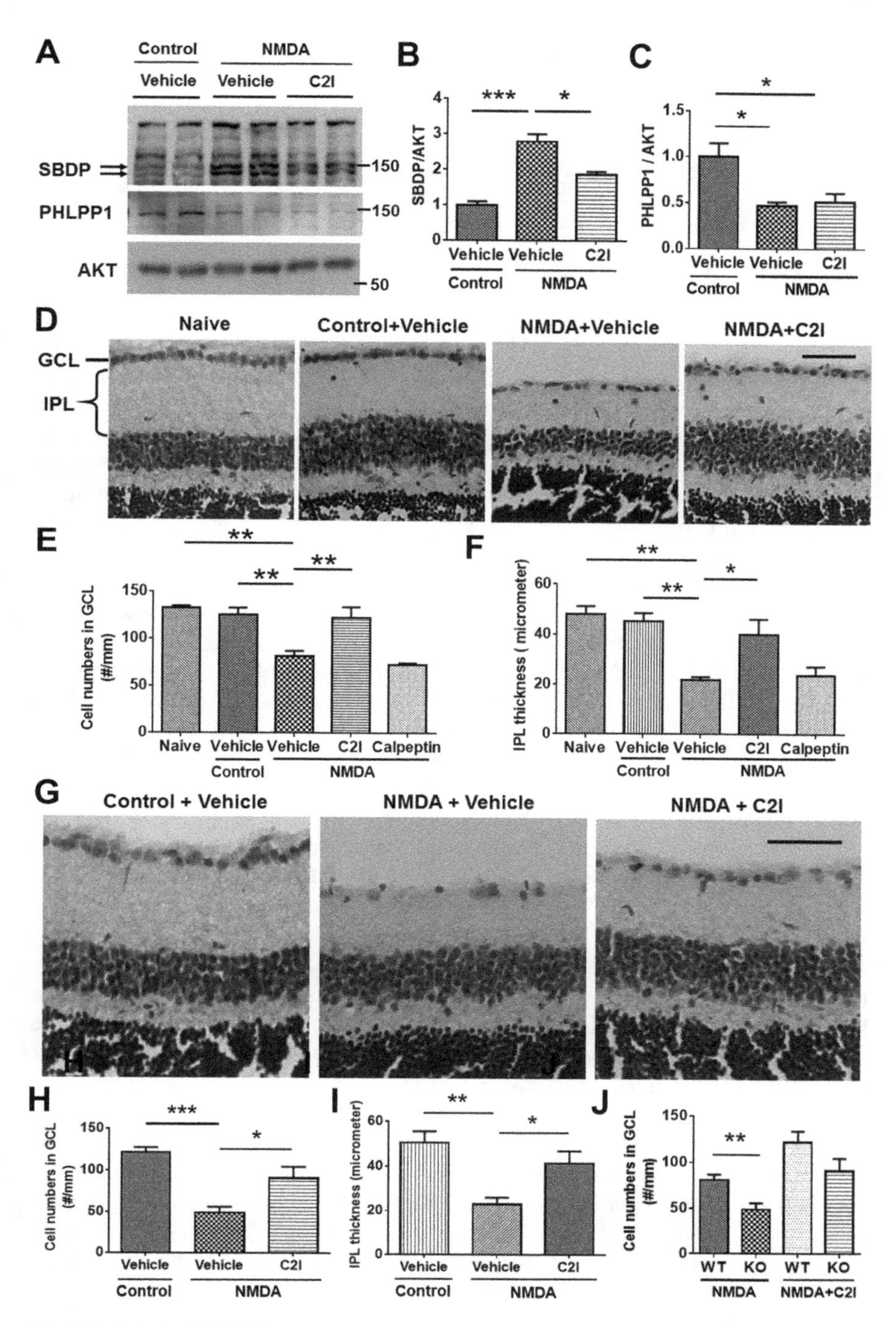

Fig. 2.3 Calpain-2 inhibition reduces, while calpain-1 knockout exacerbates cell death in ganglion cell layer induced by NMDA intravitreal injection. (**a**) Representative immunoblot of indicated proteins in mouse retinal extracts 6 h after intravitreal injection of PBS (control) or NMDA

(Fig. 2.3g–i). A very similar pattern of results was obtained in a different model of acute glaucoma, consisting in a brief period of increased intraocular pressure (Wang et al. 2016b). Furthermore, recent studies in a mouse model of TBI also support the notion that calpain-2 activation is prolonged and responsible for neuronal death, while calpain-1 activation is neuroprotective (Wang et al 2017).

2.5 Clinical Implications of Specific Calpain-2 Inhibition and Calpain-1 Activation

Our results clearly demonstrate that calpain-1 and calpain-2 have opposite functions in both synaptic plasticity and neuronal survival/death after acute insults. Thus, calpain-1 activation is required for LTP induction and for hippocampus-dependent learning and is neuroprotective both during the postnatal developmental period and in adulthood following acute insults. On the other hand, calpain-2 activation limits the extent of hippocampus-dependent learning and is neurodegenerative following acute insults, and in particular excitotoxicity. Our results have important implications for the development of new approaches for treating diseases associated with excitotoxicity, such as epilepsy, stroke, Alzheimer's and Parkinson's disease, Huntington disease and ischemia. In all these cases, it has been suggested that extra-synaptic NMDAR activation and STEP degradation are involved in neurodegeneration. Our results would, therefore, suggest that specific inhibition of calpain-2 but not calpain-1 would have neuroprotective effects under these conditions. Conversely, overexpression or activation of calpain-1, by cleaving PHLPP1 and stimulating pro-survival cascades, could also have beneficial effects. In addition, calpain-2 activation is involved in regulating the magnitude of long-term potentiation (LTP) in hippocampus, due to the existence of a molecular brake consisting in calpain-2-mediated PTEN degradation and stimulation of m-TOR dependent PHLPP1β synthesis (Wang et al. 2014). We also showed that low doses of a selective calpain-2 inhibitor facilitate learning in normal mice, while higher doses, which inhibit calpain-1, impair learning. Thus, a selective calpain-2 inhibitor could be extremely beneficial for preventing neurodegeneration, while facilitating certain forms of learning and memory. As discussed above, a selective calpain-2 inhibitor prevented death of retinal ganglion cells and maintained vision in a mouse model of acute glaucoma (Wang et al. 2016b). Calpain inhibitors have previously been proposed to represent potential treatments for a variety of eye disorders, including glaucoma and macular degeneration (Azuma and Shearer 2008; Paquet-Durand et al. 2007), and further studies are needed to assess the potential use of selective calpain-2 inhibitors for these disorders. Calpain inhibition has been proposed to represent a therapeutic approach for stroke and TBI, although this notion has not been supported by a variety of experiments. We postulate that the use of selective calpain-2 inhibitors might overcome the problems associated with that of non-selective calpain inhibitors. Our results in a mouse model of TBI supports this notion, as we have found that post-treatment with a selective calpain-2 inhibitor provides a highly significant degree of neuroprotection and facilitates behavioral recovery (Wang et al 2017). The potential use of selective calpain-2 inhibitors for chronic neurodegenerative disorders needs to be further evaluated. It is important to note that calpain has been proposed to participate in neurodegeneration associated with Parkinson's disease as well as Alzheimer's disease, and it is tempting to speculate that selective calpain-2 inhibitors might also be beneficial in these disorders.

Acknowledgements This work was supported by grant P01NS045260-01 from NINDS (PI: Dr. C.M. Gall), grant R01NS057128 from NINDS to M.B., and grant R15MH101703 from NIMH to X.B. X.B. is also supported by funds from the Daljit and Elaine Sarkaria Chair.

References

Anagli J, Han Y, Stewart L, Yang D, Movsisyan A, Abounit K, Seyfried D (2009) A novel calpastatin-based inhibitor improves postischemic neurological recovery. Biochem Biophys Res Commun 385(1):94–99

Arias E, Koga H, Diaz A, Mocholi E, Patel B, Cuervo AM (2015) Lysosomal mTORC2/PHLPP1/Akt regulate chaperone-mediated autophagy. Mol Cell 59(2):270–284. https://doi.org/10.1016/j.molcel.2015.05.030

Azuma M, Shearer T (2008) The role of calcium-activated protease calpain in experimental retinal pathology. Surv Ophthalmol 53(2):150–163

Bains M, Cebak JE, Gilmer LK, Barnes CC, Thompson SN, Geddes JW, Hall ED (2013) Pharmacological analysis of the cortical neuronal cytoskeletal protective efficacy of the calpain inhibitor SNJ-1945 in a mouse traumatic brain injury model. J Neurochem 125(1):125–132

Balazs R, Jorgensen OS, Hack N (1988) N-methyl-D-aspartate promotes the survival of cerebellar granule cells in culture. Neuroscience 27(2):437–451

Bartus RT, Baker KL, Heiser AD, Sawyer SD, Dean RL, Elliott PJ, Straub JA (1994a) Postischemic administration of AK275, a calpain inhibitor, provides substantial protection against focal ischemic brain damage. J Cereb Blood Flow Metab 14(4):537–544. https://doi.org/10.1038/jcbfm.1994.67

Bartus RT, Hayward NJ, Elliott PJ, Sawyer SD, Baker KL, Dean RL, Akiyama A, Straub JA, Harbeson SL, Li Z et al (1994b) Calpain inhibitor AK295 protects neurons from focal brain ischemia. Effects of postocclusion intra-arterial administration. Stroke 25(11):2265–2270

Baudry M, Bi X (2016) Calpain-1 and Calpain-2: the Yin and Yang of synaptic plasticity and neurodegeneration. Trends Neurosci 39(4):235–245. https://doi.org/10.1016/j.tins.2016.01.007

Baudry M, Simonson L, Dubrin R, Lynch G (1986) A comparative study of soluble calcium-dependent proteolytic activity in brain. J Neurobiol 17(1):15–28. https://doi.org/10.1002/neu.480170103

Bevers MB, Lawrence E, Maronski M, Starr N, Amesquita M, Neumar RW (2009) Knockdown of m-calpain increases survival of primary hippocampal neurons following NMDA excitotoxicity. J Neurochem 108(5):1237–1250

Boda B, Mendez P, Boury-Jamot B, Magara F, Muller D (2014) Reversal of activity-mediated spine dynamics and learning impairment in a mouse model of Fragile X syndrome. Eur J Neurosci 39(7):1130–1137. https://doi.org/10.1111/ejn.12488

Briz V, Hsu Y-T, Li Y, Lee E, Bi X, Baudry M (2013) Calpain-2-mediated PTEN degradation contributes to BDNF-induced stimulation of dendritic protein synthesis. J Neurosci 33(10):4317–4328

Cagmat EB, Guingab-Cagmat JD, Vakulenko AV, Hayes RL, Anagli J (2015) Potential use of calpain inhibitors as brain injury therapy. In: Kobeissy FH (ed) Brain neurotrauma: molecular, neuropsychological, and rehabilitation aspects. CRC Press/Taylor & Francis, Boca Raton, FL. Chapter 40. Frontiers in Neuroengineering

Chazot PL (2004) The NMDA receptor NR2B subunit: a valid therapeutic target for multiple CNS pathologies. Curr Med Chem 11(3):389–396

Chen M, Pratt CP, Zeeman ME, Schultz N, Taylor BS, O'Neill A, Castillo-Martin M, Nowak DG, Naguib A, Grace DM, Murn J, Navin N, Atwal GS, Sander C, Gerald WL, Cordon-Cardo C, Newton AC, Carver BS, Trotman LC (2011) Identification of PHLPP1 as a tumor suppressor reveals the role of feedback activation in PTEN-mutant prostate cancer progression. Cancer Cell 20(2):173–186. https://doi.org/10.1016/j.ccr.2011.07.013

Chen B, Van Winkle JA, Lyden PD, Brown JH, Purcell NH (2013) PHLPP1 gene deletion protects the brain from ischemic injury. J Cereb Blood Flow Metab 33(2):196–204

Chiu K, Lam TT, Ying Li WW, Caprioli J, Kwong Kwong JM (2005) Calpain and N-methyl-d-aspartate (NMDA)-induced excitotoxicity in rat retinas. Brain Res 1046(1–2):207–215. https://doi.org/10.1016/j.brainres.2005.04.016

Donkor IO (2011) Calpain inhibitors: a survey of compounds reported in the patent and scientific literature. Expert Opin Ther Pat 21(5):601–636

Downward J (1999) How BAD phosphorylation is good for survival. Nat Cell Biol 1(2):E33–E35

Du K, Montminy M (1998) CREB is a regulatory target for the protein kinase Akt/PKB. J Biol Chem 273(49):32377–32379

Forman OP, De Risio L, Mellersh CS (2013) Missense mutation in CAPN1 is associated with spinocerebellar ataxia in the Parson Russell Terrier dog breed. PLoS One 8(5):e64627. https://doi.org/10.1371/journal.pone.0064627

Frey U, Morris RG (1998) Synaptic tagging: implications for late maintenance of hippocampal long-term potentiation. Trends Neurosci 21(5):181–188

Gan-Or Z, Bouslam N, Birouk N, Lissouba A, Chambers DB, Veriepe J, Androschuck A, Laurent SB, Rochefort D, Spiegelman D, Dionne-Laporte A, Szuto A, Liao M, Figlewicz DA, Bouhouche A, Benomar A, Yahyaoui M, Ouazzani R, Yoon G, Dupre N, Suchowersky O, Bolduc FV, Parker JA, Dion PA, Drapeau P, Rouleau GA, Bencheikh BO (2016) Mutations in CAPN1 cause autosomal-recessive hereditary spastic paraplegia. Am J Hum Genet 98(5):1038–1046. https://doi.org/10.1016/j.ajhg.2016.04.002

Gao T, Furnari F, Newton AC (2005) PHLPP: a phosphatase that directly dephosphorylates Akt, promotes apoptosis, and suppresses tumor growth. Mol Cell 18(1):13–24

Hamakubo T, Kannagi R, Murachi T, Matus A (1986) Distribution of calpains I and II in rat brain. J Neurosci 6(11):3103–3111

Hardingham GE, Bading H (2010) Synaptic versus extrasynaptic NMDA receptor signalling: implications for neurodegenerative disorders. Nat Rev Neurosci 11(10):682–696

Hardingham GE, Arnold FJ, Bading H (2001) Nuclear calcium signaling controls CREB-mediated gene expression triggered by synaptic activity. Nat Neurosci 4(3):261–267

Hashimoto K, Fukaya M, Qiao X, Sakimura K, Watanabe M, Kano M (1999) Impairment of AMPA receptor function in cerebellar granule cells of ataxic mutant mouse stargazer. J Neurosci 19(14):6027–6036

Hong SC, Goto Y, Lanzino G, Soleau S, Kassell NF, Lee KS (1994) Neuroprotection with a calpain inhibitor in a model of focal cerebral ischemia. Stroke 25(3):663–669

Jackson TC, Verrier JD, Semple-Rowland S, Kumar A, Foster TC (2010) PHLPP1 splice variants differentially regulate AKT and PKCα signaling in hippocampal neurons: characterization of PHLPP proteins in the adult hippocampus. J Neurochem 115(4):941–955

Kim AH, Khursigara G, Sun X, Franke TF, Chao MV (2001) Akt phosphorylates and negatively regulates apoptosis signal-regulating kinase 1. Mol Cell Biol 21(3):893–901

Kim JC, Cook MN, Carey MR, Shen C, Regehr WG, Dymecki SM (2009) Linking genetically defined neurons to behavior through a broadly applicable silencing allele. Neuron 63(3):305–315. https://doi.org/10.1016/j.neuron.2009.07.010

Kobeissy FH, Liu MC, Yang Z, Zhang Z, Zheng W, Glushakova O, Mondello S, Anagli J, Hayes RL, Wang KK (2015) Degradation of βII-Spectrin protein by Calpain-2 and Caspase-3 under neurotoxic and traumatic brain injury conditions. Mol Neurobiol 52(1):696–709

Koumura A, Nonaka Y, Hyakkoku K, Oka T, Shimazawa M, Hozumi I, Inuzuka T, Hara H (2008) A novel calpain inhibitor,((1S)-1 ((((1S)-1-benzyl-3-cyclopropylamino-2, 3-di-oxopropyl) amino) carbonyl)-3-methylbutyl) carbamic acid 5-methoxy-3-oxapentyl ester, protects neuronal cells from cerebral ischemia-induced damage in mice. Neuroscience 157(2):309–318

Krapivinsky G, Krapivinsky L, Manasian Y, Ivanov A, Tyzio R, Pellegrino C, Ben-Ari Y, Clapham DE, Medina I (2003) The NMDA receptor is coupled to the ERK pathway by a direct interaction between NR2B and RasGRF1. Neuron 40(4):775–784

Li PA, Howlett W, He QP, Miyashita H, Siddiqui M, Shuaib A (1998) Postischemic treatment with calpain inhibitor MDL 28170 ameliorates brain damage in a gerbil model of global ischemia. Neurosci Lett 247(1):17–20

Li D, Qu Y, Mao M, Zhang X, Li J, Ferriero D, Mu D (2009) Involvement of the PTEN-AKT-FOXO3a pathway in neuronal apoptosis in developing rat brain after hypoxia-ischemia. J Cereb Blood Flow Metab 29(12):1903–1913. https://doi.org/10.1038/jcbfm.2009.102

Liu J, Liu MC, Wang K (2008) Calpain in the CNS: from synaptic function to neurotoxicity. Sci Signal 1(14):re1

Liu J, Weiss HL, Rychahou P, Jackson LN, Evers BM, Gao T (2009) Loss of PHLPP expression in colon cancer: role in proliferation and tumorigenesis. Oncogene 28(7):994–1004

Liu S, Yin F, Zhang J, Qian Y (2014) The role of calpains in traumatic brain injury. Brain Inj 28(2):133–137

Mao L, Jia J, Zhou X, Xiao Y, Wang Y, Mao X, Zhen X, Guan Y, Alkayed NJ, Cheng J (2013) Delayed administration of a PTEN inhibitor BPV improves functional recovery after experimental stroke. Neuroscience 231:272–281. https://doi.org/10.1016/j.neuroscience.2012.11.050

Markgraf CG, Velayo NL, Johnson MP, McCarty DR, Medhi S, Koehl JR, Chmielewski PA, Linnik MD (1998) Six-hour window of opportunity for calpain inhibition in focal cerebral ischemia in rats. Stroke 29(1):152–158

Masubuchi S, Gao T, O'Neill A, Eckel-Mahan K, Newton AC, Sassone-Corsi P (2010) Protein phosphatase PHLPP1 controls the light-induced resetting of the circadian clock. Proc Natl Acad Sci U S A 107(4):1642–1647. https://doi.org/10.1073/pnas.0910292107

Mingorance-Le Meur A, O'Connor TP (2009) Neurite consolidation is an active process requiring constant repression of protrusive activity. EMBO J 28(3):248–260

Monti B, Contestabile A (2000) Blockade of the NMDA receptor increases developmental apoptotic elimination of granule neurons and activates caspases in the rat cerebellum. Eur J Neurosci 12(9):3117–3123

Monti B, Marri L, Contestabile A (2002) NMDA receptor-dependent CREB activation in survival of cerebellar granule cells during in vivo and in vitro development. Eur J Neurosci 16(8):1490–1498

Moran J, Patel AJ (1989) Stimulation of the N-methyl-D-aspartate receptor promotes the biochemical differentiation of cerebellar granule neurons and not astrocytes. Brain Res 486(1):15–25

Papadia S, Stevenson P, Hardingham NR, Bading H, Hardingham GE (2005) Nuclear Ca2+ and the cAMP response element-binding protein family mediate a late phase of activity-dependent neuroprotection. J Neurosci 25(17):4279–4287

Papouin T, Oliet SH (2014) Organization, control and function of extrasynaptic NMDA receptors. Philos Trans R Soc B 369(1654):20130601

Paquet-Durand F, Johnson L, Ekström P (2007) Calpain activity in retinal degeneration. J Neurosci Res 85(4):693–702

Pennacchio LA, Bouley DM, Higgins KM, Scott MP, Noebels JL, Myers RM (1998) Progressive ataxia, myoclonic epilepsy and cerebellar apoptosis in cystatin B-deficient mice. Nat Genet 20(3):251–258. https://doi.org/10.1038/3059

Perkinton MS, Ip J, Wood GL, Crossthwaite AJ, Williams RJ (2002) Phosphatidylinositol 3-kinase is a central mediator of NMDA receptor signalling to MAP kinase (Erk1/2), Akt/PKB and CREB in striatal neurones. J Neurochem 80(2):239–254

Perlmutter LS, Siman R, Gall C, Seubert P, Baudry M, Lynch G (1988) The ultrastructural localization of calcium-activated protease "calpain" in rat brain. Synapse 2(1):79–88

Saavedra A, Garcia-Martinez J, Xifro X, Giralt A, Torres-Peraza J, Canals J, Diaz-Hernandez M, Lucas J, Alberch J, Perez-Navarro E (2010) PH domain leucine-rich repeat protein phosphatase 1 contributes to maintain the activation of the PI3K/Akt pro-survival pathway in Huntington's disease striatum. Cell Death Differ 17(2):324–335

Schoch KM, Evans HN, Brelsfoard JM, Madathil SK, Takano J, Saido TC, Saatman KE (2012) Calpastatin overexpression limits calpain-mediated proteolysis and behavioral deficits following traumatic brain injury. Exp Neurol 236(2):371–382

Shimazawa M, Suemori S, Inokuchi Y, Matsunaga N, Nakajima Y, Oka T, Yamamoto T, Hara H (2010) A novel calpain inhibitor, ((1S)-1-((((1S)-1-Benzyl-3-cyclopropylamino-2,3-di-oxopropyl)amino)carbonyl)-3-me thylbutyl)carbamic acid 5-methoxy-3-oxapentyl ester (SNJ-1945), reduces murine retinal cell death in vitro and in vivo. J Pharmacol Exp Ther 332(2):380–387. https://doi.org/10.1124/jpet.109.156612

Shimizu K, Okada M, Nagai K, Fukada Y (2003) Suprachiasmatic nucleus circadian oscillatory protein, a novel binding partner of K-Ras in the membrane rafts, negatively regulates MAPK pathway. J Biol Chem 278(17):14920–14925

Shimizu K, Phan T, Mansuy IM, Storm DR (2007) Proteolytic degradation of SCOP in the hippocampus contributes to activation of MAP kinase and memory. Cell 128(6):1219–1229

Shmerling D, Hegyi I, Fischer M, Blattler T, Brandner S, Gotz J, Rulicke T, Flechsig E, Cozzio A, von Mering C, Hangartner C, Aguzzi A, Weissmann C (1998) Expression of amino-terminally truncated PrP in the mouse leading to ataxia and specific cerebellar lesions. Cell 93(2):203–214

Siklos M, BenAissa M, Thatcher GR (2015) Cysteine proteases as therapeutic targets: does selectivity matter? A systematic review of calpain and cathepsin inhibitors. Acta Pharm Sin B 5(6):506–519

Simonson L, Baudry M, Siman R, Lynch G (1985) Regional distribution of soluble calcium activated proteinase activity in neonatal and adult rat brain. Brain Res 327(1–2):153–159

Soriano FX, Papadia S, Hofmann F, Hardingham NR, Bading H, Hardingham GE (2006) Preconditioning doses of NMDA promote neuroprotection by enhancing neuronal excitability. J Neurosci 26(17):4509–4518

Steward O, Wallace CS (1995) mRNA distribution within dendrites: relationship to afferent innervation. J Neurobiol 26(3):447–459

Thompson SN, Carrico KM, Mustafa AG, Bains M, Hall ED (2010) A pharmacological analysis of the neuroprotective efficacy of the brain-and cell-permeable calpain inhibitor MDL-28170 in the mouse controlled cortical impact traumatic brain injury model. J Neurotrauma 27(12):2233–2243

Tovar KR, Westbrook GL (1999) The incorporation of NMDA receptors with a distinct subunit composition at nascent hippocampal synapses in vitro. J Neurosci 19(10):4180–4188

Tsubokawa T, Solaroglu I, Yatsushige H, Cahill J, Yata K, Zhang JH (2006) Cathepsin and calpain inhibitor E64d attenuates matrix metalloproteinase-9 activity after focal cerebral ischemia in rats. Stroke 37(7):1888–1894. https://doi.org/10.1161/01.STR.0000227259.15506.24

Vosler P, Brennan C, Chen J (2008) Calpain-mediated signaling mechanisms in neuronal injury and neurodegeneration. Mol Neurobiol 38(1):78–100

Wang C-F, Huang Y-S (2012) Calpain 2 activated through N-methyl-D-aspartic acid receptor signaling cleaves CPEB3 and abrogates CPEB3-repressed translation in neurons. Mol Cell Biol 32(16):3321–3332

Wang JT, Medress ZA, Barres BA (2012a) Axon degeneration: molecular mechanisms of a self-destruction pathway. J Cell Biol 196(1):7–18

Wang Y-B, Wang J-J, Wang S-H, Liu S-S, Cao J-Y, Li X-M, Qiu S, Luo J-H (2012b) Adaptor protein APPL1 couples synaptic NMDA receptor with neuronal prosurvival phosphatidylinositol 3-kinase/Akt pathway. J Neurosci 32(35):11919–11929

Wang Y, Briz V, Chishti A, Bi X, Baudry M (2013) Distinct roles for mu-calpain and m-calpain in synaptic NMDAR-mediated neuroprotection and extrasynaptic NMDAR-mediated neurodegeneration. J Neurosci 33(48):18880–18892. https://doi.org/10.1523/JNEUROSCI.3293-13.2013

Wang Y, Zhu G, Briz V, Hsu YT, Bi X, Baudry M (2014) A molecular brake controls the magnitude of long-term potentiation. Nat Commun 5:3051. https://doi.org/10.1038/ncomms4051

Wang Y, Hersheson J, Lopez D, Hammer M, Liu Y, Lee KH, Pinto V, Seinfeld J, Wiethoff S, Sun J, Amouri R, Hentati F, Baudry N, Tran J, Singleton AB, Coutelier M, Brice A, Stevanin G, Durr A, Bi X, Houlden H, Baudry M (2016a) Defects in the CAPN1 gene result in alterations in cerebellar development and cerebellar ataxia in mice and humans. Cell Rep 16(1):79–91. https://doi.org/10.1016/j.celrep.2016.05.044

Wang Y, Lopez D, Davey PG, Cameron DJ, Nguyen K, Tran J, Marquez E, Liu Y, Bi X, Baudry M (2016b) Calpain-1 and calpain-2 play opposite roles in retinal ganglion cell degeneration induced by retinal ischemia/reperfusion injury. Neurobiol Dis 93:121–128. https://doi.org/10.1016/j.nbd.2016.05.007

Wang, Y, Liu, Y, Lopez, D, Lee, M, Dayal, S, Hirtado, A, Bi, X and Baudry, M (2017) Protection against TBI-induced neuronal death with post-treatment with a selective calpain-2 inhibitor in mice. J Neurotrauma 34:1–13

Xiong Y, Mahmood A, Chopp M (2013) Animal models of traumatic brain injury. Nat Rev Neurosci 14(2):128–142

Xu J, Kurup P, Zhang Y, Goebel-Goody SM, Wu PH, Hawasli AH, Baum ML, Bibb JA, Lombroso PJ (2009) Extrasynaptic NMDA receptors couple preferentially to excitotoxicity via calpain-mediated cleavage of STEP. J Neurosci 29(29):9330–9343. https://doi.org/10.1523/JNEUROSCI.2212-09.2009

Yamaguchi A, Tamatani M, Matsuzaki H, Namikawa K, Kiyama H, Vitek MP, Mitsuda N, Tohyama M (2001) Akt activation protects hippocampal neurons from apoptosis by inhibiting transcriptional activity of p53. J Biol Chem 276(7):5256–5264

Yan X-X, Jeromin A (2012) Spectrin breakdown products (SBDPs) as potential biomarkers for neurodegenerative diseases. Curr Trans Geriatr Exp Gerontol Rep 1(2):85–93

Yildiz-Unal A, Korulu S, Karabay A (2015) Neuroprotective strategies against calpain-mediated neurodegeneration. Neuropsychiatr Dis Treat 11:297

Zadran S, Jourdi H, Rostamiani K, Qin Q, Bi X, Baudry M (2010) Brain-derived neurotrophic factor and epidermal growth factor activate neuronal m-calpain via mitogen-activated protein kinase-dependent phosphorylation. J Neurosci 30(3):1086–1095

Zhou M, Baudry M (2006) Developmental changes in NMDA neurotoxicity reflect developmental changes in subunit composition of NMDA receptors. J Neurosci 26(11):2956–2963

Zhu G, Liu Y, Wang Y, Bi X, Baudry M (2015) Different patterns of electrical activity lead to long-term potentiation by activating different intracellular pathways. J Neurosci 35(2):621–633. https://doi.org/10.1523/JNEUROSCI.2193-14.2015

Part II
Traumatic Brain Injury

Chapter 3
Oxidative Damage Mechanisms in Traumatic Brain Injury and Antioxidant Neuroprotective Approaches

Edward D. Hall, Indrapal N. Singh, and John E. Cebak

Abstract This chapter reviews our current knowledge of the role of oxidative damage mechanisms and pharmacological antioxidant neuroprotective strategies for inhibiting reactive oxygen species (ROS) and reactive nitrogen species (RNS)-mediated secondary injury following traumatic brain injury (TBI). First of all, the chemistry of the main forms of oxidative damage: lipid peroxidation, carbonylation and nitration are presented as well as the interactions of oxidative damage with other secondary injury mechanisms including glutamate-mediated excitotoxicity, intracellular calcium overload and mitochondrial dysfunction. Secondly, the general mechanistic approaches to interrupting oxidative damage are presented: decreasing ROS/RNS formation or scavenging ROS and RNS-derived radicals, inhibition of lipid peroxidation propagation, chelation of iron, which is a potent catalyst of lipid peroxidation reactions, scavenging of neurotoxic aldehydic lipid peroxidation products ('carbonyls'), and enhancement of the expression of the pleiotopic Nrf2-antioxidant response element (ARE) pathway that controls the synthesis of several endogenous antioxidant enzymes and chemical antioxidants. Pharmacological examples of compounds that effectively inhibit oxidative damage and produce neuroprotective effects in animal TBI models by each of these various approaches are presented. Finally, the results of large phase III clinical trials with the either the radical scavenger polyethylene glycol-coupled superoxide dismutase (PEG-SOD) or the 21-aminosteroid lipid peroxidation inhibitor tirilazad are revisited in which the latter compound was found to selectively improve survival after moderate and severe TBI, particularly in male patients, suggesting that successful clinical translation of neuroprotective antioxidant compounds, or combinations of mechanistically complimentary antioxidants, should be possible.

E. D. Hall (✉) · I. N. Singh
Spinal Cord & Brain Injury Research Center (SCoBIRC) and Department of Neuroscience, University of Kentucky Medical Center, Lexington, KY, USA
e-mail: edhall@uky.edu; ising2@uky.edu

J. E. Cebak
Lincoln Memorial University DeBusk College of Osteopathic Medicine, Harrogate, TN, USA

D. G. Fujikawa (ed.), *Acute Neuronal Injury*,
https://doi.org/10.1007/978-3-319-77495-4_3

Keywords Glutamate excitotoxicity · Calcium overload · Reactive oxygen species · Lipid peroxidation · Antioxidant · Neuroprotection · Traumatic brain injury

3.1 Introduction

At present, there are no FDA-approved pharmacological therapies for acute treatment of traumatic brain injury (TBI) patients that are conclusively proven to mitigate the often devastating neurological effects of their injuries. However, the possibility of discovering and developing an effective 'neuroprotective' treatment that will limit posttraumatic brain damage and improve neurological recovery is based upon the fact that even though some of the neural injury is due to the primary mechanical injury to the parenchymal neurons, glia and vascular elements, the majority of post-traumatic neurodegeneration is due to a pathophysiological secondary injury cascade triggered initially by massive release of glutamate and its excitotoxic effects that occur during the first minutes, hours and days following the injury, which exacerbates the damaging effects of the primary injury. One of the most validated "secondary injury" mechanisms, as revealed in experimental TBI studies, that contributes to glutamate-mediated excitotoxic neurodegeneration, involves the downstream increase in reactive oxygen species (ROS) that cause oxygen radical-induced oxidative damage to brain cell lipids and proteins. This chapter outlines the key sources of reactive oxygen species (ROS), including highly reactive (i.e. rapidly oxidizing) free radicals, the pathophysiological mechanisms associated with oxidative neural damage and pharmacological antioxidants that have been shown to produce neuroprotective effects that limit excitotoxic neurodegeneration in preclinical TBI models, one of which has revealed some evidence of neuroprotective efficacy in a major pathological subset of TBI patients in a large phase III clinical trial.

3.2 Reactive Oxygen Species and Reactive Nitrogen Species

The term reactive oxygen species (ROS) includes oxygen-derived radicals such as the modestly reactive superoxide radical ($O_2^{\bullet-}$) and the highly reactive hydroxyl ($OH^{\bullet}$) radical as well as non-radicals such as hydrogen peroxide (H_2O_2) and peroxynitrite ($ONOO^-$), the latter often referred to as a reactive nitrogen species (RNS). The cascade of posttraumatic oxygen radical reactions begins in response to the primary mechanical injury triggering neuronal depolarization, due to the voltage-dependent opening of sodium (Na^+) and calcium (Ca^{2+}) channels, which causes a massive increase in intracellular Ca^{2+} that stimulates rapid elevations in extracellular glutamate levels that excessively stimulates N-methyl-aspartate (NMDA) glutamate receptors, causing a further exacerbation of the injury-induced increase in intracellular Ca^{2+}. This voltage-dependent and glutamate receptor-mediated intracellular Ca^{2+} overload initiates multiple downstream neurodegenerative processes, one of which is the increased generation of oxygen free radicals that initiate oxidative damage to brain cell phospholipid membranes and proteins. The primordial oxygen free radical that comes from several pathophysiological sources involves the single electron (e^-) reduction of an oxygen molecule (O_2) to produce superoxide ($O_2^{\bullet-}$). Superoxide can be generated from several sources; one of the main ones is $O_2^{\bullet-}$ leakage from complex I of the mitochondrial electron transport chain in Ca^{2+}-overloaded brain mitochondria. However, $O_2^{\bullet-}$ is considered by many free radical chemists and biologists to be a modestly reactive radical that can potentially react with other molecules to give rise to much more reactive, and thus more potentially damaging, radical species. The reason that $O_2^{\bullet-}$ is only modestly reactive is that it can act as either an oxidant by stealing an electron from another oxidizable molecule or it can act as a reductant by which it donates its unpaired electron to another radical species, thus acting as an antioxidant. However, if $O_2^{\bullet-}$ reacts with a proton (H^+) to form a hydroperoxyl radical ($HO\bullet_2$) this results in a superoxide form that is much more likely to cause oxidation (i.e. act as an electron stealer).

One of the most important endogenous antioxidants is the enzyme superoxide dismutase (SOD) which rapidly catalyzes the dismutation of $O_2^{\bullet-}$ into H_2O_2 and oxygen. At low pH, $O_2^{\bullet-}$ can dismutate spontaneously. The formation of highly reactive oxygen radicals, which have unpaired electron(s) in their outer molecular orbitals, and the propagation of chain reactions are fueled by non-radical ROS, which do not have unpaired electron(s), but are chemically reactive. For example, $OH^\bullet$ radicals are generated in the iron-catalyzed Fenton reaction, where ferrous iron (Fe^{2+}) is oxidized to form $OH^\bullet$ in the presence of H_2O_2 ($Fe^{2+} + H_2O_2 \rightarrow Fe^{3+} + OH^\bullet + OH^-$). Superoxide, acting as a reducing agent (i.e. an electron donor), can actually donate its unpaired electron to ferric iron (Fe^{3+}), cycling it back to the ferrous state in the Haber-Weiss reaction ($O_2^{\bullet-} + Fe^{3+} \rightarrow Fe^{2+} + O_2$), thus driving subsequent Fenton reactions and increased production of $OH^\bullet$. Under physiological conditions, iron is tightly regulated by its transport protein, transferrin and storage protein, ferritin, both of which bind the ferric (Fe^{3+}) form. This reversible bond of transferrin and ferritin with iron decreases with declining pH (below pH 7). Indeed, tissue acidosis is known to occur in the traumatized CNS that will cause the release of iron and initiation of iron-dependent oxygen radical production. A second source of iron comes from hemoglobin released into the blood during injury-induced hemorrhage.

Although $O_2^{\bullet-}$ is much less reactive than $OH^\bullet$ radical, its reaction with nitric oxide ($NO^\bullet$) radical forms the highly reactive oxidizing agent, peroxynitrite (PN: $ONOO^-$).This reaction ($O_2^{\bullet-} + NO^\bullet \rightarrow ONOO^-$) occurs at a very high rate constant that out competes SOD's ability to convert $O_2^{\bullet-}$ into H_2O_2. Subsequently, at physiological pH, $ONOO^-$ will largely undergo protonation to form peroxynitrous acid (ONOOH) or it can react with carbon dioxide (CO_2) to form nitrosoperoxycarbonate ($ONOOCO_2^-$). The ONOOH can break down to form highly reactive nitrogen dioxide ($NO\bullet_2$) and $OH^\bullet$ ($ONOOH \rightarrow NO^\bullet_2 + OH^\bullet$). Alternatively, the $ONOOCO_2^-$ can decompose into $NO^\bullet_2$ and carbonate radical ($CO_3^{\bullet-}$) ($ONOOCO_2^- \rightarrow NO\bullet_2 + CO_3^{\bullet-}$).

3.3 Lipid Peroxidation

Increased production of reactive free radicals (i.e. "oxidative stress") in the injured brain has been shown to cause "oxidative damage" to cellular lipids and proteins, leading to functional compromise and cell death in both the microvascular and brain parenchymal compartments. Extensive study shows that a major form of radical-induced oxidative damage involves oxidative attack on cell membrane polyunsaturated fatty acids, triggering the process of lipid peroxidation (LP) that is characterized by three distinct steps: initiation, propagation and termination (Gutteridge 1995), which are shown in Fig. 3.1 in the context of radical-induced LP of arachidonic acid. **Initiation:** LP is initiated when highly reactive oxygen radicals (e.g. OH•, NO_2•, $CO_3^{\bullet -}$) react with polyunsaturated fatty acids such as arachidonic acid (AA), linoleic acid (LA), eicosapentaenoic acid (EPA) or docosahexaenoic acid (DHA), resulting in disruptions in cellular and subcellular membrane integrity. Initiation of LP begins when ROS-induced hydrogen atom (H^+) and its one associated electron is abstracted from an allylic carbon. The basis for the susceptibility of the allylic carbon of the polyunsaturated fatty acid having one of its electrons stolen by a highly electrophilic free radical is that the carbon is surrounded by two relatively electronegative double bonds which tend to pull one of the electrons from the carbon. Consequently, a reactive free radical has an easy time pulling the hydrogen electron off of the carbon because the commitment of the carbon electron to staying

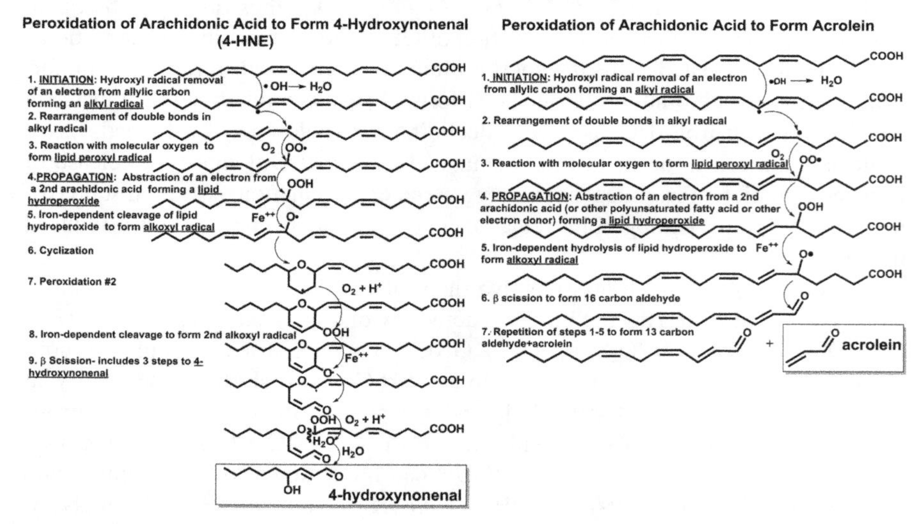

Fig. 3.1 Biochemistry involved in the initiation, propagation and termination reactions of arachidonic acid during lipid peroxidation, with the resulting formation of the aldehydic end-products 4-hydroxynonenal (4-HNE) and acrolein

paired with it has been weakened by the surrounding electronegative double bonds. This results in the original radical being quenched while the polyunsaturated fatty acid (L) becomes a lipid radical (L•) due to its having lost an electron. **Propagation:** Subsequently, in the propagation step, the unstable L• reacts with O_2 to form a lipid peroxyl radical (LOO•). The LOO• in turn extracts a hydrogen atom from an adjacent polyunsaturated fatty acid, yielding a lipid hydroperoxide (LOOH) and a second L•, which sets off a series of propagation "chain" reactions.

Termination: These propagation reactions are terminated in the third step, when the substrate becomes depleted and a lipid radical reacts with another radical to yield potentially neurotoxic non-radical aldehydic end products. One of those end-products that is often used to measure LP is the three carbon-containing malondialdehyde (MDA) which is mainly a stable non-toxic compound that when measured represents a LP 'tombstone'. In contrast, two highly neurotoxic aldehydic products of LP (commonly referred to as 'carbonyls') are 4-hydroxynonenal (4-HNE) or 2-propenal (acrolein), both of which have been well characterized in CNS injury experimental models (Bains and Hall 2012; Hall et al. 2010; Hamann and Shi 2009). These latter two aldehydic LP end products covalently bind to proteins and amino acids (lysine, histidine, arginine) by either Schiff base or Michael addition reactions altering their structure and functional properties. Immunohistochemical and immunoblotting (western, slot, dot) techniques are commonly used to measure 4-HNE or acrolein-modified proteins (i.e. 'protein carbonyls') in the injured brain (Hall and Bosken 2009).

3.4 Free Radical-Induced Protein Carbonylation and Nitration

Free radicals can cause various forms of oxidative protein damage. Firstly, a major mechanism involves carbonylation by reaction of various free radicals with susceptible amino acids such as arginine, lysine and histidine. The protein carbonyls thus formed are measurable through immunoblotting after derivatization of the carbonyl groups with diphenylhydrazine (DNPH). Indeed, the measurement of protein carbonyls by the so-called DNPH assay has long been used to measure free radical-induced protein oxidation. However, the carbonyl assay also picks up protein carbonyls that are present due to covalent binding of LP-derived 4-HNE and acrolein to cysteine residues, in addition to those resulting from direct free radical-induced amino acid oxidation. Thus, as a result, the carbonyl assay is as much an indirect index of LP as it is of direct radical-induced protein oxidation.

Secondly, $NO•_2$ can nitrate the three position of aromatic amino acids tyrosine or phenylalanine in proteins; 3-NT is a specific footprint of PN-induced cellular damage. Similarly, lipid peroxyl radicals (LOO•) can promote nitration of aromatic amino acids by producing initial oxidation (i.e. loss of an electron) which would enhance the ability of $NO•_2$ to nitrate the phenyl ring. Multiple commercially available polyclonal and monoclonal antibodies are available for immunoblot or immunohistochemical measurement of proteins that have been nitrated by PN.

3.5 Interaction of Oxidative Damage with Other Secondary Injury Mechanisms

The impact of ROS/RNS production is heightened when oxygen radicals feed back and amplify other secondary injury pathways creating a continuous cycle of ion imbalance, Ca^{2+} buffering impairment, mitochondrial dysfunction, glutamate-induced excitotoxicity and microvascular disruption. One example of ROS-induced ionic disruption arises from LP-induced damage to the plasma membrane ATP-driven Ca^{2+} pump (Ca^{2+}-ATPase) and Na^+ pump (Na^+/K^+-ATPase), which contributes to increases in intracellular Ca^{2+} concentrations, mitochondrial dysfunction and additional ROS production. Both Ca^{2+}-ATPase and Na^+/K^+-ATPase disruptions result in further increases in intracellular Ca^{2+} and Na^+ accumulation respectively (Bains and Hall 2012), the latter causing reversal of the Na^+/Ca^{2+} exchanger which further exacerbates intracellular Ca^{2+} (Rohn et al. 1993, 1996). As already noted above, PN formed from mitochondrial Ca^{2+} overload also contributes to mitochondrial dysfunction. Specifically, nitric oxide (NO•), formed from mitochondrial NOS, which in turn reacts with $O_2^{\bullet-}$ to produce the highly toxic PN, which impairs respiratory and Ca^{2+} buffering capacity via its derived free radicals (Bringold et al. 2000). Indeed increased PN-derived 3NT and 4HNE has been detected during the time of mitochondrial dysfunction and correlates with respiratory and Ca^{2+} buffering impairment (Sullivan et al. 2007). Increased synaptosomal 4-HNE content is associated with impaired synaptosomal glutamate and amino acid uptake (Carrico et al. 2009; Zhang et al. 1996). Glutamate and NMDA- induced damage in neuronal cultures is attenuated with LP inhibition, confirming LP and oxidative damage as promoters of glutamate excitotoxicity (Monyer et al. 1990; Pellegrini-Giampietro et al. 1990).

3.6 Mechanisms for Pharmacological Inhibition of Oxidative Damage

Based upon the discussion above concerning oxidative stress (increased ROS/RNS) and oxidative damage (LP, protein oxidation and nitration), a number of potential mechanisms for its inhibition are apparent which fall into five categories. The first category includes compounds that inhibit the initiation of LP and other forms of oxidative damage by **attenuating the formation of ROS or RNS species**. For instance, nitric oxide synthase (NOS) inhibitors exert an indirect antioxidant effect by limiting NO• production and thus PN formation. However, they also have the potential to interfere with the physiological roles that $NO^{\bullet}$ is responsible for, including antioxidant effects which are due to its important role as a scavenger of lipid peroxyl radicals (e.g. $LOO^{\bullet} + NO^{\bullet} \rightarrow LOONO$) (Hummel et al. 2006). Another approach to blocking posttraumatic radical formation is the inhibition of the enzymatic (e.g. cyclooxygenase, 5-lipoxygenases) arachidonic acid (AA) cascade during which $O_2^{\bullet-}$ is produced as a by-product of prostanoid and leukotriene synthesis. Kontos and colleagues (Kontos 1989; Kontos and Wei 1986) and Hall (1986) have shown that cyclooxygenase-inhibiting non-steroidal anti-inflammatory agents (e.g. indomethacin, ibuprofen) are vaso- and neuro-protective in TBI models.

Another example of an indirect approach for reducing the formation of ROS/RNS in the injured brain is via the inhibition of brain mitochondrial functional failure with the drug cyclosporine A which has been shown to reduce mitochondrial permeability transition pore (mPTP) formation by blocking cyclophilin D interaction with other components of the pathological pore which has been shown to lessen mitochondrial free radical formation and consequently attenuate LP and nitrative mitochondrial protein oxidative damage (Mbye et al. 2008; Sullivan et al. 1999).

A second indirect LP inhibitory approach involves **chemically scavenging the radical species** (e.g. $O_2^{\bullet-}$, $OH^{\bullet}$, $NO\bullet_2$, $CO_3^{\bullet-}$) before they have a chance to steal an electron from a polyunsaturated fatty acid and thus initiate LP. The use of pharmacologically-administered SOD represents an example of this strategy. Another example concerns the use of the nitroxide antioxidant tempol which has been shown to catalytically scavenge the PN-derived free radicals $NO\bullet_2$ and $CO_3\bullet^-$ (Carroll et al. 2000). In either case, a general limitation to these first two approaches and antioxidant agents that work by this mechanism is that they would be expected to have a short therapeutic window and would have to be administered rapidly in order to have a chance to interfere with the initial posttraumatic "burst" of free radical production that has been documented in TBI models (Kontos and Wei 1986; Hall et al. 1993). While it is believed that ROS, including PN production, persists several hours after injury, the major portion is an early event that peaks in the first 60 min after injury, making it clinically impractical to pharmacologically inhibit, unless the antioxidant compound is already "on board" when the TBI occurs (Fig. 3.2).

In contrast to the above indirect-acting antioxidant mechanisms, the third category involves stopping the "chain reaction" propagation of LP once it has begun. The most demonstrated way to accomplish this is by **scavenging of lipid peroxyl (LOO•) radicals**. The prototype scavenger of these lipid radicals is alpha tocopherol or vitamin E (Vit E) which can donate an electron from its phenolic hydroxyl moiety to quench a $LOO^{\bullet}$. However, the scavenging process is stoichiometric (1 Vit E can only quench 1 $LOO^{\bullet}$) and in the process vitamin E loses its antioxidant efficacy and becomes Vitamin E radical ($LOO^{\bullet}$ + Vit E $\rightarrow$ LOOH + Vit $E^{\bullet}$). Although Vit $E^{\bullet}$ is relatively unreactive (i.e. harmless), it also cannot scavenge another $LOO^{\bullet}$ until it is reduced back to its active form by receiving an electron from other endogenous antioxidant reducing agents such as ascorbic acid (Vitamin C) or glutathione (GSH). While this tripartite $LOO^{\bullet}$ antioxidant defense system (Vit E, Vit C, GSH) works fairly effectively in the absence of a major oxidative stress, numerous studies have shown that each of these antioxidants are rapidly consumed during the early

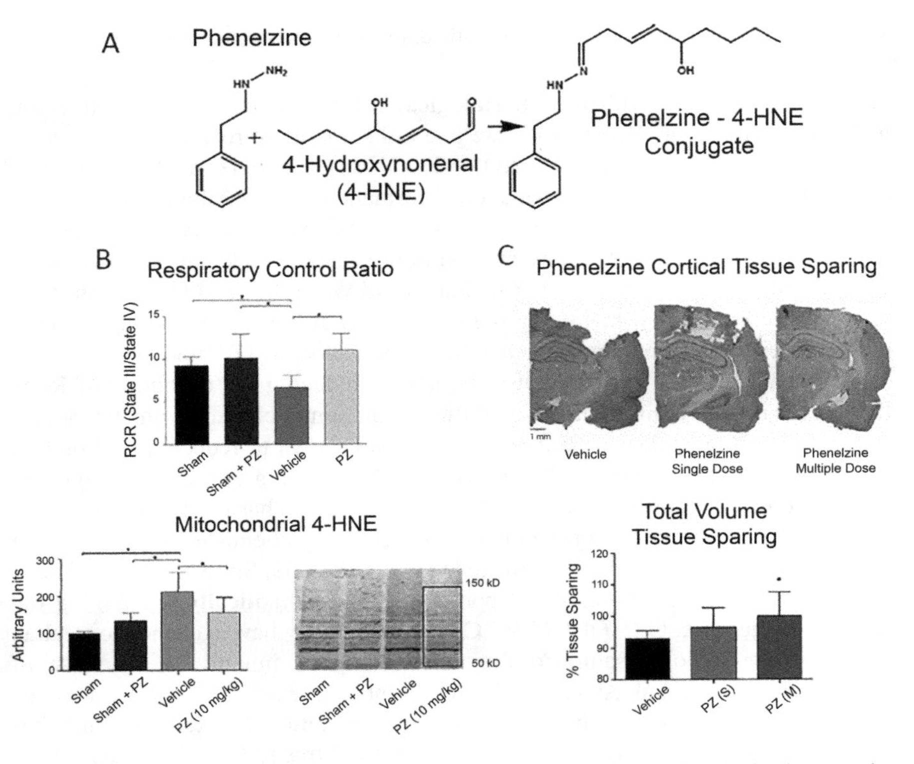

Fig. 3.2 (**a**) Chemical scavenging mechanism involved in the reactivity of the hydrazine-containing compound phenelzine with 4-HNE. (**b**) (Top): Effects of repeated phenelzine (PZ: 10 mg/kg s.c. 15 min after injury followed by maintenance dosing (5 mg/kg s.c.) every 12 h) on cortical mitochondrial bioenergetics 72 h following severe controlled cortical impact TBI. Mitochondrial respiration was measured with a Clark-type electrode expressed as respiratory control ratio (RCR). The RCR is rate of oxygen consumption during State III divided by State IV respiration. Animals received PZ rats were euthanized at 72 h. Sham and Sham + PZ groups were significantly different compared to Vehicle groups. RCR of PZ treatment was significantly increased compared to vehicle and not significantly different from either sham control variant (Sham or Sham + PZ). One-way ANOVA ($F = 7.7$, $df = 3, 24$, $P < 0.009$) followed by Student Newman-Keuls post-hoc test. $*p < 0.05$. Error bars represent ±SD; n = 8–9 rats per group except sham where n = 5 rats per group. (**b**) (Bottom): Repeated PZ reduces 4-hydroxynonenal (4-HNE) accumulation in mitochondria 72 h after TBI. As revealed by quantitative western blot (see sample blot and bar graph), 4-HNE-modified proteins were significantly elevated in the Vehicle group compared to both Sham groups. PZ treatment group exhibited significantly reduced oxidative damage compared to Vehicle group, but did not return to Sham levels. ANOVA ($F = 9.9$, $df = 3, 24$, $p < 0.0002$) followed by Student Newman-Keuls post-hoc test. $*p < 0.05$. (**c**) Repeated PZ reduces cortical neurodegeneration 72 h after TBI: Coronal sections of ipsilateral rat brains rat taken at 1.2× magnification. **Left**: Vehicle (0.9% saline) treated rat brain injected 15 min after TBI; **Center**: Phenelzine (PZs) single 10 mg/kg s.c. dose treated animal; **Right**: Rat brain of PZ-treated with a multiple dosing paradigm (PZm): single subcutaneous injection of PZ 15 min after injury, followed by maintenance dosing of 5 mg/kg every 12 h thereafter. All groups (Vehicle, PZ(S), PZ(M)) were euthanized 72 h after first injection. Black bar under the photomicrographs represents 1 mm. The graph below the photos shows percent of cortical tissue sparing followed by either Vehicle (saline), PZ(S), or PZ(M) treatment. Rats were euthanized in all treatment paradigms at 72 h after first injection. PZs did not exhibit a statistically significant amount of cortical tissue sparing when compared to Vehicle. However, PZm significantly increased the total volume of spared cortical tissue. One-way ANOVA followed by Dunnett's post-hoc test. $*p < 0.05$ compared to Vehicle. Error bars represent mean ± SD; n = 6 rats for vehicle group; n = 8 rats per group for drug-treated rats. These data are reproduced with permission from Cebak et al. (2017)

minutes and hours after CNS injury (Hall et al. 1989, 1992). Thus, it has long been recognized that more effective brain penetrable pharmacological LOO• scavengers are needed. Furthermore, compared to antioxidants that are scavengers of the initial post-TBI oxygen radical burst, it is reasonable to theorize that antioxidants that interrupt the LP process after it has begun would be able to exert a more clinically practical neuroprotective effect (i.e. possess a longer antioxidant therapeutic window).

An additional approach to inhibiting the propagation of LP reactions is to **chelate free iron**, either ferrous (Fe^{2+}) or ferric (Fe^{3+}), which potently catalyzes the breakdown of lipid hydroperoxides (LOOH), an essential event in the continuation of LP chain reactions in cellular membranes. The prototypical iron-chelating drug which chelates Fe^{3+}, is the tri-hydroxamic acid compound deferoxamine.

The fourth antioxidant category that has begun to be explored for neuroprotection following TBI concerns **pharmacological scavenging of LP-derived aldehydic (carbonyl-containing) breakdown products 4-HNE and acrolein**. As introduced earlier, these highly neurotoxic compounds have high affinity for covalently binding to basic amino acid residues including histidine, lysine, arginine and cysteine. These modifications have been shown to inhibit the activities of a variety of enzymatic proteins (Halliwell and Gutteridge 2008). Also, 4-HNE and acrolein, formed by LP oxidative damage, are also associated with stimulating additional free radical generation (i.e. oxidative stress) in injured CNS tissue (Hamann and Shi 2009). Several compounds have been identified that are able to antagonize this "carbonyl stress" by covalently binding to reactive LP-derived aldehydes. Two commercially available FDA-approved drugs that have been tested in TBI models are D-penicillamine and phenelzine, whose neuroprotective effects will be briefly discussed in the next section of this chapter.

A fifth antioxidant category that is theoretically an attractive broad spectrum mechanistic approach for achieving neuroprotection in TBI involves **pharmacologically activating the body's endogenous pleiotropic antioxidant defense system** that is largely regulated by nuclear factor E2-related factor 2/antioxidant response element (Nrf2/ARE) signaling at the transcriptional level (Kensler et al. 2007). As will be discussed below, Nrf2 activation and the up-regulation of antioxidant and anti-inflammatory genes, which has been previously described in experimental models of stroke and neurodegenerative disease (Shih et al. 2003), appears to be particularly promising in TBI models. Indeed, it has been documented that in the mouse controlled cortical impact TBI paradigm the injury itself upregulates Nrf2 and antioxidant gene expression. However, the time course of that antioxidant response occurs simultaneously with the time course of posttraumatic LP in brain tissue (Miller et al. 2014). Thus, what is needed is a compound that speeds up and increases the magnitude of the post-TBI Nrf2/ARE activation in the injured brain so that is has a chance to attenuate the peak of posttraumatic oxidative neural damage. Two such naturally occurring compounds that have been shown to be protective in TBI models are sulforaphane, found in high concentrations in broccoli, and carnosic acid, found in the herb rosemary.

Finally, a sixth strategy for achieving antioxidant neuroprotection in injured brain tissue involves **protecting the neural mitochondrion** which is essential for maintaining ATP production via the multi-complex electron transport chain as well as for its role in excessive post-TBI cytoplasmic Ca^{2+} accumulation. While mitochondrial Ca^{2+} buffering is one of the major functions of cellular mitochondria, as the intra-mitochondrial Ca^{2+} concentration increases, this leads to increases in mitochondrial $O_2^{\bullet-}$ as well as activation of a mitochondrial NOS that produces $NO^{\bullet}$. These two radicals rapidly combine to generate peroxynitrite and its derived highly reactive radicals ($OH^{\bullet}$, $NO^{\bullet}_2$, $CO_3^{\bullet-}$) which cause damage NO• leakage, causing oxidative damage to the electron transport chain (Bains and Hall 2012). Ultimately, mitochondrial dysfunction triggers the formation of the multi-component mitochondrial permeability transition pore (mPTP), which when it opens triggers mitochondrial permeability transition (mPT) loss of ionic gradients and leakage of important mitochondrial proteins (e.g. cytochrome C). Cyclosporine A (CsA), in addition to its immunosuppressive properties caused by inhibition of calcineurin, also has the ability to prevent mPTP formation by binding to one of the mPTP components, cyclophilin D, preventing the latter from joining mPTP complex, which is required in order for mPT to take place. Consequently, CsA acts to rescue the mitochondrion, preserves membrane potential and lessens additional ROS generation and oxidative damage. This has repeatedly been demonstrated in TBI models (Mbye et al. 2008; Sullivan et al. 1999). Because CsA is not an electron-donor or radical scavenger, its antioxidant action is consequently indirect. In other words, by preventing mPT from occurring it decreases mitochondrial ROS generation and thus indirectly limits oxidative damage. That this protective effect of CsA has little or nothing to do with its inhibition of calcineurin is due to the demonstration that the non-immunosuppressive CsA analog NIM811 does not inhibit calcineurin, but does bind to cyclophilin D, is just as protective as CsA in terms of mitochondrial function in the injured brain and able to reduce lesion volume (Mbye et al. 2008, 2009; Readnower et al. 2011)

3.7 Neuroprotective Effects of Pharmacological Antioxidants in TBI Patients and Models

TBI Clinical Trial Results with PEG-SOD and Tirilazad: During the past 30 years, there has been an intense effort to discover and develop pharmacological agents for acute treatment of TBI. This has included two compounds that possess free radical scavenging/antioxidant properties, including polyethylene glycol-conjugated superoxide dismutase (PEG-SOD) and the LP inhibitor tirilazad, that were tested in phase III clinical trials in a pathologically heterogeneous population of TBI patients (Langham et al. 2000; Marshall et al. 1998; Narayan et al. 2002). However, each of these trials was a therapeutic failure in that no overall benefit was documented. These failures have been hypothesized to be due to several factors (Narayan et al. 2002).

PEG-SOD: As mentioned earlier, the initial studies of free radical scavenging compounds in TBI models were carried out with Cu/Zn SOD based upon the work of Kontos and colleagues, who showed that post-traumatic microvascular dysfunction was initiated by $O_2^{\bullet-}$ generated as a by-product of the arachidonic acid cascade, which is massively activated during the first minutes and hours after TBI (Kontos 1989; Kontos and Wei 1986; Kontos and Povlishock 1986). Their work showed that administration of SOD prevented the post-traumatic microvascular dysfunction. This led to clinical trials in which the more metabolically stable polyethylene glycol (PEG)-conjugated SOD was examined in moderate and severe TBI patients when administered within the first 8 h after injury. Although an initial small phase II study showed a positive trend, subsequent multi-center phase III studies failed to show a significant benefit in terms of increased survival or improved neurological outcomes (Muizelaar et al. 1995). One theoretical reason may be that a large protein like SOD is unlikely to have much brain penetrability and therefore its radical scavenging effects may be limited to the microvasculature. A second reason may be that attempting to scavenge the short-lived inorganic radical $O_2^{\bullet-}$ may be associated with a very short therapeutic window, as suggested above. As pointed out earlier, the time course of measurable post-traumatic $OH^{\bullet}$ formation in the injured rodent brain has been shown to largely run its course by the end of the first hour after TBI (Hall et al. 1993; Smith et al. 1994). A more rational strategy would be to inhibit the LP that is triggered by the initial burst of inorganic radicals. A comparison of the time course of LP with that of post-traumatic $OH^{\bullet}$ shows that LP reactions continue to build beyond the first post-traumatic hours (Smith et al. 1994) and may continue for 3–4 days (Du et al. 2004; Miller et al. 2014; Hall et al. 2012). Despite the failure of PEG-SOD in human TBI, experimental studies have shown that transgenic mice that over-express Cu/Zn SOD are significantly protected against post-TBI pathophysiology and neurodegeneration (Chan et al. 1995; Gladstone et al. 2002; Lewen et al. 2000; Mikawa et al. 1996; Xiong et al. 2005). This fully supports the importance of post-traumatic $O_2^{\bullet-}$ in post-traumatic secondary injury, despite the fact that targeting this primordial radical is probably not the best antioxidant strategy for acute TBI compared to trying to stop the downstream LP process that is initiated by the early increases in $OH^{\bullet}$, $NO^{\bullet}_2$ and $CO_3^{\bullet-}$.

Tirilazad: Consistent with that rationale, the 21-aminosteroid LP inhibitor tirilazad was discovered, which inhibits free radical-induced LP by a combination of $LOO^{\bullet}$ scavenging and a membrane-stabilizing action that limits the propagation of LP reactions between an $LOO^{\bullet}$ and an adjacent polyunsaturated fatty acid (Hall et al. 1994). The protective efficacy of tirilazad has been demonstrated in multiple animal models of acute TBI in mice (Hall et al. 1988), rats (McIntosh et al. 1992) and cats (Dimlich et al. 1990). While the compound is largely localized in the microvascular endothelium, the post-traumatic disruption of the BBB is known to allow the successful penetration of tirilazad into the brain parenchyma as noted earlier (Hall et al. 1992). Other mechanistic data derived from the rat controlled cortical impact and the mouse diffuse concussive head injury models have definitively shown that a major effect of tirilazad is to lessen post-traumatic microvascular damage including BBB opening (Hall et al. 1992; Smith et al. 1994).

Tirilazad was taken into clinical development in the early 1990s, and following a small phase II dose-escalation study that demonstrated the drug's safety in TBI patients, it was evaluated in two phase III multi-center clinical trials for its ability to improve neurological recovery in moderately and severely injured closed TBI patients. One trial was conducted in North America and the other in Europe. In both trials, TBI patients were treated within 4 h after injury with either vehicle or tirilazad (10 mg/kg i.v. q6h for 5 days). The North American trial was never published, due to a major confounding imbalance in the randomization of the patients to placebo or tirilazad in regards to injury severity and pre-treatment neurological status. In contrast, the European trial that 1120 enrolled had much better randomization balance and was published (Marshall et al. 1998). As observed for PEG-SOD, tirilazad failed to show a significant beneficial effect of tirilazad in either moderate (GCS = 9–12) or severe (GCS = 4–8) patient categories. However, a post hoc analysis showed that moderately-injured male TBI patients with traumatic subarachnoid hemorrhage (tSAH) had a significantly lower incidence of 6 month mortality after treatment with tirilazad (6%) compared to placebo (24%, $p < 0.042$). In severely injured males with tSAH, tirilazad also lessened mortality from 43% in placebo-treated to 34% ($p < 0.026$). This result is consistent with the fact that tirilazad is also highly effective in reducing SAH-induced brain edema and vasospasm in animal models of SAH (Hall et al. 1994). Nevertheless, additional trials would have been required in order to establish the neuroprotective utility of tirilazad in tSAH patients in order to gain FDA approval. However, the sponsoring company Pharmacia & Upjohn opted not to continue the compound's development for TBI although tirilazad was successfully approved and marketed for use in aneurysmal SAH (aSAH) in several western European countries, Australia, New Zealand and South Africa, based upon its demonstrated efficacy in phase III aSAH trials conducted in those countries (Kassell et al. 1996; Lanzino and Kassell 1999). Therefore, the apparent post hoc-identified benefit in tSAH patients is consistent with tirilazad's prospectively demonstrated efficacy in aSAH patients, also mainly observed in males.

Effects of Other Direct and Indirect-Acting Lipid Peroxidation Inhibitors: In addition to tirilazad, several other LP inhibitors have been reported to be effective neuroprotectants in TBI models. These include the lipid peroxyl radical (LOO•) scavenging 2-methylaminochromans U-78517F and U-83836E (Hall et al. 1991; Mustafa et al. 2010, 2011), the pyrrolopyrimidine U-101033E (Hall et al. 1997; Xiong et al. 1997, 1998), OPC-14117 (Mori et al. 1998) and the naturally-occurring LOO• scavengers curcumin (Sharma et al. 2009; Wu et al. 2006) and resveratrol (Ates et al. 2007; Sonmez et al. 2007), the indoleamine melatonin (Beni et al. 2004; Cirak et al. 1999; Mesenge et al. 1998; Ozdemir et al. 2005a, b) and lastly, the endogenous antioxidant lipoic acid (Toklu et al. 2009). In the case of curcumin and resveratrol, these are potent LOO• scavengers due to their possession of multiple phenolic hydroxyl groups that can donate electrons to LOO• radicals. Melatonin also has LOO• scavenging capability (Longoni et al. 1998), but in addition appears to react with PN (Zhang et al. 1999). Lipoic acid may also have LOO• scavenging effects, but these are more likely to be indirect via the regeneration (i.e. re-reduction) of other endogenous electron-donating antioxidants, including vitamin E, glutathione and vitamin C.

Among these LP inhibitors, arguably the most potent and effective LOO• scavenging LP inhibitor yet discovered is the 2-methylaminochroman compound U-83836E which combines the LOO• scavenging antioxidant chroman ring structure of vitamin E with the bis-pyrrolopyrimidine moiety of tirilazad. The phenolic chroman antioxidant moiety, after it sacrifices it phenolic electron to scavenge an LOO•, can be re-reduced by endogenous ascorbic acid (vitamin C) or glutathione (GSH) making it able to quench a second and then a third LOO•, etc. The bis-pyrrolopyrimidine moiety, on the other hand, can also scavenge multiple moles of LOO• by a true catalytic mechanism (Hall et al. 1991; Hall et al. 1995). Thus, U-83836E, is a dual functionality LOO• scavenger that is understandably more effective than either vitamin E, tirilazad (Hall et al. 1991) and possibly the other naturally-occurring LOO• scavengers such as curcumin, resveratrol, melatonin and lipoic acid. Furthermore, U-83836E possesses a high degree of lipophilicity endowing it with a high affinity for membrane phospholipids where LP takes place. Studies from the authors' laboratory in the mouse CCI-TBI model have shown that U-83836E is able to reduce post-traumatic LP and protein nitration and preserve mitochondrial respiratory function, and lessen calpain-mediated neuronal cytoskeletal degradation and decrease injured tissue (Mustafa et al. 2010, 2011).

Nitroxide Antioxidants and Peroxynitrite Scavengers: In addition to the lipid peroxyl (LOO•) radical scavengers, the neuroprotective effects of a family of nitroxide-containing antioxidants have also been examined in experimental TBI models. These are sometimes referred to as "spin-trapping agents" and include α-phenyl-tert-butyl nitrone (PBN) and its thiol analog NXY-059 and tempol. Both PBN and tempol have been shown to be protective in rodent TBI paradigms (Awasthi et al. 1997; Marklund et al. 2001). As mentioned earlier, tempol has been shown by the author and colleagues to catalytically scavenge PN-derived $NO\bullet_2$ and $CO_3^{\bullet-}$ (Carroll et al. 2000; Bonini et al. 2002), and to reduce post-traumatic oxidative damage (both LP and protein nitration), preserve mitochondrial function, decrease calcium-activated, calpain-mediated cytoskeletal damage and reduce neurodegeneration in mice subjected to a severe controlled cortical impact-induced focal TBI (Deng-Bryant et al. 2008). Earlier, another laboratory reported that tempol can reduce post-traumatic brain edema and improve neurological recovery in a rat contusion injury model (Beit-Yannai et al. 1996; Zhang et al. 1998). However, the neuroprotective effect of tempol, administered alone, is associated with a therapeutic window of an hour or less in the mouse controlled cortical impact TBI (CCI-TBI) model. Moreover, tempol is not effective at directly inhibiting LP in the latter model (Deng-Bryant et al. 2008).

Effects of the Iron Chelator Deferoxamine: The prototype iron chelator deferoxamine, which binds ferric (Fe^{3+}) iron and thereby would lessen the catalytic effects of iron on LP, has also been reported to have beneficial actions in preclinical TBI or TBI-related models (Gu et al. 2009; Long et al. 1996). However, deferoxamine is hindered by its limited brain penetration and rapid plasma elimination rate. To counter the latter limitation, a dextran-coupled deferoxamine has been synthesized that has been reported to significantly improve early neurological recovery in a mouse diffuse TBI model (Panter et al. 1992). Much of this activity, however, is probably due to microvascular antioxidant protection because of limited brain penetrability. Another caveat to the iron-chelation antioxidant neuroprotective approach that is at least relevant to the ferric iron chelators such as deferoxamine is that at they can cause a pro-oxidant effect in that their binding of Fe^{3+} can actually drive the oxidation of ferrous to ferric iron which can increase superoxide radical formation in the process ($Fe^{2+} + O_2 \rightarrow Fe^{3+} + O_2^{\bullet-}$).

Effects of Carbonyl Scavengers: We have previously demonstrated that D-penicillamine is able to scavenge PN (Althaus et al. 1994) and to protect brain mitochondria from PN-induced respiratory dysfunction in isolated rat brain mitochondria (Singh et al. 2007). D-Penicillamine has also been documented to form an irreversible bond to primary aldehydes, enabling it to scavenge neurotoxic LP-derived carbonyl compounds such as 4-HNE and acrolein (Wood et al. 2008). Consistent with that mechanism of action, D-penicillamine was shown to attenuate the levels 4-HNE-modified brain mitochondrial proteins after exposure of isolated mitochondria to 4-HNE (Singh et al. 2007). The PN scavenging action of D-penicillamine along with its carbonyl scavenging capability may jointly explain our previous findings that acutely administered penicillamine can improve early neurological recovery of mice subjected to moderately severe concussive TBI (Hall et al. 1999).

More recently, it has been demonstrated that a variety of FDA-approved hydrazine ($-NH-NH_2$)-containing compounds including the anti-hypertensive agent hydralazine and the anti-depressant phenelzine can react with the carbonyl moieties of 4-HNE or acrolein, which prevents the latter from binding to susceptible amino acids in proteins (Galvani et al. 2008). Most impressive is the fact that the application of hydrazines can rescue cultured cells from 4-HNE toxicity even when administered after the 4-HNE has already covalently bound to cellular proteins (Galvani et al. 2008). Consistent with this effect being neuroprotective, others have shown that hydralazine inhibits either compression or acrolein-mediated injuries to ex vivo spinal cord (Hamann et al. 2008). However, hydralazine, which is a potent vasodilator, would be difficult to administer in vivo after either spinal cord injury or TBI in which hypotension is already a common pathophysiological problem. In contrast, another FDA-approved hydrazine-containing drug phenelzine, used for certain depressive patients, should not compromise blood pressure as readily as hydralazine. Accordingly, a recently published paper has shown that phenelzine administration to rats subjected to acute contusion SCI mitigated post-SCI neuropathic pain, reduces motor deficits and improves spinal cord tissue sparing (Chen et al. 2016). Earlier studies have demonstrated neuroprotective efficacy in a rodent ischemia-reperfusion stroke model that were attributed to reducing 'aldehyde load' in the stroke-injured brain (Wood et al. 2006).

In vitro studies in our laboratory have documented the ability of phenelzine to protect isolated rat brain mitochondria from the respiratory depressant effects of 4-HNE, together with a concentration-related attenuation of the accumulation of

4-HNE modified mitochondrial proteins. More recently, we have observed that phenelzine is able to protect isolated mitochondria from respiratory functional depression and modification of mitochondrial proteins following application of the more highly reactive aldehyde acrolein (Cebak et al. 2017). Subsequent in vivo studies in the rat controlled cortical impact TBI model have found that a single 10 mg/kg s.c. dose of phenelzine can also reduce early (3 h) posttraumatic mitochondrial respiratory failure as well as reducing cortical lesion volume at 14 days post-injury (Singh et al. 2013). To better define the optimal neuroprotective use of phenelzine, additional in vivo TBI studies have demonstrated that repeated dosing with phenelzine over a 60 h post-TBI period is capable of protecting delayed mitochondrial failure at its peak at 72 h in the same TBI model along with a reduction in cortical lesion volume that is greater than that seen with a single early dose. This makes sense in that the adequate carbonyl-scavenging drug levels logically need to be maintained during the 72 h long time course of posttraumatic generation of LP-derived neurotoxic aldehydes (Cebak et al. 2017).

Effects of Nrf2/ARE Signaling Activators: The body's endogenous antioxidant defense system is largely regulated by nuclear factor E2-related factor 2/antioxidant response element (Nrf2/ARE) signaling at the transcriptional level (Zhang 2006; Kensler et al. 2007). Nrf2 activation and the up-regulation of antioxidant and anti-inflammatory genes represents a valid neurotherapeutic intervention in CNS injury and has been previously described in various experimental models of stroke and neurodegenerative diseases (Shih et al. 2003). More recently, the role of Nrf2/ARE activation in SCI has been explored as a targeted neuroprotective strategy for both TBI and SCI. Indeed, studies in Nrf2 (−/−) mice demonstrated increased spinal cord edema and expression of inflammatory cytokines compared to wild-type Nrf2 mice following SCI (Mao et al. 2010; Mao et al. 2011). In mild rat thoracic SCI, it has been reported that Nrf2 levels increase as early as 30 min post-injury and remain elevated through 3 days. In the same study, application of the natural product sulforaphrane, a Nrf2/ARE signaling activator, significantly reduced contusion volume and increased post-SCI coordination. These positive outcomes were a result of sulforaphrane-induced increases in Nrf2, glutamine and decreases in inflammatory cytokines, IL-1β and TBFα (Wang et al. 2012).

The mRNA levels of Nrf2-regulated antioxidant enzymes, heme oxygenase (HO-1) and NADPH:quinone oxidoreductase-1 (NQO1), are up-regulated 24 h post TBI (Yan et al. 2008). In addition, Nrf2-knockout mice are susceptible to increased oxidative stress and neurologic deficits following TBI compared to their wild-type counterparts (Hong et al. 1994). Administration of sulforaphane is also neuroprotective in various animal models of TBI, specifically reducing cerebral edema and oxidative stress and improving BBB function and cognitive deficits (Dash et al. 2009). Studies by Chen et al. (2011) demonstrated increased expression of Nrf2 and HO-1 in the cortex of the rat subarachnoid hemorrhage model. Treatment with sulforaphane further increased the expression of Nrf2, HO-1, NQ01 and glutathione S-transferase-α1 (GST-α1), resulting in the reduction of brain edema, cortical neuronal death and motor deficits (Chen et al. 2011). Tert-butylhydroquinone, another activator of Nrf2, protects against TBI-induced inflammation and damage via reduction in NF-KB activation and TNFα and IL-1β production following injury in the mouse closed head injury model (Jin et al. 2011). Collectively these studies demonstrate a significant neuroprotective role of Nrf2 signaling through the activation of antioxidant enzymes and reduction of oxidative secondary injury responses following CNS injury. Thus, Nrf2 activation may be a prime candidate for the attenuation of oxidative stress and subsequent neurotoxicity in TBI via the development of small-molecule activators of the Nrf2/ARE pathway.

Recent work in our laboratory has revealed that following controlled cortical impact TBI in mice, there is indeed a progressive activation of the Nrf2-ARE system in the traumatically-injured brain as evidenced by an increase in HO-1 mRNA and protein that peaks at 72 h after TBI. However, this effect does not precede, but rather it is coincident with the post-injury increase in LP-related 4-HNE (Miller et al. 2014). Therefore, it is apparent that this endogenous neuroprotective antioxidant response needs to be pharmacologically enhanced and/or sped up if it is to be capable of exerting acute post-TBI neuroprotection. Our laboratory is currently studying another Nrf2-ARE activator, natural product carnosic acid, that has been shown by others to more effectively induce this antioxidant defense system than the prototype sulforaphane (Satoh et al. 2008). We have shown that administration of carnosic acid to non-TBI mice is able to significantly increase the resistance of cortical mitochondria harvested 48 h later to the respiratory depressant effects of the in vitro applied 4-HNE together with in decrease in 4-HNE modification of mitochondrial proteins (Miller et al. 2013). Subsequently, we have administered a single 1 mg/kg i.p. dose of carnosic acid to mice at 15 min after controlled cortical impact TBI and observed that the compound is able to significantly preserve respiratory function along with a reduction in the level of LP-mediated damage in mitochondria harvested from the injured cortex at 24 h after TBI (Miller et al. 2015). Furthermore, carnosic acid's antioxidant effects were still apparent at 48 h post-injury in terms of an attenuation of 4-HNE and 3-NT in the injured cortical tissue together with a decrease in Ca^{2+}-activated, calpain-mediated neuronal cytoskeletal degradation. In regards to the latter neuronal protective effect, a decrease in 48 h cytoskeletal degradation was also shown to occur even with a post-TBI treatment delay of 8 h. Ongoing studies are evaluating the behavioral recovery and tissue protective effects of carnosic acid whether these are achievable with a clinically practical therapeutic window.

3.8 Combination Antioxidant Treatment of Traumatic Brain Injury

Antioxidant neuroprotective therapeutic discovery directed at acute TBI has consistently been focused upon attempting to inhibit the secondary injury cascade by pharmacological targeting of a single oxidative damage mechanism. As presented above, these efforts have included either enzymatic scavenging of superoxide radicals with SOD (Muizelaar et al. 1995) or inhibition of LP with tirilazad

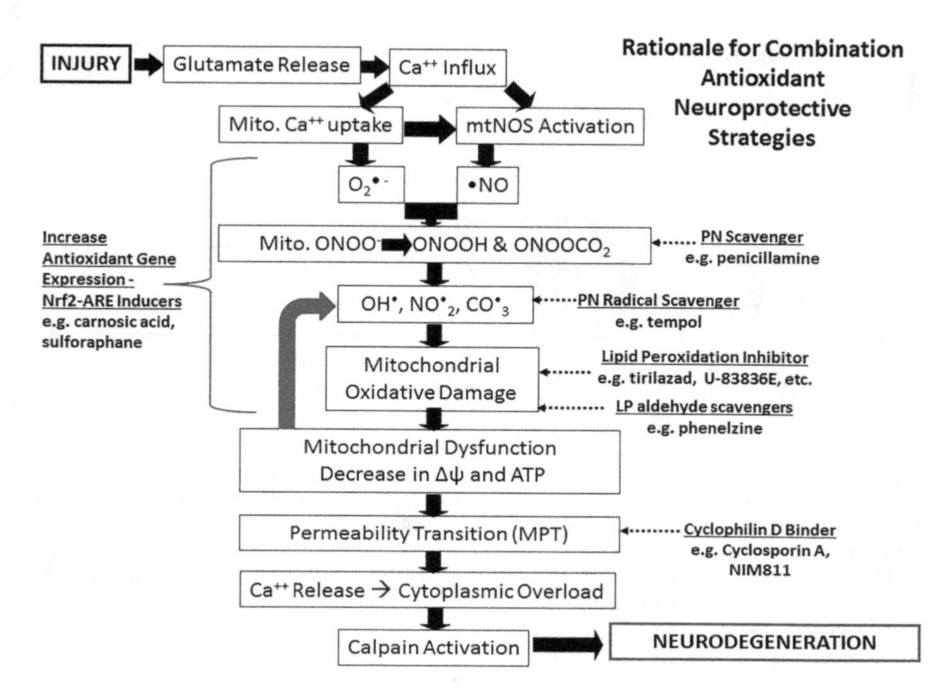

Fig. 3.3 Rationale for the combination of two or more antioxidant strategies to achieve a more effective and consistent (i.e. less variable) neuroprotective effect in the injured brain

(Marshall et al. 1998). While each of these strategies alone has shown protective efficacy in animal models of TBI, phase III clinical trials with either compound failed to demonstrate a statistically significant positive effect although post hoc subgroup analysis suggests that the microvascularly localized tirilazad may have efficacy in moderate and severe TBI patients with tSAH (Marshall et al. 1998). While many reasons have been identified as possible contributors to the failure, one logical explanation has to do with the possible need to interfere at multiple points in the oxidative damage portion of the secondary injury cascade either simultaneously or in a phased manner in order to achieve a clinically demonstrable level of neuroprotection. To begin to address this hypothesis, we are currently exploring the possibility that reducing posttraumatic oxidative damage more completely and less variably might be achievable by combined treatment with two or more mechanistically complimentary antioxidant compounds. Figure 3.3 summarizes the overall rationale for a multi-mechanistic antioxidant therapy for TBI. It is anticipated that the combination of two or three antioxidant mechanistic strategies may improve the extent of neuroprotective efficacy, lessen the variability of the effect and possibly provide a longer therapeutic window of opportunity compared to the window for the individual strategies. Figure 3.4 shows preliminary, not yet published data, suggesting that combination treatment of the PN radical scavenger tempol with the LP inhibitor U-83836E in mice subjected to controlled cortical impact TBI was more effective in

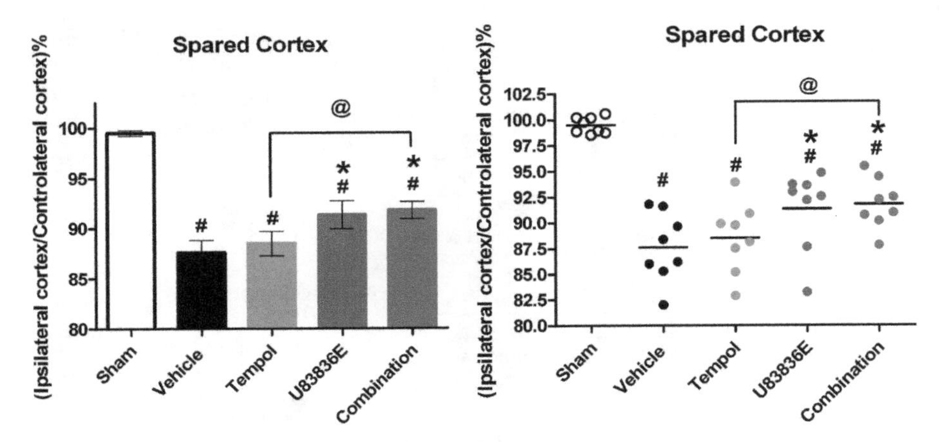

Fig. 3.4 Preliminary data on the neuroprotective effects of 15 min post-injury administration of tempol, U83836E or the combination on cortical tissue sparing. U83836E and the combination both significantly improved tissue sparing whereas tempol in this experiment did not. However, only the combination significantly out-performed tempol. As in scatter plot on the right, the variability in the combination group was considerably lower than in the single treatment groups. All values = mean ± SEM for N = 8/group; #$p < 0.05$ vs sham; *$p < 0.05$ vs. vehicle, @$p < 0.05$ tempol alone vs. combination

reducing 7 day post-TBI cortical tissue damage as well as resulting in a reduction in the variability of the data to half of that seen in the parallel groups treated with the either of the two drugs alone.

In other published studies, we have observed that combined treatment with an LP inhibitor with an inhibitor of excitotoxic glutamate release increases the neuroprotective therapeutic window. Using an infant rat model of shaken baby-induced brain damage model, we have documented that treatment with riluzole, an inhibitor of excitotoxic glutamate release, attenuated cortical neurodegeneration measured at 14 days post-TBI, but the therapeutic window for this neuroprotective effect was limited to the initial riluzole dose having to be administered during the first hour after TBI. However, when the infant rats received the LP inhibitor tirilazad at 30 min after TBI, it increased the neuroprotective therapeutic window for riluzole to 4 h after TBI (Smith and Hall 1998). Thus, combination treatments may extend the neuroprotective efficacy window significantly. Accordingly, combination neuroprotective therapy might be able to improve efficacy, reduce variability and improve the therapeutic window for achievement of clinically measurable neuroprotection in TBI patients, although additional work remains to be conducted to determine whether that neuroprotective hypothesis is correct.

Acknowledgements Portions of the work reviewed in this chapter were supported by funding from 5R01 NS046566, 5P30 NS051220, and 5P01 NS58484 and currently by 5R01 NS083405, 5R01 NS084857 and 1R01 NS100093 and from the Kentucky Spinal Cord & Head Injury Research Trust.

References

Althaus JS, Oien TT, Fici GJ, Scherch HM, Sethy VH, VonVoigtlander PF (1994) Structure activity relationships of peroxynitrite scavengers an approach to nitric oxide neurotoxicity. Res Commun Chem Pathol Pharmacol 83(3):243–254

Ates O, Cayli S, Altinoz E, Gurses I, Yucel N, Sener M, Kocak A, Yologlu S (2007) Neuroprotection by resveratrol against traumatic brain injury in rats. Mol Cell Biochem 294(1–2):137–144

Awasthi D, Church DF, Torbati D, Carey ME, Pryor WA (1997) Oxidative stress following traumatic brain injury in rats. Surg Neurol 47(6):575–581

Bains M, Hall ED (2012) Antioxidant therapies in traumatic brain and spinal cord injury. Biochim Biophys Acta 1822(5):675–684

Beit-Yannai E, Zhang R, Trembovler V, Samuni A, Shohami E (1996) Cerebroprotective effect of stable nitroxide radicals in closed head injury in the rat. Brain Res 717(1–2):22–28

Beni SM, Kohen R, Reiter RJ, Tan DX, Shohami E (2004) Melatonin-induced neuroprotection after closed head injury is associated with increased brain antioxidants and attenuated late-phase activation of NF-kappaB and AP-1. FASEB J 18(1):149–151

Bonini MG, Mason RP, Augusto O (2002) The Mechanism by which 4-hydroxy-2,2,6,6-tetramethylpiperidene-1-oxyl (tempol) diverts peroxynitrite decomposition from nitrating to nitrosating species. Chem Res Toxicol 15(4):506–511

Bringold U, Ghafourifar P, Richter C (2000) Peroxynitrite formed by mitochondrial NO synthase promotes mitochondrial Ca2+ release. Free Radic Biol Med 29(3–4):343–348

Carrico KM, Vaishnav R, Hall ED (2009) Temporal and spatial dynamics of peroxynitrite-induced oxidative damage after spinal cord contusion injury. J Neurotrauma 26(8):1369–1378

Carroll RT, Galatsis P, Borosky S, Kopec KK, Kumar V, Althaus JS, Hall ED (2000) 4-Hydroxy-2,2,6,6-tetramethylpiperidine-1-oxyl (Tempol) inhibits peroxynitrite-mediated phenol nitration. Chem Res Toxicol 13(4):294–300

Cebak JE, Singh IN, Hill RL, Wang JA, Hall ED (2017) Phenelzine protects brain mitochondrial function in vitro and in vivo following traumatic brain injury by scavenging the reactive carbonyls 4-hydroxynonenal and acrolein leading to cortical histological protection. J Neurotrauma 34(7):1302–1317

Chan PH, Epstein CJ, Li Y, Huang TT, Carlson E, Kinouchi H, Yang G, Kamii H, Mikawa S, Kondo T et al (1995) Transgenic mice and knockout mutants in the study of oxidative stress in brain injury. J Neurotrauma 12(5):815–824

Chen G, Fang Q, Zhang J, Zhou D, Wang Z (2011) Role of the Nrf2-ARE pathway in early brain injury after experimental subarachnoid hemorrhage. J Neurosci Res 89(4):515–523

Chen Z, Park J, Butler B, Acosta G, Alvarez S, Zheng L, Tang J, McCain R, Zhang W, Ouyang Z, Cao P, Shi R (2016) Mitigation of sensory and motor deficits by acrolein scavenger phenelzine in a rat model of spinal cord contusive injury. J Neurochem 138(2):328–338

Cirak B, Rousan N, Kocak A, Palaoglu O, Palaoglu S, Kilic K (1999) Melatonin as a free radical scavenger in experimental head trauma. Pediatr Neurosurg 31(6):298–301

Dash PK, Zhao J, Orsi SA, Zhang M, Moore AN (2009) Sulforaphane improves cognitive function administered following traumatic brain injury. Neurosci Lett 460(2):103–107

Deng-Bryant Y, Singh IN, Carrico KM, Hall ED (2008) Neuroprotective effects of tempol, a catalytic scavenger of peroxynitrite-derived free radicals, in a mouse traumatic brain injury model. J Cereb Blood Flow Metab 28(6):1114–1126

Dimlich RV, Tornheim PA, Kindel RM, Hall ED, Braughler JM, McCall JM (1990) Effects of a 21-aminosteroid (U-74006F) on cerebral metabolites and edema after severe experimental head trauma. Adv Neurol 52:365–375

Du L, Bayir H, Lai Y, Zhang X, Kochanek PM, Watkins SC, Graham SH, Clark RS (2004) Innate gender-based proclivity in response to cytotoxicity and programmed cell death pathway. J Biol Chem 279(37):38563–38570

Galvani S, Coatrieux C, Elbaz M, Grazide MH, Thiers JC, Parini A, Uchida K, Kamar N, Rostaing L, Baltas M, Salvayre R, Negre-Salvayre A (2008) Carbonyl scavenger and antiatherogenic effects of hydrazine derivatives. Free Radic Biol Med 45(10):1457–1467

Gladstone DJ, Black SE, Hakim AM (2002) Toward wisdom from failure: lessons from neuroprotective stroke trials and new therapeutic directions. Stroke 33(8):2123–2136

Gu Y, Hua Y, Keep RF, Morgenstern LB, Xi G (2009) Deferoxamine reduces intracerebral hematoma-induced iron accumulation and neuronal death in piglets. Stroke 40(6):2241–2243

Gutteridge JM (1995) Lipid peroxidation and antioxidants as biomarkers of tissue damage. Clin Chem 41(12 Pt 2):1819–1828

Hall E (1986) Beneficial effects of acute intravenous ibuprofen on neurological recovery of head injured mice: comparison of cyclooxygenase inhibition of thromboxane A2 synthetase or 5-lipoxygenase. CNS. Trauma 2:75–83

Hall ED, Bosken JM (2009) Measurement of oxygen radicals and lipid peroxidation in neural tissues. Curr Protoc Neurosci Chapter 7:Unit 7 17. 11–51

Hall ED, Yonkers PA, McCall JM, Braughler JM (1988) Effects of the 21-aminosteroid U74006F on experimental head injury in mice. J Neurosurg 68(3):456–461

Hall ED, Yonkers PA, Horan KL, Braughler JM (1989) Correlation between attenuation of post-traumatic spinal cord ischemia and preservation of tissue vitamin E by the 21-aminosteroid U74006F: evidence for an in vivo antioxidant mechanism. J Neurotrauma 6(3):169–176

Hall ED, Braughler JM, Yonkers PA, Smith SL, Linseman KL, Means ED, Scherch HM, Von Voigtlander PF, Lahti RA, Jacobsen EJ (1991) U-78517F: a potent inhibitor of lipid peroxidation with activity in experimental brain injury and ischemia. J Pharmacol Exp Ther 258(2):688–694

Hall ED, Yonkers PA, Andrus PK, Cox JW, Anderson DK (1992) Biochemistry and pharmacology of lipid antioxidants in acute brain and spinal cord injury. J Neurotrauma 9(Suppl 2): S425–S442

Hall ED, Andrus PK, Yonkers PA (1993) Brain hydroxyl radical generation in acute experimental head injury. J Neurochem 60(2):588–594

Hall ED, McCall JM, Means ED (1994) Therapeutic potential of the lazaroids (21-aminosteroids) in acute central nervous system trauma, ischemia and subarachnoid hemorrhage. Adv Pharmacol 28:221–268

Hall ED, Andrus PK, Smith SL, Oostveen JA, Scherch HM, Lutzke BS, Raub TJ, Sawada GA, Palmer JR, Banitt LS, Tustin JM, Belonga KL, Ayer DE, Bundy GL (1995) Neuroprotective efficacy of microvascularly-localized versus brain-penetraiting antioxidants. Acta Neurochir (Suppl) 66:107–113

Hall ED, Andrus PK, Smith SL, Fleck TJ, Scherch HM, Lutzke BS, Sawada GA, Althaus JS, Vonvoigtlander PF, Padbury GE, Larson PG, Palmer JR, Bundy GL (1997) Pyrrolopyrimidines: novel brain-penetrating antioxidants with neuroprotective activity in brain injury and ischemia models. J Pharmacol Exp Ther 281(2):895–904

Hall ED, Kupina NC, Althaus JS (1999) Peroxynitrite scavengers for the acute treatment of traumatic brain injury. Ann N Y Acad Sci 890:462–468

Hall ED, Vaishnav RA, Mustafa AG (2010) Antioxidant therapies for traumatic brain injury. Neurotherapeutics 7(1):51–61

Hall ED, Wang JA, Miller DM (2012) Relationship of nitric oxide synthase induction to peroxynitrite-mediated oxidative damage during the first week after experimental traumatic brain injury. Exp Neurol 238(2):176–182

Halliwell B, Gutteridge J (2008) Free radicals in biology and medicine, 3rd edn. Oxford University Press, New York

Hamann K, Shi R (2009) Acrolein scavenging: a potential novel mechanism of attenuating oxidative stress following spinal cord injury. J Neurochem 111(6):1348–1356

Hamann K, Nehrt G, Ouyang H, Duerstock B, Shi R (2008) Hydralazine inhibits compression and acrolein-mediated injuries in ex vivo spinal cord. J Neurochem 104(3):708–718

Hong SC, Goto Y, Lanzino G, Soleau S, Kassell NF, Lee KS (1994) Neuroprotection with a calpain inhibitor in a model of focal cerebral ischemia. Stroke 25(3):663–669

Hummel SG, Fischer AJ, Martin SM, Schafer FQ, Buettner GR (2006) Nitric oxide as a cellular antioxidant: a little goes a long way. Free Radic Biol Med 40(3):501–506

Jin W, Kong J, Wang H, Wu J, Lu T, Jiang J, Ni H, Liang W (2011) Protective effect of tert-butylhydroquinone on cerebral inflammatory response following traumatic brain injury in mice. Injury 42(7):714–718

Kassell NF, Haley EC Jr, Apperson-Hansen C, Alves WM (1996) Randomized, double-blind, vehicle-controlled trial of tirilazad mesylate in patients with aneurysmal subarachnoid hemorrhage: a cooperative study in Europe, Australia, and New Zealand. J Neurosurg 84(2):221–228

Kensler TW, Wakabayashi N, Biswal S (2007) Cell survival responses to environmental stresses via the Keap1-Nrf2-ARE pathway. Annu Rev Pharmacol Toxicol 47:89–116

Kontos HA (1989) Oxygen radicals in CNS damage. Chem Biol Interact 72(3):229–255

Kontos HA, Povlishock JT (1986) Oxygen radicals in brain injury. Cent Nerv Syst Trauma 3(4):257–263

Kontos HA, Wei EP (1986) Superoxide production in experimental brain injury. J Neurosurg 64(5):803–807

Langham J, Goldfrad C, Teasdale G, Shaw D, Rowan K (2000) Calcium channel blockers for acute traumatic brain injury. Cochrane Database Syst Rev 2:CD000565

Lanzino G, Kassell NF (1999) Double-blind, randomized, vehicle-controlled study of high-dose tirilazad mesylate in women with aneurysmal subarachnoid hemorrhage. Part II. A cooperative study in North America. J Neurosurg 90(6):1018–1024

Lewen A, Matz P, Chan PH (2000) Free radical pathways in CNS injury. J Neurotrauma 17(10):871–890

Long DA, Ghosh K, Moore AN, Dixon CE, Dash PK (1996) Deferoxamine improves spatial memory performance following experimental brain injury in rats. Brain Res 717(1–2):109–117

Longoni B, Salgo MG, Pryor WA, Marchiafava PL (1998) Effects of melatonin on lipid peroxidation induced by oxygen radicals. Life Sci 62(10):853–859

Mao L, Wang H, Qiao L, Wang X (2010) Disruption of Nrf2 enhances the upregulation of nuclear factor-kappaB activity, tumor necrosis factor-alpha, and matrix metalloproteinase-9 after spinal cord injury in mice. Mediat Inflamm 2010:238321

Mao L, Wang H, Wang X, Liao H, Zhao X (2011) Transcription factor Nrf2 protects the spinal cord from inflammation produced by spinal cord injury. J Surg Res 170(1):e105–e115

Marklund N, Clausen F, Lewen A, Hovda DA, Olsson Y, Hillered L (2001) Alpha-phenyl-tert-N-butyl nitrone (PBN) improves functional and morphological outcome after cortical contusion injury in the rat. Acta Neurochir 143(1):73–81

Marshall LF, Maas AI, Marshall SB, Bricolo A, Fearnside M, Iannotti F, Klauber MR, Lagarrigue J, Lobato R, Persson L, Pickard JD, Piek J, Servadei F, Wellis GN, Morris GF, Means ED, Musch B (1998) A multicenter trial on the efficacy of using tirilazad mesylate in cases of head injury. J Neurosurg 89(4):519–525

Mbye LH, Singh IN, Sullivan PG, Springer JE, Hall ED (2008) Attenuation of acute mitochondrial dysfunction after traumatic brain injury in mice by NIM811, a non-immunosuppressive cyclosporin A analog. Exp Neurol 209(1):243–253

Mbye LH, Singh IN, Carrico KM, Saatman KE, Hall ED (2009) Comparative neuroprotective effects of cyclosporin A and NIM811, a nonimmunosuppressive cyclosporin A analog, following traumatic brain injury. J Cereb Blood Flow Metab 29(1):87–97

McIntosh TK, Thomas M, Smith D, Banbury M (1992) The novel 21-aminosteroid U74006F attenuates cerebral edema and improves survival after brain injury in the rat. J Neurotrauma 9(1):33–46

Mesenge C, Margaill I, Verrecchia C, Allix M, Boulu RG, Plotkine M (1998) Protective effect of melatonin in a model of traumatic brain injury in mice. J Pineal Res 25(1):41–46

Mikawa S, Kinouchi H, Kamii H, Gobbel GT, Chen SF, Carlson E, Epstein CJ, Chan PH (1996) Attenuation of acute and chronic damage following traumatic brain injury in copper, zinc-superoxide dismutase transgenic mice. J Neurosurg 85(5):885–891

Miller DM, Singh IN, Wang JA, Hall ED (2013) Administration of the Nrf2-ARE activators sulforaphane and carnosic acid attenuates 4-hydroxy-2-nonenal-induced mitochondrial dysfunction ex vivo. Free Radic Biol Med 57:1–9

Miller D, Wang J, Buchanan A, Hall E (2014) Temporal and spatial dynamics of Nrf2-ARE-mediated gene targets in cortex and hippocampus following controlled cortical impact traumatic brain injury in mice. J Neurotrauma 31:1194–1201

Miller DM, Singh IN, Wang JA, Hall ED (2015) Nrf2-ARE activator carnosic acid decreases mitochondrial dysfunction, oxidative damage and neuronal cytoskeletal degradation following traumatic brain injury in mice. Exp Neurol 264:103–110

Monyer H, Hartley DM, Choi DW (1990) 21-Aminosteroids attenuate excitotoxic neuronal injury in cortical cell cultures. Neuron 5(2):121–126

Mori T, Kawamata T, Katayama Y, Maeda T, Aoyama N, Kikuchi T, Uwahodo Y (1998) Antioxidant, OPC-14117, attenuates edema formation, and subsequent tissue damage following cortical contusion in rats. Acta Neurochir Suppl (Wien) 71:120–122

Muizelaar JP, Kupiec JW, Rapp LA (1995) PEG-SOD after head injury. J Neurosurg 83(5):942

Mustafa AG, Singh IN, Carrico KM, Hall ED (2010) Mitochondrial protection after traumatic brain injury by scavenging lipid peroxyl radicals. J Neurochem 114(1):271–280

Mustafa AG, Wang JA, Carrico KM, Hall ED (2011) Pharmacological inhibition of lipid peroxidation attenuates calpain-mediated cytoskeletal degradation after traumatic brain injury. J Neurochem 117(3):579–588

Narayan RK, Michel ME, Ansell B, Baethmann A, Biegon A, Bracken MB, Bullock MR, Choi SC, Clifton GL, Contant CF, Coplin WM, Dietrich WD, Ghajar J, Grady SM, Grossman RG, Hall ED, Heetderks W, Hovda DA, Jallo J, Katz RL, Knoller N, Kochanek PM, Maas AI, Majde J, Marion DW, Marmarou A, Marshall LF, McIntosh TK, Miller E, Mohberg N, Muizelaar JP, Pitts LH, Quinn P, Riesenfeld G, Robertson CS, Strauss KI, Teasdale G, Temkin N, Tuma R, Wade C, Walker MD, Weinrich M, Whyte J, Wilberger J, Young AB, Yurkewicz L (2002) Clinical trials in head injury. J Neurotrauma 19(5):503–557

Ozdemir D, Tugyan K, Uysal N, Sonmez U, Sonmez A, Acikgoz O, Ozdemir N, Duman M, Ozkan H (2005a) Protective effect of melatonin against head trauma-induced hippocampal damage and spatial memory deficits in immature rats. Neurosci Lett 385(3):234–239

Ozdemir D, Uysal N, Gonenc S, Acikgoz O, Sonmez A, Topcu A, Ozdemir N, Duman M, Semin I, Ozkan H (2005b) Effect of melatonin on brain oxidative damage induced by traumatic brain injury in immature rats. Physiol Res 54(6):631–637

Panter SS, Braughler JM, Hall ED (1992) Dextran-coupled deferoxamine improves outcome in a murine model of head injury. J Neurotrauma 9:47–53

Pellegrini-Giampietro DE, Cherici G, Alesiani M, Carla V, Moroni F (1990) Excitatory amino acid release and free radical formation may cooperate in the genesis of ischemia-induced neuronal damage. J Neurosci 10(3):1035–1041

Readnower RD, Pandya JD, McEwen ML, Pauly JR, Springer JE, Sullivan PG (2011) Post-injury administration of the mitochondrial permeability transition pore inhibitor, NIM811, is neuroprotective and improves cognition after traumatic brain injury in rats. J Neurotrauma 28(9):1845–1853

Rohn TT, Hinds TR, Vincenzi FF (1993) Ion transport ATPases as targets for free radical damage. Protection by an aminosteroid of the Ca2+ pump ATPase and Na+/K+ pump ATPase of human red blood cell membranes. Biochem Pharmacol 46(3):525–534

Rohn TT, Hinds TR, Vincenzi FF (1996) Inhibition of Ca2+-pump ATPase and the Na+/K+-pump ATPase by iron-generated free radicals. Protection by 6,7-dimethyl-2,4-DI-1-pyrrolidinyl-7H-pyrrolo[2,3-d] pyrimidine sulfate (U-89843D), a potent, novel, antioxidant/free radical scavenger. Biochem Pharmacol 51(4):471–476

Satoh T, Kosaka K, Itoh K, Kobayashi A, Yamamoto M, Shimojo Y, Kitajima C, Cui J, Kamins J, Okamoto S, Izumi M, Shirasawa T, Lipton SA (2008) Carnosic acid, a catechol-type electrophilic compound, protects neurons both in vitro and in vivo through activation of the Keap1/Nrf2 pathway via S-alkylation of targeted cysteines on Keap1. J Neurochem 104(4):1116–1131

Sharma S, Zhuang Y, Ying Z, Wu A, Gomez-Pinilla F (2009) Dietary curcumin supplementation counteracts reduction in levels of molecules involved in energy homeostasis after brain trauma. Neuroscience 161(4):1037–1044

Shih AY, Johnson DA, Wong G, Kraft AD, Jiang L, Erb H, Johnson JA, Murphy TH (2003) Coordinate regulation of glutathione biosynthesis and release by Nrf2-expressing glia potently protects neurons from oxidative stress. J Neurosci 23(8):3394–3406

Singh IN, Sullivan PG, Hall ED (2007) Peroxynitrite-mediated oxidative damage to brain mitochondria: Protective effects of peroxynitrite scavengers. J Neurosci Res 85(10):2216–2223

Singh IN, Gilmer LK, Miller DM, Cebak JE, Wang JA, Hall ED (2013) Phenelzine mitochondrial functional preservation and neuroprotection after traumatic brain injury related to scavenging of the lipid peroxidation-derived aldehyde 4-hydroxy-2-nonenal. J Cereb Blood Flow Metab 33(4):593–599

Smith SL, Hall ED (1998) Tirilazad widens the therapeutic window for riluzole-induced attenuation of progressive cortical degeneration in an infant rat model of the shaken baby syndrome. J Neurotrauma 15(9):707–719

Smith SL, Andrus PK, Zhang JR, Hall ED (1994) Direct measurement of hydroxyl radicals, lipid peroxidation, and blood-brain barrier disruption following unilateral cortical impact head injury in the rat. J Neurotrauma 11(4):393–404

Sonmez U, Sonmez A, Erbil G, Tekmen I, Baykara B (2007) Neuroprotective effects of resveratrol against traumatic brain injury in immature rats. Neurosci Lett 420(2):133–137

Sullivan PG, Thompson MB, Scheff SW (1999) Cyclosporin A attenuates acute mitochondrial dysfunction following traumatic brain injury. Exp Neurol 160(1):226–234

Sullivan PG, Krishnamurthy S, Patel SP, Pandya JD, Rabchevsky AG (2007) Temporal characterization of mitochondrial bioenergetics after spinal cord injury. J Neurotrauma 24(6):991–999

Toklu HZ, Hakan T, Biber N, Solakoglu S, Ogunc AV, Sener G (2009) The protective effect of alpha lipoic acid against traumatic brain injury in rats. Free Radic Res 43(7):658–667

Wang X, de Rivero Vaccari JP, Wang H, Diaz P, German R, Marcillo AE, Keane RW (2012) Activation of the nuclear factor E2-related factor 2/antioxidant response element pathway is neuroprotective after spinal cord injury. J Neurotrauma 29(5):936–945

Wood PL, Khan MA, Moskal JR, Todd KG, Tanay VA, Baker G (2006) Aldehyde load in ischemia-reperfusion brain injury: neuroprotection by neutralization of reactive aldehydes with phenelzine. Brain Res 1122(1):184–190

Wood PL, Khan MA, Moskal JR (2008) Mechanism of action of the disease-modifying anti-arthritic thiol agents D-penicillamine and sodium aurothiomalate: restoration of cellular free thiols and sequestration of reactive aldehydes. Eur J Pharmacol 580(1–2):48–54

Wu A, Ying Z, Gomez-Pinilla F (2006) Dietary curcumin counteracts the outcome of traumatic brain injury on oxidative stress, synaptic plasticity, and cognition. Exp Neurol 197(2):309–317

Xiong Y, Peterson PL, Muizelaar JP, Lee CP (1997) Amelioration of mitochondrial function by a novel antioxidant U-101033E following traumatic brain injury in rats. J Neurotrauma 14(12):907–917

Xiong Y, Peterson PL, Verweij BH, Vinas FC, Muizelaar JP, Lee CP (1998) Mitochondrial dysfunction after experimental traumatic brain injury: combined efficacy of SNX-111 and U-101033E. J Neurotrauma 15(7):531–544

Xiong Y, Shie FS, Zhang J, Lee CP, Ho YS (2005) Prevention of mitochondrial dysfunction in post-traumatic mouse brain by superoxide dismutase. J Neurochem 95(3):732–744

Yan W, Wang HD, Hu ZG, Wang QF, Yin HX (2008) Activation of Nrf2-ARE pathway in brain after traumatic brain injury. Neurosci Lett 431(2):150–154

Zhang DD (2006) Mechanistic studies of the Nrf2-Keap1 signaling pathway. Drug Metab Rev 38(4):769–789

Zhang JR, Scherch HM, Hall ED (1996) Direct measurement of lipid hydroperoxides in iron-dependent spinal neuronal injury. J Neurochem 66(1):355–361

Zhang R, Shohami E, Beit-Yannai E, Bass R, Trembovler V, Samuni A (1998) Mechanism of brain protection by nitroxide radicals in experimental model of closed-head injury. Free Radic Biol Med 24(2):332–340

Zhang H, Squadrito GL, Uppu R, Pryor WA (1999) Reaction of peroxynitrite with melatonin: a mechanistic study. Chem Res Toxicol 12(6):526–534

Chapter 4
Mitochondrial Damage in Traumatic CNS Injury

W. Brad Hubbard, Laurie M. Davis, and Patrick G. Sullivan

Abstract Traumatic brain injury (TBI) is ever-present in societal issues and puts a tremendous economic burden on U.S. healthcare and affected individuals. Symptomology after TBI is sometimes subtle without outward signs of injury and neurological impairments can be long-lasting. Unfortunately, the underlying cellular mechanisms that occur in response to TBI have not fully been deduced and as such there is no FDA-approved treatment for the resulting neurological pathology. In this chapter, we will review the essential role of mitochondria in the CNS while highlighting numerous functions of this crucial organelle under basal conditions. Furthermore, we review the bevy of literature showing the dysfunction of mitochondria after TBI. After TBI, these normal mitochondrial functions become dysregulated in response to elevated Ca^{2+} mitochondrial buffering and further cause downstream damage. The loss of adequate membrane potential can lead to lower ATP levels and necrotic cell death. Ca^{2+} overload also leads to the formation of the mitochondrial permeability transition pore and cytochrome c-mediated apoptosis. Aberrant levels of reactive oxygen species resultant from mitochondrial dysfunction cause increased lipid peroxidation and presence of 4-hydroxynonenal. All of these secondary cascades involved in mitochondrial dysfunction present unique therapeutic targets that investigators throughout the years have evaluated. The development of new pharmacological tools, with translatable potential, targeting mitochondrial pathways will further validate the critical role of mitochondria in outcomes after TBI and hopefully advance research to find an effective treatment.

Keywords Mitochondria · Traumatic brain injury · Electron transport chain · Cell death · Reactive oxygen species

---

W. B. Hubbard · L. M. Davis · P. G. Sullivan (✉)
Department of Neuroscience & The Spinal Cord & Brain Injury Research Center (SCoBIRC), The University of Kentucky Chandler College of Medicine, Lexington, KY, USA
e-mail: patsullivan@uky.edu

D. G. Fujikawa (ed.), *Acute Neuronal Injury*,
https://doi.org/10.1007/978-3-319-77495-4_4

4.1 Traumatic Brain Injury

With over 1.5 million injuries every year, traumatic brain injury (TBI) has become an almost ubiquitous phenomenon in our country. As the underlying neural damage can present without any outward sign of physical damage and the victims are usually cognitively impaired, it has become a largely 'silent epidemic' (Jennett 1998; Thurman et al. 1999; Jager et al. 2000; Langlois et al. 2006; CDC 2015). Current advancements in medicine have allowed patients who would have previously succumbed to their wounds to survive long after their initial injury. Also, there is a growing population of individuals who sustain a mild to moderate TBI who do not seek treatment and often develop prolonged and chronic neurological symptoms (CDC 2015). The growing injured population has presented our society with an enormous economic and social burden, as these patients are commonly unable to properly reintegrate into their previous professional and social networks. An estimated 43% of hospitalized TBI patients sustain long-term disability, highlighting the need for rehabilitative care and support for these patients throughout their lives (Selassie et al. 2008). There is a substantial economic impact with initial health care costs and long term support for TBI totaling tens of billions dollars per year nationally (Langlois et al. 2006). There are some clinical options designed to allow people to survive their injuries, which include assuaging acute brain edema, decreasing intracranial pressure, and preventing of peripheral complications; however there is a void of treatment to mitigate neuropathology or to attenuate and recover the loss of neural tissue (Hatton 2001).

Perhaps the most insidious aspect of TBI is that it can occur without obvious signs of injury to the patient's body. There have been recorded medical incidences of mysterious neurological disorders dating back to World War I (WWI). Physicians in the British armed forces had given it the somewhat enigmatic label of "shell shock" (SS) (Jones et al. 2007). Although some cases could be attributed to psychosis, by 1917 SS was responsible for 14% of all discharges from the armed forces, and accounted for 33% of all discharges of non-wounded soldiers. Actual brain injuries that were labeled under the umbrella term of SS were potentially the result of combat conducted in "trenches," where explosives fragments as well as reflected blast overpressure could impact soldiers (Mott 1916). Symptoms of SS ranged from headaches to depression with lack of standard symptom scoring. It had become so prevalent throughout the armed forces and had such a wide array of presenting symptoms that it was highly debated whether or not it was a real condition, and the etiology and management was highly disputed during the early twentieth century. By the end of WWI the prevalence of SS began to incur a large financial burden upon the British armed forces, primarily due to the 32,000 pensions that had been awarded to "neurasthenic" soldiers suffering from SS with no obvious cerebral injury. The controversial definition of the disorder and its method of treatment, in addition to the development of public controversy and stigma over diagnosis, delayed the development of a treatment protocol and even caused the British army to ban the use of the term "shell shock" from medical reports.

During World War II (WWII) the British army banned the SS terminology in hopes of avoiding another epidemic of these cases, which they may or may not have viewed as physical disorders. However, with the start of the war it became readily apparent that disavowing the existence of this disorder did not prevent another epidemic. In response to the army regulations regarding this disorder, alternative terminology arose in its place, such as post-concussion syndrome (PS) or post-trauma concussion state (coined by Shaller). Eventually, physicians began to realize that many of the soldiers that suffered from this concussed state had been in close proximity of an explosion during battle. This led them to speculate that some force, that had no perceptible outward effect on the body, had a substantial effect on fragile neural tissues. In an attempt to, once again, clarify the etiology of this disorder, Denny-Brown suggested that it was the timeline of symptom presentation within the individual patient instead of the symptom type that was the key factor between severe head injury and PS. His etiological account indicated that severe head injury would present with immediate neurologic symptoms that would trend toward recovery, whereas PS would have delayed onset of neurologic symptoms with a trend toward worsening symptoms (Jones et al. 2007). It has been estimated that 50% of patients with a mild TBI can develop PS, consisting of dizziness, headaches, cognitive dysfunction, sleep disorders, and depression (Langlois et al. 2006; Rapoport et al. 2006; CDC 2015). This delayed development of symptoms in the mild to moderate patient populations is perhaps the most unfortunate aspect of this condition, as soldiers and civilians can often suffer immense psychiatric morbidity without realizing that they require medical treatment for a physical injury. It has been observed clinically that even mild or moderate TBI can require neurosurgical intervention, and any delay in treatment could prove to be costly in terms of cognitive and functional recovery (Setnik and Bazarian 2007; Losoi et al. 2016).

Figure 4.1 shows the "iceberg" effect of TBI incidence according to the CDC. TBI accounts for over 50,000 deaths per year, though this is only the tip of the iceberg in terms of those whose lives are impacted by TBI and have to receive medical care. The alarming statistic is that the number of individuals who sustain TBIs that are not treated or diagnosed is inestimable. Persons in this category can often lead lives with complications that go unresolved. Although great strides have been made in TBI education and diagnosis (ED visits for TBI-related injures doubled from 2003 to 2010), these efforts need to continue to provide the ability for diagnosis for all TBIs (CDC 2015). This can range from informing the general public about brain injury as well as clinical head trauma management. As advances in medicine continue, patients are able to survive injuries that would be fatal if treated several decades ago in the battlefield and clinic (Okie 2005; Warden 2006). For multiple organ injury sustained from motor vehicle accidents or severe falls, necessary steps must be taken to observe brain damage early on, despite other high priority injuries to the patient (Leong et al. 2013; Sharma et al. 2014). It has also been shown that multi/polytrauma coupled with TBI can change the inflammatory dynamics in the prognosis of TBI, so interaction with other injuries should be accounted for in treatment (McDonald et al. 2016).

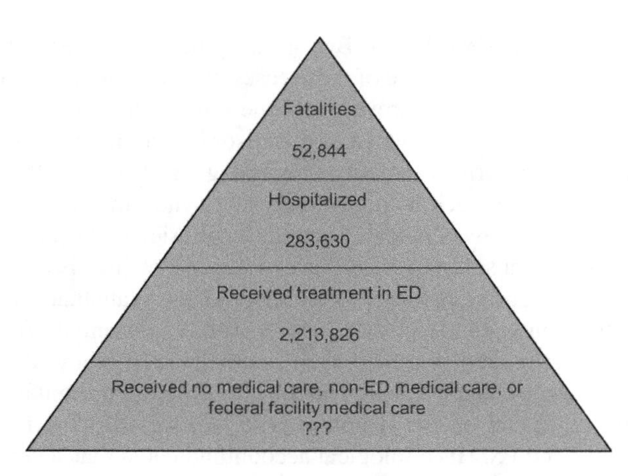

Fig. 4.1 The iceberg effect of TBI. Although there are over 50,000 fatalities yearly in the US, this is only a small fraction of the people who are affected each year by TBIs. With respect to TBI-related fatalities, over five times more people are hospitalized and over 40 times more people are treated in an Emergency Department (ED). These numbers greatly underestimate the total amount of people who sustain a TBI

In recent years, mild TBI has become a major concern of military medicine, as over 80% of TBI are classified as mild by the Department of Defense. Since 2000, the number of TBIs diagnosed in U.S. service members steadily increased until 2011, where the peak reached around 34,000 diagnoses (DoD 2014). Blast has caused the majority of injuries and TBIs in recent war zones, and has been labeled as a signature injury of recent U.S. involvement in Middle Eastern conflicts, such as Operation Enduring Freedom (OEF) and Operation Iraqi Freedom (OIF) (Warden 2006; Hoge et al. 2008; Snell and Halter 2010). Mild and moderate injury have also become a long-term problem coupled with prolonged cognitive dysfunction within the armed forces population, with approximately 18% prevalence in various reports (Hoge et al. 2008). In addition to prolonged cognitive deficits, this injury population also has an increased predisposition to the development of post-traumatic stress disorder (PTSD). Several studies have found that Veterans from recent military conflicts (OIF/OEF) predominantly present with a combination of conditions, namely chronic pain, PTSD, and persistent postconcussive symptoms (PPCS) (Clark et al. 2007a, 2009; Lew et al. 2009; Sayer et al. 2009). For clinical reports, TBI refers to the initial insult sustained and its pathology, whereas PPCS relates to the expression of symptoms following TBI (Marshall et al. 2015). There is also a problem of failure to report due to a perceived stigma concerning psychological problems within the armed forces population, which could contribute to the development of chronic neurologic dysfunction due to physical injury within this population of injured patients (Hoge et al. 2004).

Within the civilian population of the United States ~2% of the population (5.3 million) is currently living with long-term disabilities resulting from TBI (Langlois et al. 2006). Mild TBI is a growing concern in the US civilian population as the

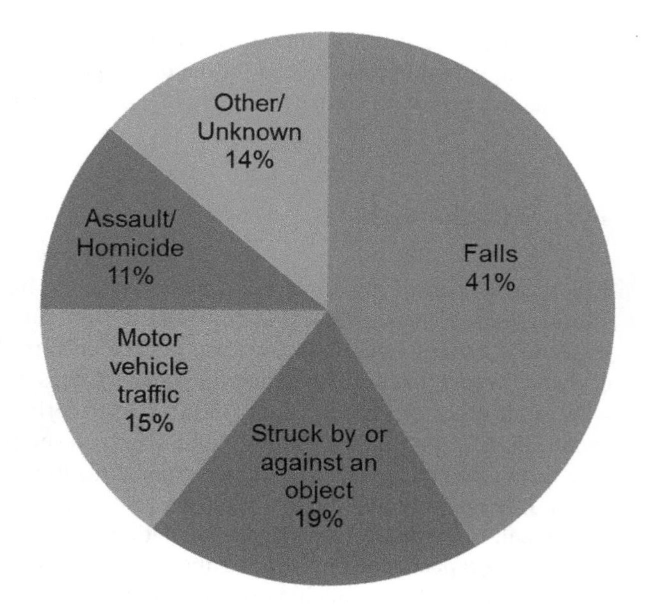

Fig. 4.2 Mechanism of TBI for Emergency Department (ED) visits. Falls and being struck by or against an object are the two leading causes of ED visits for TBI-related incidents. These type of insults often results in "mild" TBIs that do not result in fatalities

leading causes of emergency department (ED) visits due to TBI (Fig. 4.2) are falls and "struck by/against" incidents, which often are non-severe impacts (CDC 2015). There has also been an increasing population of pediatric (5–18 years) TBI cases resulting from sports-related injuries, which can often be misdiagnosed as the symptoms manifest as lethargy, irritability or fatigue (Yang et al. 2008; CDC 2015). As knowledge is gained, it has become apparent that special precautions are needed after pediatric head trauma and return to play/return to classroom initiatives and protocols garner more attention. However, implementation of these novel protocols is limited due to the unique challenges they present (Halstead et al. 2013; Dettmer et al. 2014). TBI has a biphasic age distribution of incidence; occurring in young (<25) and elderly (>75) populations (Langlois et al. 2006; Rutland-Brown et al. 2006). Unique considerations are required for these populations; children and adolescents can face setbacks in cognitive and functional development after sustained TBI, while older adults have a higher mortality risk and slower rates of functional recovery (Cifu et al. 1996; Susman et al. 2002; Anderson et al. 2006; Gerrard-Morris et al. 2010). Over 50,000 deaths are attributed to TBI each year, as well as 283,000 hospitalizations and 2,200,000 ED visits (Fig. 4.1). With such a high incidence and great propensity for the development of chronic symptoms, the total medical costs incurred by individuals currently living with TBI within the U.S. can reach $50 billion dollars per year. This figure increases to $60 billion when lost societal productivity of these individuals is factored in; however, these figures do not factor in how this disorder impacts social and family dynamics (Langlois et al. 2006; Rutland-Brown et al. 2006). Despite life satisfaction for some individuals after sustaining

TBI, numbers are still low, reportedly at 30% 2 years after injury, for the general population, stressing a clear need for the development of neuroprotective therapies and effective protocols for the treatment of TBI (Corrigan et al. 2001).

4.2 Mitochondria

The development of mitochondrial function was the basis of the development of multi-cellular organisms. It was at this evolutionary crossroads that the cell was able to produce enough energy, in the form of ATP, to form highly complex interconnected networks that developed into the organ systems we see in the human body as well as all other organisms (Lane 2006). Underscoring the dependence on mitochondrial ATP production is the evolutionary development of all multicellular organisms upon this planet to require oxygen utilization through some sort of respiration. It is essential that mitochondria are provided with adequate oxygen in order for the cell to maintain homeostatic regulation of its intercellular processes (Lane 2006). This dependence on oxygen supply for mitochondria is further exaggerated in neural tissue where mitochondria play an indispensable role in supplying energy for CNS function. For example, the brain makes up ~2% of body mass but the brain consumes ~20% of oxygen taken into the body. The importance of oxygen consumption is highly evident when we examine any pathological disease in which tissues become oxygen-deprived (ischemic) for even the shortest time period. These regions undergo massive cellular loss as a result of mitochondrial damage and dysfunction, leading to the initiation of cell death pathways, such as necrosis and apoptosis (Obrenovitch 2008).

Mitochondria have been studied since the end of the nineteenth century, and these organelles have proven to be one of the most important discoveries in the history of cellular research. Mitochondrial function has been the root of many discoveries ever since Kolliker, Altman, and Benda described their presence in cells in the later part of the nineteenth century. The first Nobel Prize for mitochondrial research was awarded to Meyerhof in 1922 for the discovery of the connection between substrate oxidation and oxygen consumption in relation to glycolysis. Next to be awarded in 1931 was the work done by Warburg on the nature and mode of action of the respiratory enzyme, indicating that ATP production was coupled with enzymatic oxidation of glyceraldehyde phosphate. Szent-Gyorgyi was awarded the Nobel Prize in 1937 for the discovery of the connection with biological combustion process of dicarboxylic acids within respiration. In 1997, the Nobel Prize in Chemistry was awarded to Boyer and Walker for their discovery of the enzymatic mechanism underlying the synthesis of ATP. The role mitochondria play in intercellular cascades continues to expand for normal and disease states of cell function (Pagliarini and Rutter 2013). Although we have learned much over the past century about mitochondrial bioenergetics, there remains a great deal to be discovered.

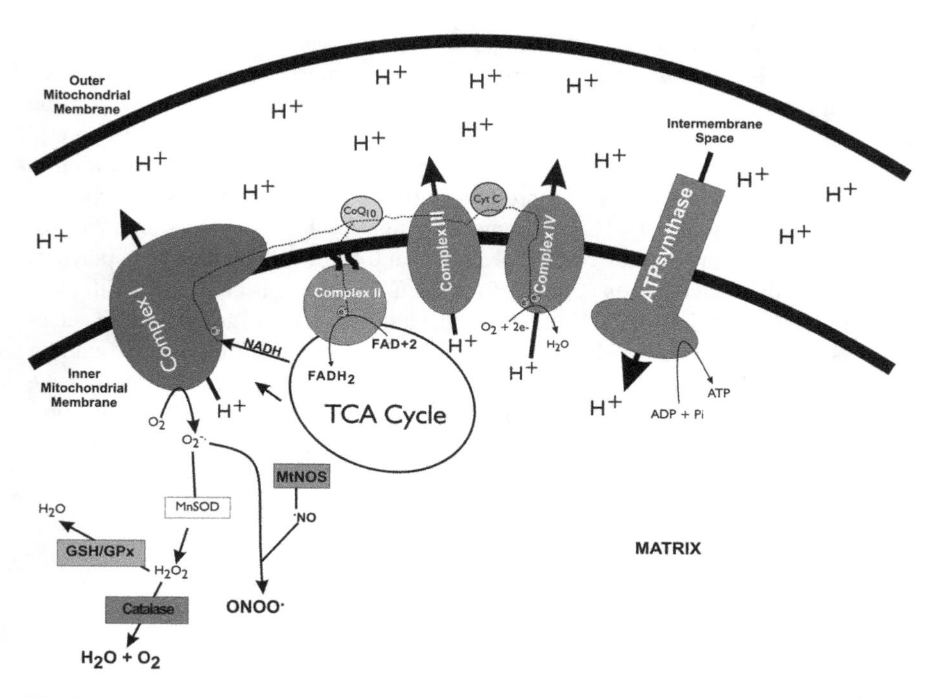

Fig. 4.3 Normal mitochondrial function, utilizing five protein complexes which make up the electron transport chain (ETC). The proton gradient established by the ETC is what drives Complex V to create ATP, the energy source for cellular processes

In order to discuss mitochondrial dysfunction we must first discuss normal mitochondrial function (Fig. 4.3). Mitochondria are intracellular organelles with a dual (inner and outer) membrane system, each of which is responsible for specific functions. The outer membrane (OM) contains many transporter proteins which import and export many ions and proteins necessary for mitochondrial function (Nicholls and Ferguson 2002). The inner membrane (IM) exhibits many folds, termed cristae, which increase the surface area available for mitochondrial respiration. The space which is enclosed by the IM is called the matrix and contains enzymes involved in cellular metabolism and calcium regulation. Within the IM there are a series of five protein complexes that comprise the electron transport chain (ETC), which is the primary site of ATP production within the cell (Fig. 4.3). Complex I (NADH-Ubiquinone Oxidoreductase), which is embedded within the IM, converts NADH to NAD^+ by accepting an electron into the Fe-S center of the protein. As a byproduct of this electron donation, a proton is translocated from the matrix to the intermembrane space (IMS), which is located between the inner and outer membranes. Complex II (Succinate Dehydrogenase), in addition to its function as an ETC protein, is also a key component of the Krebs Cycle, which converts the glycolytic product pyruvate into different molecules in order to produce substrates for the ETC. This complex utilizes the conversion of succinate to accept electrons from $FADH_2$ into the ETC, and as it is only anchored to the inner half of the IM, there is no translocation of protons from the matrix to the IMS. Complex I and II transfer their electrons to ubiquinone (Coenzyme Q_{10}) located within the IM (Fig. 4.3). These electrons are then passed to Complex III (Ubiquinone-Cytochrome-C Oxidoreductase) via the Q-cycle, resulting in proton translocation to the IMS. Another electron transfer protein, Cytochrome c, accepts this electron and transports it to Complex IV (Cytochrome-C Oxidase) which displaces another proton into the IMS via complex IV (Fig. 4.3). It is at Complex IV that oxygen plays its vital role as the final electron acceptor for the ECT, where it is combined with electrons to form H_2O (Fig. 4.3). Without the presence of the oxygen molecule, electrons become backed up within the ETC, resulting in damage to the surrounding structures. Meanwhile, all of the protons that have been pumped into the IMS create a proton concentration gradient ($\Delta\Psi$) which is utilized by Complex V (ATP synthase) to facilitate phosphorylation of ADP into ATP for use as an energy source for cellular processes (Fig. 4.3).

4.3 Role of Ca^{2+} in Neurons and Mitochondrial Ca^{2+} Sequestration

Although the complex mechanisms of secondary neuronal injury following TBI are poorly understood, it is clear that excitatory amino acid (EAA) neurotoxicity plays an important upstream role (Faden et al. 1989). Elevated EAAs increase the levels of intracellular Ca^{2+} ($[Ca^{2+}]_i$) by activation of N-methyl-D-asparate (NMDA) receptor/ion channels, α-amino-3-hydroxy-5-methyl-4-isoxazole propionic acid (AMPA) receptors and voltage-gated Ca^{2+} channels. This results in excessive entry of Ca^{2+} into the cell, leading to a loss of homeostasis and subsequent neuronal Ca^{2+} overload. Calcium is the most common signal transduction element in cells, but unlike other second-messenger molecules, it is required for life. Paradoxically, prolonged high levels of $[Ca^{2+}]_i$ lead to cell death. Excessive $[Ca^{2+}]_i$ can damage the structure of nucleic acids and proteins, interfere with kinase activity, and activate proteases or phospholipases, causing cellular damage. Therefore, maintenance of low $[Ca^{2+}]_i$ is necessary for proper cell function and to allow brief pulses of it to initiate second-messenger pathways, allowing intracellular communication. Since Ca^{2+} cannot be metabolized like other second-messenger molecules, it must be tightly regulated by cells. Numerous intracellular proteins and some organelles have adapted to bind or sequester Ca^{2+} to ensure that homeostasis is maintained. Mitochondria are one such organelle.

The mitochondrial membrane potential ($\Delta\Psi$) is generated by the translocation of protons across the inner mitochondria membrane via the ETC, culminating in the reduction of O_2 to H_2O. This store of potential energy (the electrochemical gradient) can then be coupled to ATP production as protons flow back through ATP synthase and complete the proton circuit. The potential can also be used to drive Ca^{2+} into the mitochondrial matrix via the electrogenic uniporter when cytosolic levels increase. When cytosolic levels of Ca^{2+} decrease, mitochondria pump Ca^{2+} out to precisely regulate cytosolic Ca^{2+} homeostasis. During excitotoxic insult, Ca^{2+} uptake into mitochondria has been shown to increase ROS production, inhibit ATP synthesis, and induce mitochondrial permeability transition.

4.4 Free Radicals, ΔΨ, Mitochondria and Neuronal Cell Death

Free radical production is a byproduct of ATP generation in mitochondria via the ETC. Electrons escape from the chain and reduce oxygen (O_2) to superoxide ($\bullet O_2^-$) when kinetic factors favor one-electron reduction (Murphy 2009). Normally cells convert $\bullet O_2^-$ to hydrogen peroxide (H_2O_2), utilizing both manganese superoxide dismutase (MnSOD), which is localized to the mitochondria, and copper-zinc superoxide dismutase (Cu-ZnSOD), found in the cytosol. With normal mitochondria function, H_2O_2 is rapidly converted to H_2O via catalase and glutathione peroxidase, but extensive H_2O_2 efflux can occur when there is a high NADH/NAD^+ in the matrix or a highly reduced coenzyme Q (CoQ) pool accompanied by high proton motive force in isolated mitochondria (Murphy 2009). H_2O_2 has the potential to be converted to the highly reactive hydroxyl radical (•OH) via the Fenton reaction, underlying ROS neurotoxicity (Thomas et al. 2009). •OH rapidly attacks unsaturated fatty acids in membranes, causing lipid peroxidation and the production of 4-hydroxynonenal (HNE) that conjugates to membrane proteins, impairing their function (Fig. 4.4). Such oxidative injury results in significant alterations in cellular function. Free radicals can also damage DNA, leading to activation of DNA repair enzymes, such as poly-ADP ribose polymerase (PARP) (Prins et al. 2013). PARP activation has been shown to deplete NAD^+ pools in the cytoplasm and lead to eventual cell death (Cipriani et al. 2005; Alano et al. 2010). In particular, ROS induction of lipid peroxidation and protein oxidation products may be particularly important in neurodegeneration and TBI (Hall and Sullivan 2004; Lifshitz et al. 2004).

Mitochondrial ROS production is intimately linked to ΔΨ such that hyperpolarization (high ΔΨ) increases and promotes ROS production. The underlying mechanism is the altered redox potential of ETC carriers (reduced) and an increase in semiquinone anion half-life time (high ΔΨ prevents oxidation of cytochrome b1 in the Q cycle). In other words, at a high ΔΨ, protons can no longer be pumped out of the matrix (against the electrochemical proton gradient) by the chain resulting in electron transport stalling. This leads to a sustained low level of intermediates, which increases the chance that electrons escape from these intermediates, thereby reducing oxygen and increasing ROS production. Since the magnitude of ROS production is largely dependent on- and correlates with—ΔΨ, even a modest reduction via increased proton conductance (decreases ΔΨ, the electrochemical proton gradient) across the mitochondrial inner membrane (uncoupling) reduces ROS formation (Sullivan et al. 2004b; Pandya et al. 2007).

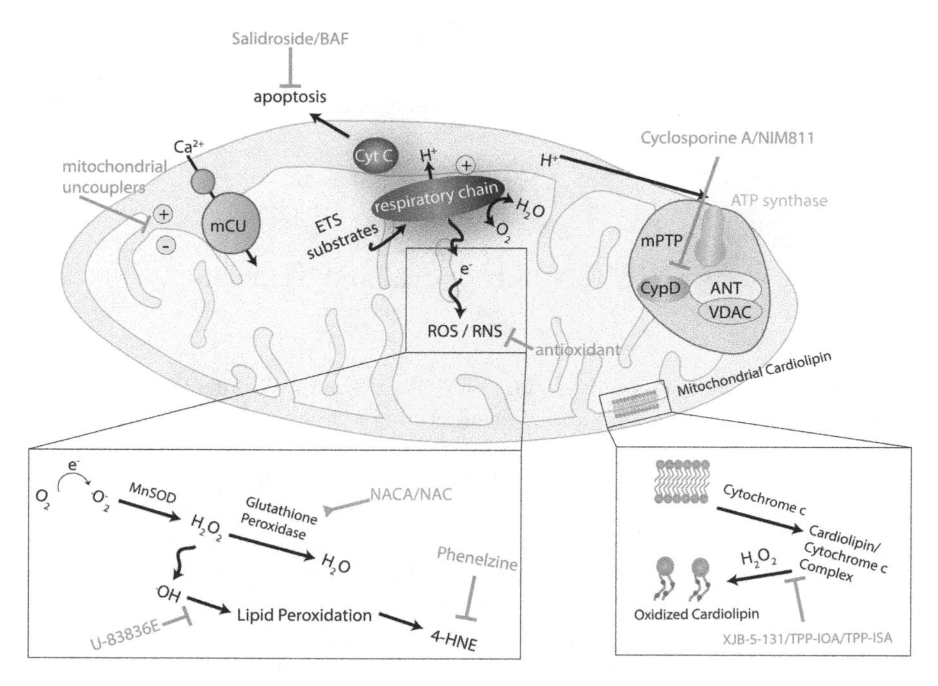

Fig. 4.4 Mitochondrial-targeted therapeutic approaches to combat eventual ROS overload and cell death. *BAF* boc-aspartyl(OMe)-fluoromethylketone, *Cyt C* cytochrome c; *mCU* mitochondrial calcium uniporter, *ETS* electron transport chain, *ROS* reactive oxygen species, *RNS* reactive nitrogen species, *NIM811* N-methyl-4-isoleucine cyclosporin, *mPTP* mitochondrial permeability transition pore, *CypD* cyclophilin D, *ANT* adenine nucleotide translocase, *VDAC* voltage-dependent anion channel, *MnSOD* manganese superoxide dismutase, *NACA* N-acetylcysteineamide, *NAC* N-acetylcysteine, *4-HNE* 4-hydroxynonenal, *TPP-IOA* triphenylphosphonium conjugated imidazole oleic acid, *TPP-ISA* triphenylphosphonium conjugated imidazole stearic acid. Adapted from Yonutas et al. (2016)

4.5 Mitochondria and TBI

It has become increasingly clear that TBI, as well as other neurological disorders, either cause or are the result of mitochondrial dysfunction (Hovda et al. 1992; Sullivan et al. 1998, 2002, 2005; Nicholls and Budd 2000; Hatton 2001; Pellock et al. 2001; Schurr 2002; Tieu et al. 2003; Lifshitz et al. 2004; Sullivan 2005). Unfortunately there is very little we can do to prevent the initial blunt force trauma that is caused by TBI; however, we may be able to intervene within the massive secondary signaling cascade that can last for hours to weeks following the primary insult. Secondary injury is initiated by a massive depolarization of the plasma membrane by voltage-dependent Na^+ channels. Along with glutamate release, this depolarization causes the removal of the Mg^+ block within NMDA channels, causing a massive Ca^{2+} influx into the cell (Nicholls et al. 1999; Nicholls and Budd 2000;

Gunter et al. 2004). This Ca^{2+} can activate many damaging cellular enzymes within the cytosol, and as such must be sequestered by intracellular organelles, mainly the mitochondria. After Ca^{2+} is imported into mitochondria via membrane potential-driven transporters, it is stored as a Ca^{2+} phosphate compound within the matrix, causing the matrix to have an almost gel-like consistency (Nicholls and Budd 2000).

However, the Ca^{2+} buffering capacity of mitochondria is finite and eventually the Ca^{2+} influx becomes too great, resulting in mitochondrial dysfunction and subsequent initiation of cell death pathways (Brookes et al. 2004; Lifshitz et al. 2004; Sullivan et al. 2004a, 2005). Calcium seems to affect primarily complex-driven respiration; damage to this major site of electron acceptance can significantly hinder the ability of the mitochondria to produce ATP (Tieu et al. 2003; Gunter et al. 2004; Sleven et al. 2006; Maalouf et al. 2007). The loss of adequate membrane potential will cause the ATP synthase to run in reverse, thereby dephosphorylating ATP and pumping protons into the IMS in an attempt to restore membrane potential and preserve mitochondrial homeostasis. However, by depleting ATP stores, membrane channels that require energy to maintain ionic balances will be unable to sustain operations. This causes the mitochondria and the cell to swell and eventually burst, which are characteristic signs of necrotic cell death (Nicholls and Budd 2000; Sullivan et al. 2005). Calcium overload can also activate intramitochondrial proteins, such as μ-calpain (calpain I), that contribute to the formation of the mitochondrial permeability transition pore (mPTP) and release of IM proteins (Scheff and Sullivan 1999; Nicholls and Budd 2000; Garcia et al. 2005; Sullivan et al. 2005).

It is also important to note that inhibition of mitochondrial Ca^{2+} uptake by reducing $\Delta\Psi$ (chemical uncoupling) following excitotoxic insults is neuroprotective, emphasizing the pivotal role of mitochondrial Ca^{2+} uptake in EAA-mediated neuronal cell death (Sullivan et al. 2004b; Pandya et al. 2007, 2009). Membrane uncouplers, such as 2,4-dinitrophenol (2,4-DNP) and carbonilcyanide p-triflouromethoxyphenylhydrazone (FCCP) ("mitochondrial uncouplers" in Fig. 4.4), have been shown to decrease Ca^{2+} burden and lower oxidative damage, but optimal dosage is extremely important (Pandya et al. 2007, 2009).

The formation of the mPTP results in mitochondrial dysfunction and has been shown to occur after acute TBI (Nicholls and Budd 2000; Sullivan et al. 2000b, 2005). This structure spans both inner and outer membrane, and causes a massive efflux of calcium into the cytosol and the release of death-inducing proteins, ultimately leading to cellular loss and cognitive dysfunction. The mPTP is a pore comprised of multiple mitochondrial proteins within the inner and outer membranes, including the adenine nucleotide translocase (ANT), inner and outer protein transporters (Tim/Tom), voltage-dependent anion channel (VDAC), and cyclophilin D. Recent data has also implicated a role for the ATP synthase as either a component or modulator of the mPTP, but it is still controversial (Jonas et al. 2015; Richardson and Halestrap 2016). The mPTP allows nonspecific conductance of matrix and intermembrane space components to the cytosol, where they can activate detrimental signaling cascades leading to cell death. It has been repeatedly shown that cyclosporine A (CsA) (and CsA analogs) inhibits opening of the mPTP (Fig. 4.4), leading to stable mitochondrial membrane potential and calcium homeostasis

(Sullivan et al. 1999, 2000a, b, 2011; Mbye et al. 2008; Readnower et al. 2011). Specifically, NIM811, a non-immunosuppressive CsA analog, at 10 mg/kg improves tissue sparing, mitochondrial functionality, and reduces 4-HNE levels (Readnower et al. 2011). Recently, it has also been shown that CsA improves respiration by targeting the more damaged synaptic mitochondria, relative to non-synaptic mitochondria (Kulbe et al. 2016).

One protein that is highly involved in both normal mitochondrial respiration and cell death cascades is cytochrome *c*. It is normally found in the IMS electrostatically attached to the inner membrane where it shuttles electrons from complex III to complex IV. However, in the presence of increased Ca^{2+} levels it is cleaved from the inner membrane by the Ca^{2+}-activated cysteine protease μ-calpain (Sullivan et al. 2002; Nasr et al. 2003; Garcia et al. 2005). After mPTP opening it is released into the cytosol where it binds to apoptosis activation factor-1 (Apaf-1), which is also bound to pro-caspase 9. This complex, known as the apoptosome, initiates the activation of caspase 3 and subsequent cleavage of apoptotic substrates ultimately resulting in cellular loss. The opening of the mPTP also releases apoptosis inducing factor (AIF) and endonuclease G (Endo G), both of which are responsible for nuclear DNA degradation. Cytochrome c translocation and subsequent DNA fragmentation has been reported 1 h and 4 h, respectively, after controlled cortical impact (CCI); cytochrome c cytosolic fractions were higher after injury in animals lacking MnSOD, suggesting an oxidative-dependent mechanism (Lewen et al. 2001). It has been show that cytochrome c translocation induces apoptotic pathways (Fig. 4.4), potentially even necrosis (Lewen et al. 2001; Sullivan et al. 2002).

As described previously, a common byproduct of normal mitochondrial function is the production of ROS. However, TBI-induced mitochondrial damage leads to Ca^{2+} loading which significantly increases mitochondrial ROS production primarily at complex I and III of the ETC. Recently, it has become increasingly apparent that the key to maintaining cellular and mitochondrial function is to decrease the levels of oxidative stress- induced damage after this TBI-induced excitotoxic Ca^{2+} influx (Hatton 2001; Sullivan et al. 2005; Singh et al. 2006). However, because the production of oxidative stress molecules is a normal byproduct of mitochondrial function and mitochondrial function is required for proper cellular function, there must be a balance between preserving mitochondrial function and reducing oxidative damage. Therefore, the metabolic pathways involving mitochondria can become a critical component of the treatment of TBI (Robertson et al. 1991; Davis et al. 2008; Prins 2008).

ROS accumulation can also activate PARP, which leads to cell death (Duan et al. 2007). PARP1 is elevated 8 h after CCI and inhibition of PARP activity has been shown to decrease lesion size, maintain NAD^+ balance and improve functional performance in mice (LaPlaca et al. 2001; Satchell et al. 2003). N-Acetylcysteineamide (NACA), as a pre-cursor, can increase levels of glutathione in order to reduce ROS accumulation. NACA, which has more CNS bioavailability compared to N-acetylcysteine (NAC), administration, has been shown to improve cortical tissue sparing and reduce oxidative damage (HNE levels) in the brain, as well as restoring glutathione content (Fig. 4.4) (Pandya et al. 2014).

Pharmacological studies have been performed to target lipid peroxidation (Fig. 4.4) (Mustafa et al. 2011; Ji et al. 2012; Singh et al. 2013). Phenelzine, a scavenger of lipid peroxidation-derived reactive aldehyde 4-hydroxynonenal (HNE), given 15 min after CCI, prevented uncoupling after energy in contused brain tissue and also increased spared tissue (Singh et al. 2013). Another lipid peroxidation inhibitor, U-83836E, has been reported to attenuate oxidative damage, preserve Ca^{2+} buffering capacity, and reduce calpain-mediated cytoskeletal damage after CCI (Mustafa et al. 2010; Mustafa et al. 2011). Ji et al. has demonstrated that mitochondria-targeted small molecule inhibitors are needed to target cardiolipin oxidation, which is central to neuronal death pathways (Ji et al. 2012). Cardiolipin oxidation is a major pathway in lipid peroxidation and provides a target for therapeutic inhibitors of mitochondrial-related pathology after TBI (Anthonymuthu et al. 2016). Among these inhibitors are XJB-5-131, which prevents $\bullet O_2^-$ and H_2O_2 formation, and triphenylphosphonium conjugated imidazole oleic acid (TPP-IOA) and stearic acid (TPP-ISA), which prevent cytochrome c/cardiolipin complex peroxidase activity (Fig. 4.4) (Atkinson et al. 2011; Ji et al. 2012, 2015).

Finally, antiapoptotic approaches have been explored to preserve mitochondria after TBI (Fig. 4.4) (Clark et al. 2007b; Chen et al. 2012). Salidroside sparks an antioxidant mechanisms and increases survival signaling via PI3K/Akt pathway, aiding mitochondrial protection (Chen et al. 2012). Boc-aspartyl(OMe)-fluoromethylketone (BAF) is a caspase inhibitor that has been reported to lower cytochrome c release and brain tissue loss (Clark et al. 2007b).

Many different models of TBI have been employed in rodents to study the effects of mitochondrial dysfunction, such as controlled cortical impact (CCI) and non-impact rotational head injury. Mitochondrial dysfunction is established and easily observed in these models of moderate to severe TBI, though it is unknown whether aberrant mitochondrial function is present after mild TBI. Fluid percussion injury (FPI) has been employed to study mitochondrial function after a "mild" insult. mPTP opening, deficits in Ca^{2+} uptake in surviving mitochondria, and ETC alterations have been reported in various groups after FPI (Xing et al. 2013; Murugan et al. 2016; Sun and Jacobs 2016). While these groups label low-level FPI as "mild" TBI, it is not representative of mild closed head injury. Few studies on mitochondrial-related decreased metabolic parameters (NAA and ATP-ADP ratio) after closed head injuries (weight drop model) have been reported (Vagnozzi et al. 2005, 2007; Di Pietro et al. 2014). These models also show ROS accumulation after mild TBI (Tavazzi et al. 2007). Recently, mild TBI has risen to prominence in the preclinical research community due to its prevalence and inconspicuous nature. While there have been limited mitochondrial dysfunction studies performed in mild TBI, findings of ROS accumulation and cell death suggest that mitochondrial play a central role in mild injury as well.

4.6 Closing Remarks

Strategies that target specific mitochondrial mechanisms have proven beneficial in preclinical studies as therapeutic interventions following central nervous system (CNS) injury (Yonutas et al. 2016). The combination of all experimental data presented demonstrates that mitochondrial function is severely impaired following TBI and that this dysfunction is related to cell death pathways known to be activated in these distinct models. The loss of mitochondrial homeostasis that occurs following CNS injury implies that mPTP activation may be a common link in several models of TBI and spinal cord injury (SCI). The pathophysiological role of mPT in CNS injury is also supported by several lines of scientific work that have utilized inhibitors (e.g. CsA and its derivatives) of the mitochondrial permeability transition (mPT) in vivo to test this hypothesis. In addition, targeting of ROS accumulation, cardiolipin oxidation, lipid peroxidation and apoptosis has proven successful in preclinical studies. The development of new pharmacological tools, with translatable potential, that specifically target the mPT and other mitochondrial pathological pathways will provide further support for the vital role of mitochondrial regulation in CNS injury and may prove beneficial as possible treatments for this and other neurodegenerative conditions.

References

Alano CC, Garnier P, Ying W, Higashi Y, Kauppinen TM, Swanson RA (2010) NAD+ depletion is necessary and sufficient for poly(ADP-ribose) polymerase-1-mediated neuronal death. J Neurosci 30:2967–2978. https://doi.org/10.1523/JNEUROSCI.5552-09.2010

Anderson VA, Catroppa C, Dudgeon P, Morse SA, Haritou F, Rosenfeld JV (2006) Understanding predictors of functional recovery and outcome 30 months following early childhood head injury. Neuropsychology 20:42–57. https://doi.org/10.1037/0894-4105.20.1.42

Anthonymuthu TS, Kenny EM, Bayir H (2016) Therapies targeting lipid peroxidation in traumatic brain injury. Brain Res 1640:57–76. https://doi.org/10.1016/j.brainres.2016.02.006

Atkinson J, Kapralov AA, Yanamala N et al (2011) A mitochondria-targeted inhibitor of cytochrome c peroxidase mitigates radiation-induced death. Nat Commun 2:497. https://doi.org/10.1038/ncomms1499

Brookes PS, Yoon Y, Robotham JL, Anders MW, Sheu SS (2004) Calcium, ATP, and ROS: a mitochondrial love-hate triangle. Am J Physiol Cell Physiol 287:C817–C833

CDC (2015) Report to congress on traumatic brain injury in the United States: epidemiology and rehabilitation. National Center for Injury Prevention and Control; Division of Unintentional Injury Prevention, Atlanta, GA

Chen SF, Tsai HJ, Hung TH et al (2012) Salidroside improves behavioral and histological outcomes and reduces apoptosis via PI3K/Akt signaling after experimental traumatic brain injury. PLoS One 7:e45763. https://doi.org/10.1371/journal.pone.0045763

Cifu DX, Kreutzer JS, Marwitz JH, Rosenthal M, Englander J, High W (1996) Functional outcomes of older adults with traumatic brain injury: a prospective, multicenter analysis. Arch Phys Med Rehabil 77:883–888

Cipriani G, Rapizzi E, Vannacci A, Rizzuto R, Moroni F, Chiarugi A (2005) Nuclear poly(ADP-ribose) polymerase-1 rapidly triggers mitochondrial dysfunction. J Biol Chem 280:17227–17234. https://doi.org/10.1074/jbc.M414526200

Clark ME, Bair MJ, Buckenmaier CC 3rd, Gironda RJ, Walker RL (2007a) Pain and combat injuries in soldiers returning from Operations Enduring Freedom and Iraqi Freedom: implications for research and practice. J Rehabil Res Dev 44:179–194

Clark ME, Scholten JD, Walker RL, Gironda RJ (2009) Assessment and treatment of pain associated with combat-related polytrauma. Pain Med 10:456–469. https://doi.org/10.1111/j.1526-4637.2009.00589.x

Clark RS, Nathaniel PD, Zhang X et al (2007b) boc-Aspartyl(OMe)-fluoromethylketone attenuates mitochondrial release of cytochrome c and delays brain tissue loss after traumatic brain injury in rats. J Cereb Blood Flow Metab 27:316–326. https://doi.org/10.1038/sj.jcbfm.9600338

Corrigan JD, Bogner JA, Mysiw WJ, Clinchot D, Fugate L (2001) Life satisfaction after traumatic brain injury. J Head Trauma Rehabil 16:543–555

Davis LM, Pauly JR, Readnower RD, Rho JM, Sullivan PG (2008) Fasting is neuroprotective following traumatic brain injury. J Neurosci Res 86(8):1812–1822

Dettmer J, Ettel D, Glang A, McAvoy K (2014) Building statewide infrastructure for effective educational services for students with TBI: promising practices and recommendations. J Head Trauma Rehabil 29:224–232. https://doi.org/10.1097/HTR.0b013e3182a1cd68

Di Pietro V, Amorini AM, Tavazzi B et al (2014) The molecular mechanisms affecting N-acetylaspartate homeostasis following experimental graded traumatic brain injury. Mol Med 20:147–157. https://doi.org/10.2119/molmed.2013.00153

DoD BIRPCO, U.S. Army Medical Research & Materiel Command (2014) Prevention, mitigation, and treatment of blast injuries. FY14 report to the executive agent

Duan Y, Gross RA, Sheu SS (2007) Ca2+-dependent generation of mitochondrial reactive oxygen species serves as a signal for poly(ADP-ribose) polymerase-1 activation during glutamate excitotoxicity. J Physiol 585:741–758. https://doi.org/10.1113/jphysiol.2007.145409

Faden AI, Demediuk P, Panter SS, Vink R (1989) The role of excitatory amino acids and NMDA receptors in traumatic brain injury. Science 244:798–800

Garcia M, Bondada V, Geddes JW (2005) Mitochondrial localization of mu-calpain. Biochem Biophys Res Commun 338:1241–1247

Gerrard-Morris A, Taylor HG, Yeates KO, Walz NC, Stancin T, Minich N, Wade SL (2010) Cognitive development after traumatic brain injury in young children. J Int Neuropsychol Soc 16:157–168. https://doi.org/10.1017/S1355617709991135

Gunter TE, Yule DI, Gunter KK, Eliseev RA, Salter JD (2004) Calcium and mitochondria. FEBS Lett 567:96–102

Hall ED, Sullivan PG (2004) Preserving function in acute nervous system injury. In: Waxman SG (ed) Neuroscience, molecular medicine and the therapeutic transfomation of neurology. Elsevier/Academic, San Diego

Halstead ME, McAvoy K, Devore CD, Carl R, Lee M, Logan K (2013) Returning to learning following a concussion. Pediatrics 132:948–957. https://doi.org/10.1542/peds.2013-2867

Hatton J (2001) Pharmacological treatment of traumatic brain injury: a review of agents in development. CNS Drugs 15:553–581

Hoge CW, Castro CA, Messer SC, McGurk D, Cotting DI, Koffman RL (2004) Combat duty in Iraq and Afghanistan, mental health problems, and barriers to care. N Engl J Med 351:13–22

Hoge CW, McGurk D, Thomas JL, Cox AL, Engel CC, Castro CA (2008) Mild traumatic brain injury in U.S. Soldiers returning from Iraq. N Engl J Med 358:453–463

Hovda DA, Becker DP, Katayama Y (1992) Secondary injury and acidosis. J Neurotrauma 9(Suppl 1):S47–S60

Jager TE, Weiss HB, Coben JH, Pepe PE (2000) Traumatic brain injuries evaluated in U.S. emergency departments, 1992-1994. Acad Emerg Med 7:134–140

Jennett B (1998) Epidemiology of head injury. Arch Dis Child 78:403–406

Ji J, Baart S, Vikulina AS et al (2015) Deciphering of mitochondrial cardiolipin oxidative signaling in cerebral ischemia-reperfusion. J Cereb Blood Flow Metab 35:319–328. https://doi.org/10.1038/jcbfm.2014.204

Ji J, Kline AE, Amoscato A et al (2012) Lipidomics identifies cardiolipin oxidation as a mitochondrial target for redox therapy of brain injury. Nat Neurosci 15:1407–1413. https://doi.org/10.1038/nn.3195

Jonas EA, Porter GA Jr, Beutner G, Mnatsakanyan N, Alavian KN (2015) Cell death disguised: the mitochondrial permeability transition pore as the c-subunit of the F(1)F(O) ATP synthase. Pharmacol Res 99:382–392. https://doi.org/10.1016/j.phrs.2015.04.013

Jones E, Fear NT, Wessely S (2007) Shell shock and mild traumatic brain injury: a historical review. Am J Psychiatry 164:1641–1645

Kulbe JR, Hill RL, Singh IN, Wang JA, Hall ED (2016) Synaptic mitochondria sustain more damage than non-synaptic mitochondria after traumatic brain injury and are protected by cyclosporine A. J Neurotrauma 34:1291–1301. https://doi.org/10.1089/neu.2016.4628

Lane N (2006) Power, sex, suicide : mitochondria and the meaning of life. Oxford University Press, Oxford, NY

Langlois JA, Rutland-Brown W, Wald MM (2006) The epidemiology and impact of traumatic brain injury: a brief overview. J Head Trauma Rehabil 21:375–378

LaPlaca MC, Zhang J, Raghupathi R et al (2001) Pharmacologic inhibition of poly(ADP-ribose) polymerase is neuroprotective following traumatic brain injury in rats. J Neurotrauma 18:369–376. https://doi.org/10.1089/089771501750170912

Leong BK, Mazlan M, Abd Rahim RB, Ganesan D (2013) Concomitant injuries and its influence on functional outcome after traumatic brain injury. Disabil Rehabil 35:1546–1551. https://doi.org/10.3109/09638288.2012.748832

Lew HL, Otis JD, Tun C, Kerns RD, Clark ME, Cifu DX (2009) Prevalence of chronic pain, posttraumatic stress disorder, and persistent postconcussive symptoms in OIF/OEF veterans: polytrauma clinical triad. J Rehabil Res Dev 46:697–702

Lewen A, Fujimura M, Sugawara T, Matz P, Copin JC, Chan PH (2001) Oxidative stress-dependent release of mitochondrial cytochrome c after traumatic brain injury. J Cereb Blood Flow Metab 21:914–920. https://doi.org/10.1097/00004647-200108000-00003

Lifshitz J, Sullivan PG, Hovda DA, Wieloch T, McIntosh TK (2004) Mitochondrial damage and dysfunction in traumatic brain injury. Mitochondria 1:705–713

Losoi H, Silverberg ND, Waljas M et al (2016) Recovery from mild traumatic brain injury in previously healthy adults. J Neurotrauma 33:766–776. https://doi.org/10.1089/neu.2015.4070

Maalouf M, Sullivan PG, Davis L, Kim DY, Rho JM (2007) Ketones inhibit mitochondrial production of reactive oxygen species production following glutamate excitotoxicity by increasing NADH oxidation. Neuroscience 145:256–264

Marshall S, Bayley M, McCullagh S, Velikonja D, Berrigan L, Ouchterlony D, Weegar K (2015) Updated clinical practice guidelines for concussion/mild traumatic brain injury and persistent symptoms. Brain Inj 29:688–700. https://doi.org/10.3109/02699052.2015.1004755

Mbye LH, Singh IN, Sullivan PG, Springer JE, Hall ED (2008) Attenuation of acute mitochondrial dysfunction after traumatic brain injury in mice by NIM811, a non-immunosuppressive cyclosporin A analog. Exp Neurol 209:243–253. https://doi.org/10.1016/j.expneurol.2007.09.025

McDonald SJ, Sun M, Agoston DV, Shultz SR (2016) The effect of concomitant peripheral injury on traumatic brain injury pathobiology and outcome. J Neuroinflammation 13:90. https://doi.org/10.1186/s12974-016-0555-1

Mott FW (1916) The effects of high explsives upon the central nervous system. Lancet 48:331–338

Murphy MP (2009) How mitochondria produce reactive oxygen species. Biochem J 417:1–13. https://doi.org/10.1042/BJ20081386

Murugan M, Santhakumar V, Kannurpatti SS (2016) Facilitating mitochondrial calcium uptake improves activation-induced cerebral blood flow and behavior after mTBI. Front Syst Neurosci 10:19. https://doi.org/10.3389/fnsys.2016.00019

Mustafa AG, Singh IN, Wang J, Carrico KM, Hall ED (2010) Mitochondrial protection after traumatic brain injury by scavenging lipid peroxyl radicals. J Neurochem 114:271–280. https://doi.org/10.1111/j.1471-4159.2010.06749.x

Mustafa AG, Wang JA, Carrico KM, Hall ED (2011) Pharmacological inhibition of lipid peroxidation attenuates calpain-mediated cytoskeletal degradation after traumatic brain injury. J Neurochem 117:579–588. https://doi.org/10.1111/j.1471-4159.2011.07228.x

Nasr P, Gursahani HI, Pang Z, Bondada V, Lee J, Hadley RW, Geddes JW (2003) Influence of cytosolic and mitochondrial Ca2+, ATP, mitochondrial membrane potential, and calpain activity on the mechanism of neuron death induced by 3-nitropropionic acid. Neurochem Int 43:89–99

Nicholls DG, Budd SL (2000) Mitochondria and neuronal survival. Physiol Rev 80:315–360

Nicholls DG, Budd SL, Castilho RF, Ward MW (1999) Glutamate excitotoxicity and neuronal energy metabolism. Ann N Y Acad Sci 893:1–12

Nicholls DG, Ferguson SJ (2002) Bioenergetics3. Academic, London

Obrenovitch TP (2008) Molecular physiology of preconditioning-induced brain tolerance to ischemia. Physiol Rev 88:211–247

Okie S (2005) Traumatic brain injury in the war zone. N Engl J Med 352:2043–2047. https://doi.org/10.1056/NEJMp058102

Pagliarini DJ, Rutter J (2013) Hallmarks of a new era in mitochondrial biochemistry. Genes Dev 27:2615–2627. https://doi.org/10.1101/gad.229724.113

Pandya JD, Pauly JR, Nukala VN et al (2007) Post-injury administration of mitochondrial uncouplers increases tissue sparing and improves behavioral outcome following traumatic brain injury in rodents. J Neurotrauma 24:798–811

Pandya JD, Pauly JR, Sullivan PG (2009) The optimal dosage and window of opportunity to maintain mitochondrial homeostasis following traumatic brain injury using the uncoupler FCCP. Exp Neurol 218:381–389. https://doi.org/10.1016/j.expneurol.2009.05.023

Pandya JD, Readnower RD, Patel SP et al (2014) N-acetylcysteine amide confers neuroprotection, improves bioenergetics and behavioral outcome following TBI. Exp Neurol 257:106–113. https://doi.org/10.1016/j.expneurol.2014.04.020

Pellock JM, Dodson WE, Bourgeois BFD (2001) Pediatric epilepsy: diagnosis and therapy. DEMOS, New York

Prins M, Greco T, Alexander D, Giza CC (2013) The pathophysiology of traumatic brain injury at a glance. Dis Model Mech 6:1307–1315. https://doi.org/10.1242/dmm.011585

Prins ML (2008) Cerebral metabolic adaptation and ketone metabolism after brain injury. J Cereb Blood Flow Metab 28:1–16

Rapoport MJ, Kiss A, Feinstein A (2006) The impact of major depression on outcome following mild-to-moderate traumatic brain injury in older adults. J Affect Disord 92:273–276

Readnower RD, Pandya JD, McEwen ML, Pauly JR, Springer JE, Sullivan PG (2011) Post-injury administration of the mitochondrial permeability transition pore inhibitor, NIM811, is neuroprotective and improves cognition after traumatic brain injury in rats. J Neurotrauma 28:1845–1853. https://doi.org/10.1089/neu.2011.1755

Richardson AP, Halestrap AP (2016) Quantification of active mitochondrial permeability transition pores using GNX-4975 inhibitor titrations provides insights into molecular identity. Biochem J 473:1129–1140. https://doi.org/10.1042/BCJ20160070

Robertson CS, Goodman JC, Narayan RK, Contant CF, Grossman RG (1991) The effect of glucose administration on carbohydrate metabolism after head injury. J Neurosurg 74:43–50

Rutland-Brown W, Langlois JA, Thomas KE, Xi YL (2006) Incidence of traumatic brain injury in the United States, 2003. J Head Trauma Rehabil 21:544–548

Satchell MA, Zhang X, Kochanek PM et al (2003) A dual role for poly-ADP-ribosylation in spatial memory acquisition after traumatic brain injury in mice involving NAD+ depletion and ribosylation of 14-3-3gamma. J Neurochem 85:697–708

Sayer NA, Cifu DX, McNamee S, Chiros CE, Sigford BJ, Scott S, Lew HL (2009) Rehabilitation needs of combat-injured service members admitted to the VA Polytrauma Rehabilitation Centers: the role of PM&R in the care of wounded warriors. PM R 1:23–28. https://doi.org/10.1016/j.pmrj.2008.10.003

Scheff SW, Sullivan PG (1999) Cyclosporin A significantly ameliorates cortical damage following experimental traumatic brain injury in rodents. J Neurotrauma 16:783–792

Schurr A (2002) Energy metabolism, stress hormones and neural recovery from cerebral ischemia/hypoxia. Neurochem Int 41:1–8

Selassie AW, Zaloshnja E, Langlois JA, Miller T, Jones P, Steiner C (2008) Incidence of long-term disability following traumatic brain injury hospitalization, United States, 2003. J Head Trauma Rehabil 23:123–131. https://doi.org/10.1097/01.HTR.0000314531.30401.39

Setnik L, Bazarian JJ (2007) The characteristics of patients who do not seek medical treatment for traumatic brain injury. Brain Inj 21:1–9

Sharma B, Bradbury C, Mikulis D, Green R (2014) Missed diagnosis of traumatic brain injury in patients with traumatic spinal cord injury. J Rehabil Med 46:370–373. https://doi.org/10.2340/16501977-1261

Singh IN, Gilmer LK, Miller DM, Cebak JE, Wang JA, Hall ED (2013) Phenelzine mitochondrial functional preservation and neuroprotection after traumatic brain injury related to scavenging of the lipid peroxidation-derived aldehyde 4-hydroxy-2-nonenal. J Cereb Blood Flow Metab 33:593–599. https://doi.org/10.1038/jcbfm.2012.211

Singh IN, Sullivan PG, Deng Y, Mbye LH, Hall ED (2006) Time course of post-traumatic mitochondrial oxidative damage and dysfunction in a mouse model of focal traumatic brain injury: implications for neuroprotective therapy. J Cereb Blood Flow Metab 26:1407–1418

Sleven H, Gibbs JE, Heales S, Thom M, Cock HR (2006) Depletion of reduced glutathione precedes inactivation of mitochondrial enzymes following limbic status epilepticus in the rat hippocampus. Neurochem Int 48:75–82

Snell FI, Halter MJ (2010) A signature wound of war: mild traumatic brain injury. J Psychosoc Nurs Ment Health Serv 48:22–28. https://doi.org/10.3928/02793695-20100107-01

Sullivan PG (2005) Interventions with neuroprotective agents: novel targets and opportunities. Epilepsy Behav 7(Suppl 3):S12–S17

Sullivan PG, Keller JN, Bussen WL, Scheff SW (2002) Cytochrome c release and caspase activation after traumatic brain injury. Brain Res 949:88–96

Sullivan PG, Keller JN, Mattson MP, Scheff SW (1998) Traumatic brain injury alters synaptic homeostasis: implications for impaired mitochondrial and transport function. J Neurotrauma 15:789–798

Sullivan PG, Rabchevsky AG, Hicks RR, Gibson TR, Fletcher-Turner A, Scheff SW (2000a) Dose-response curve and optimal dosing regimen of cyclosporin A after traumatic brain injury in rats. Neuroscience 101:289–295

Sullivan PG, Rabchevsky AG, Keller JN, Lovell M, Sodhi A, Hart RP, Scheff SW (2004a) Intrinsic differences in brain and spinal cord mitochondria: implication for therapeutic interventions. J Comp Neurol 474:524–534

Sullivan PG, Rabchevsky AG, Waldmeier PC, Springer JE (2005) Mitochondrial permeability transition in CNS trauma: cause or effect of neuronal cell death? J Neurosci Res 79:231–239

Sullivan PG, Sebastian AH, Hall ED (2011) Therapeutic window analysis of the neuroprotective effects of cyclosporine A after traumatic brain injury. J Neurotrauma 28:311–318. https://doi.org/10.1089/neu.2010.1646

Sullivan PG, Springer JE, Hall ED, Scheff SW (2004b) Mitochondrial uncoupling as a therapeutic target following neuronal injury. J Bioenerg Biomembr 36:353–356

Sullivan PG, Thompson M, Scheff SW (2000b) Continuous infusion of cyclosporin A postinjury significantly ameliorates cortical damage following traumatic brain injury. Exp Neurol 161:631–637

Sullivan PG, Thompson MB, Scheff SW (1999) Cyclosporin A attenuates acute mitochondrial dysfunction following traumatic brain injury. Exp Neurol 160:226–234. https://doi.org/10.1006/exnr.1999.7197

Sun J, Jacobs KM (2016) Knockout of cyclophilin-D provides partial amelioration of intrinsic and synaptic properties altered by mild traumatic brain injury. Front Syst Neurosci 10:63. https://doi.org/10.3389/fnsys.2016.00063

Susman M, DiRusso SM, Sullivan T et al (2002) Traumatic brain injury in the elderly: increased mortality and worse functional outcome at discharge despite lower injury severity. J Trauma 53:219–223. discussion 223–214

Tavazzi B, Vagnozzi R, Signoretti S et al (2007) Temporal window of metabolic brain vulnerability to concussions: oxidative and nitrosative stresses—part II. Neurosurgery 61:390–395. discussion 395–396

Thomas C, Mackey MM, Diaz AA, Cox DP (2009) Hydroxyl radical is produced via the Fenton reaction in submitochondrial particles under oxidative stress: implications for diseases associated with iron accumulation. Redox Rep 14:102–108. https://doi.org/10.1179/135100009X392566

Thurman DJ, Alverson C, Dunn KA, Guerrero J, Sniezek JE (1999) Traumatic brain injury in the United States: a public health perspective. J Head Trauma Rehabil 14:602–615

Tieu K, Perier C, Caspersen C et al (2003) D-beta-hydroxybutyrate rescues mitochondrial respiration and mitigates features of Parkinson disease. J Clin Invest 112:892–901

Vagnozzi R, Signoretti S, Tavazzi B et al (2005) Hypothesis of the postconcussive vulnerable brain: experimental evidence of its metabolic occurrence. Neurosurgery 57:164–171. discussion 164–171

Vagnozzi R, Tavazzi B, Signoretti S et al (2007) Temporal window of metabolic brain vulnerability to concussions: mitochondrial-related impairment—part I. Neurosurgery 61:379–388.; discussion 388–379. https://doi.org/10.1227/01.NEU.0000280002.41696.D8

Warden D (2006) Military TBI during the Iraq and Afghanistan wars. J Head Trauma Rehabil 21:398–402

Xing G, Barry ES, Benford B, Grunberg NE, Li H, Watson WD, Sharma P (2013) Impact of repeated stress on traumatic brain injury-induced mitochondrial electron transport chain expression and behavioral responses in rats. Front Neurol 4:196. https://doi.org/10.3389/fneur.2013.00196

Yang J, Phillips G, Xiang H, Allareddy V, Heiden E, Peek-Asa C (2008) Hospitalizations for sport-related concussions in US children aged 5 to 18 years during 2000-2004. Br J Sports Med 42:664–669

Yonutas HM, Vekaria HJ, Sullivan PG (2016) Mitochondrial specific therapeutic targets following brain injury. Brain Res 1640:77–93. https://doi.org/10.1016/j.brainres.2016.02.007

Chapter 5
Neuroprotective Agents Target Molecular Mechanisms of Programmed Cell Death After Traumatic Brain Injury

Lu-Yang Tao

Abstract The review is to update the current state of knowledge in post-TBI pathophysiological mechanisms, including programmed cell death mechanisms and mechanism-based preclinical pharmacological intervention used in animal models. Their effects on cell death, inflammatory events, and prolonged motor and cognitive deficits will be summarized, and their potential success for clinical application will be evaluated. Many of the above-mentioned mechanisms may be important targets for limiting the consequences of TBI.

Keywords Traumatic brain injury · Cell death · Neural protection · Plasmalemma

5.1 Introduction

Traumatic brain injury (TBI), a major cause of morbidity and mortality (Jin et al. 2015), represents the quintessential neuropsychiatric paradigm with a combination of effects in cognition, personality, and the risk for psychiatric disorders (Santopietro et al. 2015). However, the hope for an effective treatment is derived from the fact that much of the post- traumatic damage to the injured brain is caused by a secondary injury cascade of pathochemical and pathophysiological events that exacerbates the primary mechanical TBI (Mustafa et al. 2010). Therefore, TBI is a significant clinical problem with few therapeutic interventions successfully translated to the clinic.

The goal of this chapter is to update the current state of knowledge in post-TBI pathophysiological mechanisms, including programmed cell death mechanisms and mechanism-based preclinical pharmacological intervention used in animal models. Their effects on cell death, inflammatory events, prolonged motor and cognitive deficits will be summarized, and their potential success for clinical application will be evaluated. Many of the above mentioned mechanisms may be important targets for limiting the consequences of TBI.

L.-Y. Tao (✉)
Department of Forensic Medicine, Soochow University, Suzhou, China
e-mail: taoluyang@suda.edu.cn

D. G. Fujikawa (ed.), *Acute Neuronal Injury*,
https://doi.org/10.1007/978-3-319-77495-4_5

5.2 Programmed Cell Death

Cell death is broadly classified into three types: necrosis, apoptosis (type 1 programmed cell death [PCD]) and autophagy (type 2 PCD) (Edinger and Thompson 2004). It has been identified that necrosis is the majority and apoptosis is the minority (Krysko et al. 2008), and necrosis can also be programmed, which has been called necroptosis (Degterev et al. 2005).

Traumatic brain injury (TBI) results in neuronal apoptosis, autophagic cell death and necroptosis (Wang et al. 2012). A greater understanding of the time course of cell death following traumatic brain injury (TBI) is important for clinical treatment. Our *in vivo* study evaluated the time course of TBI-induced cell death through dUTP nick-end labeling (TUNEL), Fluoro-Jade B, and propidium iodide (PI) staining using a fluorescent staining double-labeling method. Our results indicated PI labeling is more sensitive and reliable than TUNEL and Fluoro-Jade B staining for detecting cell death following traumatic brain injury. Moreover, PI labeling can function as a reliable marker to estimate the entire time course of cell death (Luo et al. 2010a, b). TBI elicited a significant increase in the number of PI-positive cells from 10 min to 21 days, and the count peaked in the 24 h and 48 h group (Luo et al. 2010a, b, 2011).

5.2.1 Apoptosis

Apoptosis has been attributed to programmed cell death in TBI (Tehranian et al. 2008). In mammals, two major molecular pathways can initiate the apoptotic cascade, the cell death-receptor pathway and the mitochondrial pathway (Hengartner 2000). The two pathways can both lead to caspase-3 activation and apoptotic cell death. The former is triggered by members of the death receptor superfamily such as Fas (CD95) or TNF-R1 (Siegel et al. 2000), whereas the latter is activated by stimuli involving mitochondrial membrane permeabilization (MMP) and then release of proapoptotic mitochondrial proteins (cytochrome c, etc.) into the cytosol, leading to activation of caspases and cell death (Halestrap et al. 2000; Jiang and Wang 2000; Lin et al. 2009).

Apoptosis is a caspase-dependent cell death modality. The proteolytic activation of caspases in apoptotic cells drives cell rounding, retraction of pseudopodes, reduction of cellular volume (pyknosis), chromatin condensation, nuclear fragmentation (karyorrhexis), and plasma membrane blebbing (Vanden Berghe et al. 2015).

Caspase activation by the extrinsic pathway involves the binding of extracellular death ligands (such as FasL or tumor necrosis factor-a (TNFa)) to transmembrane death receptors. Attachment of death receptors to their cognate ligands provokes the recruitment of adaptor proteins, such as the Fas-associated death domain protein (FADD), which in turn recruits and aggregates several molecules of caspase-8, thereby promoting its autoprocessing and activation (Song et al. 2016). Active caspase-8 then proteolytically processes and activates caspase-3 and caspase-7, provoking further caspase activation events that culminate in substrate proteolysis and cell death. In some situations, extrinsic death signals can cross talk with the intrinsic pathway through caspase-8-mediated proteolysis of the BH3-only protein BID (BH3-interacting domain death agonist). Truncated BID (tBID) can promote release of mitochondrial cytochrome c and assembly of the apoptosome (comprising ~7 molecules of apoptotic protease-activating factor-1 (APAF1) and the same number of caspase-9 homodimers). In the intrinsic pathway, diverse stimuli that provoke cell stress or damage typically activate one or more members of the BH3-only protein family. BH3-only proteins act as pathway-specific sensors for various stimuli and are regulated in distinct ways. BH3-only protein activation above a crucial threshold overcomes the inhibitory effect of the anti-apoptotic B cell lymphoma-2 (BCL-2) family members and promotes the assembly of BAK–BAX oligomers within mitochondrial outer membranes. These oligomers permit the efflux of intermembrane space proteins, such as cytochrome c, into the cytosol. On release from mitochondria, cytochrome c can seed apoptosome assembly (Song et al. 2016). Active caspase-9 then propagates a proteolytic cascade of caspase activation events. The granzyme B-dependent caspase activation involves the delivery of this protease into the target cell through specialized granules that are released from cytotoxic T lymphocytes (CTL) or natural killer (NK) cells. CTL and NK granules contain numerous granzymes as well as a pore-forming protein, perforin, which oligomerizes in the membranes of target cells to permit entry of the granzymes. Granzyme B, similar to the caspases, also cleaves its substrates after Asp residues and can process BID as well as caspase-3 and caspase-7 to initiate apoptosis (Taylor et al. 2008; Song et al. 2016).

Similar to the above research, our study also showed that TBI induced a marked increase in number of TUNEL positive cells, the release of cytochrome c from mitochondria to cytosol, activation of caspase-3, up-regulation of Bax, and down-regulation of Bcl-2, indicating TBI results in apoptosis (Luo et al. 2010a, b, 2011).

5.2.2 Autophagy

Autophagy is an evolutionarily conserved pathway that leads to degradation of proteins and entire organelles in cells undergoing stress (Pozuelo-Rubio 2011). There are three types of autophagy: macroautophagy, microautophagy, and chaperone-mediated autophagy (Klionsky 2005). Microautophagy involves the direct engulfment of the cargo by the lysosomal membrane. Chaperone-mediated autophagy is characterized by transfer of cytosolic proteins with a KFERQ motif to the lysosome by chaperone proteins, followed by their direct import via the lysosomal-associated membrane protein type 2A (LAMP2A) translocation complex. Finally, macroautophagy (hereafter simply called autophagy), is the most studied type of autophagy and involves the formation around the cargo of a double-membrane vesicle named the autophagosome that subsequently fuses with the lysosome to form autolysosomes (Wesselborg and Stork 2015).

Currently, p62 is identified as one of the specific substrates that are degraded through the autophagy pathway (Ichimura et al. 2008; Komatsu and Ichimura 2010). This degradation is mediated by interaction with LC3, a mammalian homologue of Atg8, which is recruited to the phagophore/isolation membrane and remains associated with the completed autophagosome (Pankiv et al. 2007; Zheng et al. 2009). In this autophagy pathway, 3-methyladenine (3-MA) is a relatively selective inhibitor of the Class III phoshatidylinositol-3-kinase (PI3K); the latter can interact with Beclin-1 to participate in the formation of autophagosomes (Petiot et al. 2000). Moreover, bafliomycin A1 (BFA) inhibits autophagy through inhibiting vacuolar H + -ATPase (Boya et al. 2005).

Recent studies have shown that autophagy is increased after TBI (Lai et al. 2008; Zhang et al. 2008). The increased LC3 immunostaining is found mainly in neurons at 24 h post-TBI (Liu et al. 2008). However, few experimental studies have addressed the role of autophagy in traumatic damage and neurologic outcome. By using autophagy inhibitors 3-MA and bafliomycin A1 (BFA), our study suggests that the autophagy pathway is involved in the pathophysiologic responses after TBI, and inhibition of this pathway may help attenuate traumatic damage and functional outcome deficits (Luo et al. 2011).

Autophagic cell death has been identified as a mechanism of programmed cell death. However, whether this is death due to autophagy or not remains controversial (Kroemer and Levine 2008). During conditions of nutrient limitation, autophagy is used to generate amino acids and energy to maintain cell viability through the bulk degradation of cytoplasmic material (Messer 2016). Accordingly, the presence of autophagy in dying cells has been proposed to be a stress response mechanism to prolong cell viability. Nevertheless, recent studies strongly support autophagy as a process that can promote programmed cell death (Lin and Baehrecke 2015).

Many of the original studies describing autophagy-induced death relied on the observation of autophagy in dying cells and did not examine autophagic flux (Messer 2016). In autophagic flux studies, while increased autophagic flux may be protective after TBI, after more severe trauma inhibition of autophagic flux may contribute to neuronal cell death, indicating disruption of autophagy as a part of the secondary injury mechanism (Sarkar et al. 2014; Lipinski et al. 2015). The concept of autophagic cell death is based on observations of increased morphological features (e.g., accumulation of autophagic vesicles) in dying cells (Galluzzi et al. 2012). Recently, a novel form of autophagy-dependent cell death has been described, autosis, which not only meets the criteria in claim (i.e., blocked by autophagy inhibition, independent of apoptosis or necrosis), but also demonstrates unique morphological features and a unique ability to be suppressed by pharmacological or genetic inhibition of Na^+/K^+-ATPase (Liu et al. 2013).

5.2.3 *Necroptosis*

In recent years, the traditional concept of necrosis has been seriously challenged. Tumor necrosis factor (TNF) receptor (TNFR) superfamily are strong regulators of apoptosis (Chan et al. 2003). However, TNFR-1, Fas and TNF-related apoptosis-inducing ligand receptor (TRAILR) can also trigger an alternative form of cell death as "programmed necrosis", namely TNF-induced programmed necrosis was facilitated by TNFR-2 signaling and caspase inhibition. It is conceivable that when the apoptotic caspases fail to be activated, then cells undergo necroptosis as an alternative death pathway (Han et al. 2011).

Necroptosis is a newly discovered caspases-independent programmed necrosis pathway which can be triggered by activation of death receptor. Necroptosis is currently the best-characterized form of regulated necrosis and is mediated by the action of receptor interacting protein kinase 1 (RIPK1) and RIPK3, and mixed lineage kinase domain-like (MLKL) in response to death receptors, Toll- and NOD-like receptors, T-cell receptor, genotoxic stress, and viruses (Linkermann and Green 2014; Zhou and Yuan 2014). As a caspase-independent cell death mode, necroptosis exists as an alternative form of programmed cell death when caspase dependent apoptosis is blocked (Christofferson and Yuan 2010). However, definite interactions between necroptosis and apoptosis are far from obvious.

Previous studies showed that TBI initiates physiopathologic cascades of cell death signals and induces multiple cell death modes (Luo et al. 2010a, b, 2011; Werner and Engelhard 2007). Beyond the classical programmed cell death pathways, apoptosis (type1) and autophagy (type2), necroptosis is a newly discovered caspase-independent programmed necrotic mode. It has been demonstrated that the occurrence of these cell death modes participate in neural injury in TBI and ischemia models (Luo et al. 2010a, b, 2011; Meloni et al. 2011; Xu et al. 2010a, b; You et al. 2008).

Being a specific inhibitor of RIPK1, necrostatin-1 (NEC-1) was found to potently depress necroptotic cell death (Degterev et al. 2005). Previous work showed that NEC-1, a specific necroptosis inhibitor, could reduce tissue damage and functional impairment through inhibiting of necroptosis following TBI (You et al. 2008). However, several lines of evidence indicate that necroptosis has indeterminate relationships with the other two types of programmed cell death (Zhang et al. 2011). We have shown that multiple cell death pathways participate in the development of TBI, and NEC-1 inhibits apoptosis and autophagy simultaneously. These interactions may further explain how NEC-1 reduces TBI-induced tissue damage and functional deficits and reflect the interrelationship among necrosis, apoptosis and autophagy (Wang et al. 2012).

5.2.4 *Interactions Between Autophagy, Apoptosis and Necroptosis*

A breakthough has been made in the understanding of the role of Bcl-2 in controlling autophagy via its interaction with Beclin1 (Erlich et al. 2006; Sadasivan et al. 2008). Autophagy is activated, and can coexist or occur sequentially with apoptosis (Uchiyama et al. 2008). The process of autophagy acts to both increase and decrease apoptosis. Autophagy limits apoptosis through degradation of pro-apoptotic stimuli such as damaged mitochondria or cytotoxic protein aggregates (Xue et al. 2001).

As a caspase-independent cell death mode, necroptosis exists as an alternative form of programmed cell death when caspase-dependent apoptosis is blocked (Christofferson and Yuan 2010). However, definite interactions between necroptosis and apoptosis are far from obvious. Caspase-3, defined as an 'executioner' caspase, can be activated via amplification of extrinsic or intrinsic apoptotic signals (Cheema et al. 1999; Zou et al. 1997). On the contrary, Bcl-2 is an anti-apoptotic member of B-cell lymphoma-2 (Bcl-2) family of proteins and plays an important role in regulation of both caspase-dependent and caspase-independent apoptosis (Graham et al. 2000). To date, the effects of NEC-1 on apoptosis are still controversial. NEC-1 could not change the activation of caspase-3 or number of TUNEL-positive cell in the ischemic brain (Degterev et al. 2005). Moreover, NEC-1 reversed shikonin-induced necroptosis to apoptosis (Han et al. 2009). On the other hand, activation of caspase-3 induced by 11′-deoxyverticillin-A in human colon carcinoma cell death was also partially inhibited by NEC-1 (Zhang et al. 2011). The causes of these differences are probably associated with the different cell types or different strategies in particular protocols. Here, in our TBI protocol, we validated an anti-apoptotic role of NEC-1 through inhibiting Bcl-2 decline and caspase-3 activation after TBI.

Autophagy is a probable downstream consequence of necroptosis rather than a contributing factor to necroptosis and is activated as a clean-up mechanism for cell death (Degterev et al. 2005; Bell et al. 2008). Treatment with autophagy inhibitors also inhibits Z-VAD-FMK induced cell death, and knockdown of autophagy related genes such as beclin-1 and atg7 were shown to inhibit necroptosis in L929 cells (Yu et al. 2004). These results suggest an interaction between necroptosis and autophagy.

5.3 Mechanism-Based Preclinical Pharmacological Intervention

The mechanisms underlying TBI are very complex, including membrane integrity, apoptosis, autophagy, necroptosis, mitochondrial dysfunction, inflammation, oxidative stress, and excitotoxicity. Drug treatments have been classified according to which categories of pathogenic mechanisms they target. Currently, there are no Food and Drug Administration (FDA)-approved pharmacologic agents for the treatment of those with TBI (Kulbe et al. 2016).

5.3.1 Membrane-Resealing Agents

Plasmalemma permeability plays an important role in the secondary neuronal death induced by TBI (Cullen et al. 2011; Serbest et al. 2005). Poloxamer 188 (P188) is an amphiphilic copolymer of polyoxyethylene and polyoxypropylene and is used as a pharmaceutical excipient. P188 can restore plasma membrane integrity, help to seal damaged cell membranes, and can have a cytoprotective action in many types of cells (Serbest et al. 2005; Greenebaum et al. 2004). We have found that plasmalemma permeability contributes to TBI-induced blood-brain barrier (BBB) disruption, brain edema and neural cell apoptosis. Maintaining plasmalemma integrity is important for TBI, and P188 may be a potential drug in clinical applications (Bao et al. 2012).

Our recent study investigated the effects of plasmalemmal resealing by P188 on neuronal autophagy in TBI. The results revealed that plasma membranes were resealed after TBI, in which P188 aggravated autophagy in vivo (Bao et al. 2016).

In addition, acute membrane damage due to traumatic brain injury (TBI) is a critical precipitating event (Lenz et al. 2007). However, the subsequent effects of the mechanical trauma, including mitochondrial and lysosomal membrane permeability, remain elusive. We have shown that injured neurons have undergone mitochondrial and lysosomal membrane permeability damage, and the mechanism can be exploited with pharmacological interventions. P188's neuroprotection appears to involve a relationship between cathepsin B and tBid-mediated mitochondrial initiation of cell death (Luo et al. 2013a, b).

5.3.2 Anti-Necroptosis Agents

Necroptosis is a newly discovered caspase-independent programmed necrosis pathway which can be triggered by activation of death receptor. NEC-1, identified as a specific inhibitor of RIPK1, effectively inhibits necroptosis (Degterev et al. 2005, 2008).

Recently, studies found that NEC-1 could change the activities of other types of cell death- associated factors. For example, (1) NEC-1 prevented glutamate-induced nuclear translocation of apoptosis inducing factor (AIF), production of reactive oxygen species (ROS) and poly ADP-ribose polymerase (PARP) activation (Xu et al. 2007, 2010a, b); (2) NEC-1 efficiently reduced arachidonic acid (AA)-induced cell death via blocking reactive oxygen species (ROS) production and c-Jun N-terminal kinases (JNKs) activation (Kim et al. 2010); and (3) NEC-1 inhibited Z-VAD-FMK- induced phosphorylation of extracellular signal-regulated kinase 1/2 (ERK1/2) in a mouse model of Huntington's disease (Zhu et al. 2011). These findings showed that NEC-1 could alter other cell death modes besides necroptosis, at least in certain conditions. These accessory effects of NEC-1 coincide with the potential role of RIPK-1, a multifunctional protein, which is involved in different pathways of cell death and survival (Degterev et al. 2008; Festjens et al. 2007).

Besides reducing the amount of cell injury and tissue damage, NEC-1 improves functional outcome after TBI (Degterev et al. 2005). We found that multiple cell death pathways participate in the development of TBI, and NEC-1 inhibits apoptosis and autophagy simultaneously. These coactions may further explain how NEC-1 can reduce TBI-induced tissue damage and functional deficits and reflect the interrelationship among necrosis, apoptosis and autophagy (Wang et al. 2012).

5.3.3 Anti-Inflammatory Agents

Mechanical disruption of brain triggers a cascade of events leading to BBB breakdown, brain edema and inflammation after TBI (Lenzlinger et al. 2001; Uryu et al. 2002). Activation of pro-inflammatory cytokines is an important neuroinflammatory response. Among proinflammatory cytokines, TNF-a, IL-1b and IL-6 appear to play a determinant role in disrupting blood–brain barrier, and accelerating the formation of cerebral edema (Wang et al. 2007). Elevated TNF-a, IL-1b and IL-6 have been detected in the brain parenchyma within the early hours after brain injury in both humans and rodents (Chao et al. 1995; Winter et al. 2002). Furthermore, previous studies have shown that inhibiting TNF-a, IL-1b and IL-6 induced by TBI is neuroprotective (Jones et al. 2005; Shohami et al. 1997). Our study indicates that TBI activates pro-inflammatory cytokines (TNF-α, IL-1β, IL-6), the MAPK pathways together with lipoxin A4 (LXA4) receptor in astrocytes, and these mechanisms may be inhibited by pharmacological interventions (Luo et al. 2013a, b).

In addition, NF-κB upregulation has been demonstrated in neurons and glial cells in response to experimental injury and neuropathological disorders, where it has been related to both neurodegenerative and neuroprotective activities (Sanz et al. 2002). It has been generally recognized that NF-κB plays important roles in the regulation of apoptosis and inflammation as well as innate and adaptive immunity. We sought first to investigate the effect of the NF-κB inhibitor SN50, which inhibits NF-κB nuclear translocation, cell death, inflammatory pathways and behavioral deficits in our mouse TBI models. Our results imply that through NF-κB/TNF-α/cathepsin networks SN50 may contribute to regulation of TBI-induced extrinsic and intrinsic apoptosis, and inflammatory pathways, which partly determined the fate of injured cells in our TBI model. The protective effects of SN50 preconditioning may be partly related to suppression of apoptosis as well as reduction of inflammatory cytokines like TNF-α. However, further study is needed to completely elucidate the protective mechanism of SN50 on TBI-induced cell death and to establish its clinical utility in the treatment of TBI.

5.3.4 Mitochondrial Protective Agents

Mitochondria, the primary energy-generating system in most eukaryotic cells, have been shown to be a crucial participant in TBI pathophysiology (Gajavelli et al. 2015), undergoing constant changes in size and shape. As central mediators of the secondary injury cascade, mitochondria are promising therapeutic targets for prevention of cellular death and dysfunction after TBI. One of the most promising and extensively studied mitochondrial targeted TBI therapies is inhibition of the mitochondrial permeability transition pore (mPTP) by the FDA-approved drug, cyclosporine A (CsA) (Kulbe et al. 2016). Kulbe et al. report that synaptic mitochondria sustain more damage than non-synaptic mitochondria 24 h after severe controlled cortical impact injury (CCI), and that intraperitoneal administration of CsA (20 mg/kg) 15 min after injury improves synaptic and non-synaptic respiration, with a significant improvement being seen in the more severely impaired synaptic population (Kulbe et al. 2016).

Furthermore, mitochondrial morphology is orchestrated by a well- conserved cellular machinery, comprised of dynamin-related GTPases, dynamin-related protein 1 (Drp1) for fission and mitofusions (Mfn1 and Mfn2) and optic atrophy-1 (OPA1) for fusion (Purnell and Fox 2013; Wang et al. 2013a, b). Drp1, which is targeted to the outer mitochondrial membrane, is primarily found in the cytosol, and it localizes to discrete spots on mitochondrial surfaces to initiate fission by interaction with Fis1 (Chan 2006; Qi et al. 2013; Sharp et al. 2015). Mdivi-1, a mitochondrial division inhibitor, is a highly efficacious small molecule acting as a selective inhibitor of Drp1 (Zhang et al. 2013a, b, c).

Disturbed regulation of mitochondrial dynamics, the balance of mitochondrial fusion and fission, has been implicated in neurodegenerative diseases, such as Parkinson's disease and cerebral ischemia/reperfusion (Knott and Bossy-Wetzel 2008; Lackner and Nunnari 2009; Ong et al. 2010). However, the role of mitochondrial dynamics in traumatic brain injury has not been illuminated. Our study was to investigate the role of Mdivi-1, a small molecule inhibitor of a key mitochondrial fission protein Drp1, in TBI-induced cell death and functional outcome. Protein expression of Drp1 was first investigated. Our data imply that inhibition of Drp1 may help attenuate TBI-induced functional outcome and cell death through maintaining normal mitochondrial morphology and inhibiting activation of apoptosis (Wu et al. 2016).

5.3.5 Hydrogen Sulfide (H_2S)

Hydrogen sulfide (H_2S) has been known as a toxic gas characterized by its offensive odor, described as the smell of rotten eggs, and its being an environment pollutant (Abe and Kimura 1996; Martelli et al. 2012). It had long been assumed that H_2S

exists in animal tissues at very low concentrations because of its toxicity, although it could be produced endogenously (Kimura 2010). H_2S is a lipid-soluble, endogenously produced gaseous messenger molecule collectively known as gasotransmitters (Kimura 2010). Over the last several decades, gasotransmitters have emerged as potent cytoprotective mediators in various models of tissue and cellular injury. Current evidence suggests that endogenous H_2S in the brain is produced from L-cysteine by the pyridoxal 50-phosphate-dependent enzyme, cystathionine-beta-synthase (CBS) (Enokido et al. 2005).

In our study, we investigated changes of H_2S and its possible role in the pathogenesis after TBI. We found that: (1) down-regulation of endogenous H_2S pathway after brain injury and the level of H_2S was in parallel with the expression of CBS in the cortex and hippocampus (Zhang et al. 2013a, b, c); (2) CBS may be implicated in neuronal death and the pathophysiology of brain after TBI (Zhang et al. 2013a, b, c); (3) a protective effect and therapeutic potential of H_2S in the treatment of brain injury and the protective effect against TBI may be associated with regulating apoptosis and autophagy (Zhang et al. 2014). Our data may provide a novel pathway to learn the underlying molecular and cellular mechanisms of CNS after TBI and a novel strategy for the treatment of CNS trauma. Future studies attempting to characterize the functional consequences of H_2S on cellular apoptosis, inflammation and the identification of their substrates and downstream signaling targets are now possible.

3-mercaptopyruvate sulfurtransferase (3-MST) is a novel hydrogen sulfide (H_2S)-synthesizing enzyme that may be involved in cyanide degradation and in thiosulfate biosynthesis (Frasdorf et al. 2014). In recent years, considerable attention has been focused on the biochemistry and molecular biology of H_2S-synthesizing enzymes. In contrast, there have been few concerted attempts to investigate the changes in the expression of the H_2S-synthesizing enzymes with disease states (Shibuya et al. 2009). We found that 3-MST is mainly located in living neurons and may be implicated in the autophagy of neurons and in the pathophysiology of brain after TBI (Zhang et al. 2016).

5.3.6 Others: Cathepsin B Inhibitor (CBI), and Humanin

Cathepsins B is one of the major lysosomal cysteine proteases that might be important in the intracellular protein catabolism in neurons and also plays an important role in apoptotic and necrotic programmed cell death (Guicciardi et al. 2004; Chwieralski et al. 2006; Krysko et al. 2008). Ellis et al. (2005) reported that in areas adjacent to the injury epicenter following spinal cord injury in the rat, cathepsin B enzymatic activity was significantly increased, and cathepsin B immunoreactivity appeared to be elevated in neurons. A lysosomal-mitochondrial axis theory of cell death has also been proposed (Kim et al. 2006). Specific inhibitors of cathepsin B, such as cystatin A and CBI (a selective cathepsin B inhibitor), can protect cells against ischemic hippocampal neuronal death (Yamashima et al. 1998; Tsuchiya et al. 1999) and excitotoxic striatal cell death (Wang et al. 2008). We found that CBI inhibits TBI-induced cell death through the programmed cell necrosis and mitochondria-mediated apoptotic pathways (Luo et al. 2010a, b).

Humanin (HN) is as an endogenous peptide that inhibits Alzheimer disease (AD)-relevant neuronal cell death (Hashimoto et al. 2001). HNG, a variant of HN in which the 14th amino acid serine was replaced with glycine, can reduce infarct volume and improve neurological deficits after ischemia/reperfusion injury (Xu et al. 2008; Zhao et al. 2011). We have found that HNG treatment improves morphological and functional outcomes after TBI in mice, and the protective effect of HNG against TBI may be associated with down-regulating apoptosis and autophagy (Wang et al. 2013a, b).

5.4 Conclusions and Perspective

Traumatic brain injury (TBI), a common cause of disability and death worldwide, causes cell death and behavioral deficits (Al Nimer et al. 2015; Jin et al. 2015). TBI-induced programmed cell death includes apoptosis, autophagic cell death and necroptosis (Wang et al. 2012). There have been limited advances in the therapeutic strategies to counter brain injury. Except for conservative management, neuroprotection and neural recovery are still the main therapeutic strategies.

Our review updates the current state of knowledge in post-TBI pathophysiological mechanisms, mainly including programmed cell death mechanisms. Mechanism-based preclinical pharmacological intervention is summarized, including membrane-resealing agents, anti-autophagy agents, anti-necroptosis agents, anti-inflammatory agents, mitochondrial protective agents and others.

Novel therapeutic agents may even have differential effects at different points in the progression of TBI. There may be numerous pathologically and/or genetically distinct forms of TBI, which might be best targeted with different therapies. Furthermore, biomarker discovery may help us to recognize the earliest symptoms of TBI, ultimately improving prognosis and contributing to clinical trials with TBI patients.

Acknowledgments This work was supported by the National Natural Science Foundation of China (No. 81530062), and Dr. Cheng-liang Luo did the mainly work of this chapter.

References

Abe K, Kimura H (1996) The possible role of hydrogen sulfide as an endogenous neuromodulator. J Neurosci 16:1066–1071

Al Nimer F, Thelin E, Nystrom H, Dring AM, Svenningsson A, Piehl F, Nelson DW, Bellander BM (2015) Comparative assessment of the prognostic value of biomarkers in trau- matic brain injury reveals an independent role for serum levels of neurofilament light. PLoS One 10:e0132177

Bao H, Yang X, Zhuang Y, Huang Y, Wang T, Zhang M, Dai D, Wang S, Xiao H, Huang G, Kuai J, Tao L (2016) The effects of poloxamer 188 on the autophagy induced by traumatic brain injury. Neurosci Lett 634:7–12

Bao HJ, Wang T, Zhang MY, Liu R, Dai DK, Wang YQ, Wang L, Zhang L, Gao YZ, Qin ZH, Chen XP, Tao LY (2012) Poloxamer-188 attenuates TBI-induced blood-brain barrier damage leading to decreased brain edema and reduced cellular death. Neurochem Res 37(12):2856–2867

Bell BD, Leverrier S, Weist BM, Newton RH, Arechiga AF, Luhrs KA, Morrissette NS, Walsh CM (2008) FADD and caspase-8 control the outcome of autophagic signaling in proliferating T cells. Proc Natl Acad Sci U S A 105(43):16677–16682

Boya P, Gonzalez-Polo RA, Casares N, Perfettini JL, Dessen P, Larochette N, Metivier D, Meley D, Souquere S, Yoshimori T, Pierron G, Codogno P, Kroemer G (2005) Inhibition of macroautophagy triggers apoptosis. Mol Cell Biol 25:1025–1040

Chan DC (2006) Mitochondria: dynamic organelles in disease, aging, and development. Cell 125:1241–1252

Chan FK, Shisler J, Bixby JG, Felices M, Zheng L, Appel M et al (2003) A role for tumor necrosis factor receptor-2 and receptorinteracting protein in programmed necrosis and antiviral responses. J Biol Chem 278:51613–51621

Chao CC, Hu S, Ehrlich L, Peterson PK (1995) Interleukin-1 and tumornecrosis factor-alpha synergistically mediate neurotoxicity: involvement of nitricoxide and of N-methyl-D-aspartate receptors. Brain Behav Immun 9:355–365

Cheema ZF, Wade SB, Sata M, Walsh K, Sohrabji F, Miranda RC (1999) Fas/Apo [apoptosis]-1 and associated proteins in the differentiating cerebral cortex: induction of caspase-dependent cell death and activation of NF-kappaB. J Neurosci 19(5):1754–1770

Christofferson DE, Yuan J (2010) Necroptosis as an alternative form of programmed cell death. Curr Opin Cell Biol 22(2):263–268

Chwieralski CE, Wehe T, Buhling F (2006) Cathepsin-regulated apoptosis. Apoptosis 11:143–149

Cullen DK, Vernekar VN, LaPlaca MC (2011) Trauma-induced plasmalemma disruptions in three-dimensional neural cultures are dependent on strain modality and rate. J Neurotrauma 28:2219–2233

Degterev A, Huang Z, Boyce M, Li Y, Jagtap P, Mizushima N, Cuny GD, Mitchison TJ, Moskowitz MA, Yuan J (2005) Chemical inhibitor of nonapoptotic cell death with therapeutic potential for ischemic brain injury. Nat Chem Biol 1(2):112–119

Degterev A, Hitomi J, Germscheid M, Ch'en IL, Korkina O, Teng X, Abbott D, Cuny GD, Yuan C, Wagner G, Hedrick SM, Gerber SA, Lugovskoy A, Yuan J (2008) Identification of RIP1 kinase as a specific cellular target of necrostatins. Nat Chem Biol 4(5):313–321

Edinger AL, Thompson CB (2004) Death by design: apoptosis, necrosis and autophagy. Curr Opin Cell Biol 16:663–669

Ellis RC, O'steen WA, Hayes RL, Nick HS, Wang KK, Anderson DK (2005) Cellular localization and enzymatic activity of cathepsin B after spinal cord injury in the rat. Exp Neurol 193:19–28

Enokido Y, Suzuki E, Iwasawa K, Namekata K, Okazawa H, Kimura H (2005) Cystathionine beta-synthase, a key enzyme for homocysteine metabolism, is preferentially expressed in the radial glia/astrocyte lineage of developing mouse CNS. FASEB J 19:1854–1856

Erlich S, Shohami E, Pinkas-Kramarski R (2006) Neurodegeneration induces upregulation of beclin 1. Autophagy 2:49–51

Festjens N, Vanden Berghe T, Cornelis S, Vandenabeele P (2007) RIP1, a kinase on the crossroads of a cell's decision to live or die. Cell Death Differ 14(3):400–410

Frasdorf B, Radon C, Leimkuhler S (2014) Characterization and interaction studies of two isoforms of the dual localized 3-mercaptopyruvate sulfurtransferase TUM1 from humans. J Biol Chem 289:34543–34556

Gajavelli S, Sinha VK, Mazzeo AT, Spurlock MS, Lee SW, Ahmed AI, Yokobori S, Bullock RM (2015) Evidence to support mitochondrial neuroprotection, in severe traumatic brain injury. J Bioenerg Biomembr 47:133–148

Galluzzi L, Vitale I, Abrams JM, Alnemri ES, Baehrecke EH, Blagosklonny MV, Dawson TM, Dawson VL, El-Deiry WS, Fulda S, Gottlieb E, Green DR, Hengartner MO, Kepp O, Knight RA, Kumar S, Lipton SA, Lu X, Madeo F, Malorni W, Mehlen P, Nuñez G, Peter ME, Piacentini M, Rubinsztein DC, Shi Y, Simon HU, Vandenabeele P, White E, Yuan J, Zhivotovsky B, Melino G, Kroemer G (2012) Molecular definitions of cell death subroutines: recommendations of the nomenclature committee on cell death. Cell Death Differ 19(1):107–120

Graham SH, Chen J, Clark RS (2000) Bcl-2 family gene products in cerebral ischemia and traumatic brain injury. J Neurotrauma 17(10):831–841

Greenebaum B, Blossfield K, Hannig J, Carrillo CS, Beckett MA, Weichselbaum RR, Lee RC (2004) Poloxamer 188 prevents acute necrosis of adult skeletal muscle cells following high-dose irradiation. Burns 30:539–547

Guicciardi ME, Leist M, Gores GJ (2004) Lysosomes in cell death. Oncogene 23:2881–2890

Halestrap AP, Doran E, Gillespie JP, O'Toole A (2000) Mitochondria and cell death. Biochem Soc Trans 278:170–177

Han J, Zhong CQ, Zhang DW (2011) Programmed necrosis: backup to and competitor with apoptosis in the immune system. Nat Immunol 12:1143–1149

Han W, Xie J, Li L, Liu Z, Hu X (2009) Necrostatin-1 reverts shikonin-induced necroptosis to apoptosis. Apoptosis 14(5):674–686

Hashimoto Y, Ito Y, Niikura T, Shao Z, Hata M, Oyama F, Nishimoto I (2001) Mechanisms of neuroprotection by a novel rescue factor humanin from Swedish mutant amyloid precursor protein. Biochem Biophys Res Commun 283:460–468

Hengartner MO (2000) The biochemistry of apoptosis. Nature 407:770–776

Ichimura Y, Kumanomidou T, Sou YS, Mizushima T, Ezaki J, Ueno T, Kominami E, Yamane T, Tanaka K, Komatsu M (2008) Structural basis for sorting mechanism of p62 in selective autophagy. J Biol Chem 283:22847–22857

Jiang X, Wang X (2000) Cytochrome c promotes caspase-9 activation by inducing nucleotide binding to Apaf-1. J Biol Chem 275:159–163

Jin Y, Lin Y, Feng JF, Jia F, Gao G, Jiang JY (2015) Attenuation of cell death in injured cortex following post-traumatic brain injury moderate hypothermia: possible involvement of autophagy pathway. World Neurosurg 84:420–430

Jones NC, Prior MJ, Burden-The E, Marsden CA, Morris PG, Murphy S (2005) Antagonism of the interleukin-1 receptor following traumatic brain injury in the mouse reduces the number of nitric oxide synthase-2-positive cells and improves anatomical and functional outcomes. Eur J Neurosci 22:72–78

Kim R, Emi M, Tanabe K (2006) Role of mitochondria as the gardens of cell death. Cancer Chemother Pharmacol 57:545–553

Kim S, Dayani L, Rosenberg PA, Li J (2010) RIP1 kinase mediates arachidonic acid-induced oxidative death of oligodendrocyte precursors. Int J Physiol Pathophysiol Pharmacol 2(2):137–147

Kimura H (2010) Hydrogen sulfide: from brain to gut. Antioxid Redox Signal 12:1111–1123

Klionsky DJ (2005) The molecular machinery of autophagy: unanswered questions. J Cell Sci 118:7–18

Knott AB, Bossy-Wetzel E (2008) Impairing the mitochondrial fission and fusion balance: a new mechanism of neurode- generation. Ann N Y Acad Sci 1147:283–292

Komatsu M, Ichimura Y (2010) Physiological significance of selective degradation of p62 by autophagy. FEBS Lett 584:1374–1378

Kroemer G, Levine B (2008) Autophagic cell death: the story of a misnomer. Nat Rev Mol Cell Biol 9:1004–1010

Krysko DV, Vanden-Berghe T, Herdek D, Vandenabeele P (2008) Apoptosis and necrosis: detection, discrimination and phagocytosis. Methods 44:205–221

Kulbe JR, Hill RL, Singh IN, Wang JA, Hall ED (2016) Synaptic mitochondria sustain more damage than non-synaptic mitochondria after traumatic brain injury and are protected by cyclosporine A. J Neurotrauma. [Epub ahead of print]

Lackner LL, Nunnari JM (2009) The molecular mechanism and cellular functions of mitochondrial division. Biochim Bio- Phys Acta 1792:1138–1144

Lai Y, Hickey RW, Chen Y, Bayir H, Sullivan ML, Chu CT, Kochanek PM, Dixon CE, Jenkins LW, Graham SH, Watkins SC, Clark RS (2008) Autophagy is increased after traumatic brain injury in mice and is partially inhibited by the antioxidant gamma-glutamylcysteinyl ethyl ester. J Cereb Blood Flow Metab 28:540–550

Lenz A, Franklin GA, Cheadle WG (2007) Systemic inflammation after trauma. Injury 38:1336–1345

Lenzlinger PM, Hans VH, Jo ller-Jemelka HI, Trentz O, Morganti-Kossmann MC, Kossmann T (2001) Markers for cell-mediated immune response are elevated in cerebrospinal fluid and serum after severe traumatic brain injury in humans. J Neurotrauma 18:479–489

Linkermann A, Green DR (2014) Necroptosis. N Engl J Med 370:455–465

Lin KM, Hsiao G, Shih CM, Chou DS, Sheu JR (2009) Mechanism of resveratrol-induced platelet apoptosis. Cardiovasc Res 83:575–585

Lin L, Baehrecke EH (2015) Autophagy cell death, and cancer. Mol Cell Oncol 2(3):e985913

Lipinski MM, Wu J, Faden AI, Sarkar C (2015) Function and mechanisms of autophagy in brain and spinal cord trauma. Antioxid Redox Signal 23(6):565–577

Liu CL, Chen S, Dietrich D, Hu BR (2008) Changes in autophagy after traumatic brain injury. J Cereb Blood Flow Metab 28:674–683

Liu Y, Shoji-Kawata S, Sumpter RM Jr, Wei Y, Ginet V, Zhang L, Posner B, Tran KA, Green DR, Xavier RJ et al (2013) Autosis is a NaC, KC-ATPase-regulated form of cell death triggered by autophagyinducing peptides, starvation, and hypoxia-ischemia. Proc Natl Acad Sci U S A 110:20364–20371

Luo CL, Chen XP, Li LL, Li QQ, Li BX, Xue AM, Xu HF, Dai DK, Shen YW, LY T, ZQ Z (2013a) Poloxamer 188 attenuates in vitro traumatic brain injury-induced mitochondrial and lysosomal membrane permeabilization damage in cultured primary neurons. J Neurotrauma 30:597–607

Luo CL, Li QQ, Chen XP, Zhang XM, Li LL, Li BX, Zhao ZQ, Tao LY (2013b) Lipoxin A4 attenuates brain damage and downregulates the production of pro-inflammatory cytokines and phosphorylated mitogen-activated protein kinases in a mouse model of traumatic brain injury. Brain Res 1502:1–10

Luo CL, Chen XP, Ni H, Li QQ, Yang R, Sun YX, Tao LY, Zhu GY (2010a) Comparison of labeling methods and time course of traumatic brain injury-induced cell death in mice. Neural Regen Res 5(9):706–709

Luo CL, Chen XP, Yang R, Sun YX, Li QQ, Bao HJ, Cao QQ, Ni H, Qin ZH, Tao LY (2010b) Cathepsin B contributes to traumatic brain injury-induced cell death through a mitochondria-mediated apoptotic pathway. J Neurosci Res 88:2847–2858

Luo CL, Li BX, Li QQ, Chen XP, Sun YX, Bao HJ, Dai DK, Shen YW, Xu HF, Ni H, Wan L, Qin ZH, Tao LY, Zhao ZQ (2011) Autophagy is involved in traumatic brain injury-induced cell death and partially contributes to functional outcome deficits in mice. Neuroscience 184:54–63. See comment in PubMed commons below.

Martelli A, Testai L, Breschi MC, Blandizzi C, Virdis A et al (2012) Hydrogen sulphide: novel opportunity for drug discovery. Med Res Rev 32:1093–1130

Meloni BP, Meade AJ, Kitikomolsuk D, Knuckey NW (2011) Characterisation of neuronal cell death in acute and delayed in vitro ischemia (oxygen-glucose deprivation) models. J Neurosci Methods 195(1):67–74

Messer JS (2016) The cellular autophagy/apoptosis checkpoint during inflammation. Cell Mol Life Sci. [Epub ahead of print]

Moquin DM, McQuade T, Chan FK-M, Harhaj EW (2013) CYLD Deubiquitinates RIP1 in the TNFÎ±-induced necrosome to facilitate kinase activation and programmed necrosis. PLoS One 8(10):e76841

Mustafa AG, Singh IN, Wang J, Carrico KM, Hall ED (2010) Mitochondrial protection after traumatic brain injury by scavenging lipid peroxyl radicals. J Neurochem 114:271–280

Ong SB, Subrayan S, Lim SY, Yellon DM, Davidson SM, Hausenloy DJ (2010) Inhibiting mitochondrial fission protects the heart against ischemia/reperfusion injury. Circulation 121:2012–2022

Pankiv S, Clausen TH, Lamark T, Brech A, Bruun JA, Outzen H, Øvervatn A, Bjørkøy G, Johansen T (2007) p62/SQSTM1 binds directly to Atg8/LC3 to facilitate degradation of ubiquitinated protein aggregates by autophagy. J Biol Chem 282:24131–24145

Petiot A, Ogier-Denis E, Blommaart EF, Meijer AJ, Codogno P (2000) Distinct classes of phosphatidylinositol 3′-kinases are involved in signaling pathways that control macroautophagy in HT-29 cells. J Biol Chem 275:992–998

Pozuelo-Rubio M (2011) 14-3-3_ binds class III phosphatidylinositol-3-kinase and inhibits autophagy. Autophagy 7:240–242

Purnell PR, Fox HS (2013) Autophagy-mediated turnover ofdynamin-related protein 1. BMC Neurosci 14:86

Qi X, Qvit N, Su YC, Mochly-Rosen D (2013) A novel Drp1 inhibitor diminishes aberrant mitochondrial fission and neu- rotoxicity. J Cell Sci 126:789–802

Sadasivan S, Dunn WA Jr, Hayes RL, Wang KK (2008) Changes in autophagy proteins in a rat model of controlled cortical impact induced brain injury. Biochem Biophys Res Commun 373:478–481

Santopietro J, Yeomans JA, Niemeier JP, White JK, Coughlin CM (2015) Traumatic brain injury and behavioral health: the state of treatment and policy. N C Med J 76:96–100

Sarkar C, Zhao Z, Aungst S, Sabirzhanov B, Faden AI, Lipinski MM (2014) Impaired autophagy flux is associated with neuronal cell death after traumatic brain injury. Autophagy 10(12):2208–2222

Sanz O, Acarin L, Gonza'lez B, Castellano B (2002) NF-jB and IjBa expression following traumatic brain injury to the immature rat brain. J Neurosci Res 67:772–780

Serbest G, Horwitz J, Barbee K (2005) The effect of poloxamer-188 on neuronal cell recovery from mechanical injury. J Neurotrauma 22:119–132

Sharp WW, Beiser DG, Fang YH, Han M, Piao L, Varughese J, Archer SL (2015) Inhibition of the mitochondrial fission protein dynamin-related protein 1 improves survival in a murine cardiac arrest model. Crit Care Med 43:e38–e47

Shibuya N, Tanaka M, Yoshida M, Ogasawara Y, Togawa T, Ishii K, Kimura H (2009) 3-Mercaptopyruvate sulfurtransferase produces hydrogen sulfide and bound sulfane sulfur in the brain. Antioxid Redox Signal 11:703–714

Shohami E, Gallily R, Mechoulam R, Bass R, Ben-Hur T (1997) Cytokine production in the brain following closed head injury: dexanabinol (HU-211) is a novel TNF-alpha inhibitor and an effective neuroprotectant. J Neuroimmunol 72:169–177

Siegel RM, Chan FK, Chun HJ, Lenardo MJ (2000) The multifaceted role of Fas signaling in immune cell homeostasis and autoimmunity. Nat Immunol 1:469–474

Song B, Zhou T, Yang WL, Liu J, Shao LQ (2016) Programmed cell death in periodontitis: recent advances and future perspectives. Oral Dis. https://doi.org/10.1111/odi.12574. [Epub ahead of print]

Taylor RC, Cullen SP, Martin SJ (2008) Apoptosis: controlled demolition at the cellular level. Nat Rev Mol Cell Biol 9:231–241

Tehranian R, Rose ME, Vagni V, Pickrell AM, Griffith RP, Liu H, Clark RS, Dixon CE, Kochanek PM, Graham SH (2008) Disruption of Bax protein prevents neuronal cell death but produces cognitive impairment in mice following traumatic brain injury. J Neurotrauma 25:755–767

Tsuchiya K, Kohda Y, Yoshida M, Zhao L, Ueno T, Yamashita J, Yoshioka T, Kominami E, Yamashima T (1999) Postictal blockade of ischemic hippocampal neuronal death in primates using selective cathepsin inhibitors. Exp Neurol 155:187–194

Uchiyama Y, Koike M, Shibata M (2008) Autophagic neuron death in neonatal brain ischemia/hypoxia. Autophagy 4:404–408

Vanden Berghe T, Kaiser WJ, Bertrand MJ, Vandenabeele P (2015) Molecular crosstalk between apoptosis, necroptosis, and survival signaling. Mol Cell Oncol 2(4):e975093

Uryu K, Laurer H, McIntosh T, Pratic OD, Martinez D, Leight S, Lee VM, Trojanowski JQ (2002) Repetitive mild brain trauma accelerates A beta deposition, lipid peroxidation, and cognitive impairment in a transgenicmouse model of Alzheimer amyloidosis. J Neurosci 22:446–454

Wang DB, Garden GA, Kinoshita C, Wyles C, Babazadeh N, Sopher B, Kinoshita Y, Morrison RS (2013a) Declines in Drp1 and parkin expression underlie DNA damage-induced changes in mitochondrial length and neuronal death. J Neurosci 33:1357–1365

Wang Q, Tang XN, Yenari MA (2007) The inflammatory response in stroke. J Neuroimmunol 184:53–56

Wang T, Zhang L, Zhang M, Bao H, Liu W, Wang Y, Wang L, Dai D, Chang P, Dong W, Chen X, Tao L (2013b) [Gly14]-Humanin reduces histopathology and improves functional outcome after traumatic brain injury in mice. Neuroscience 231:70–81

Wang Y, Han R, Liang ZQ, Wu JC, Zhang XD, Gu ZL, Qin ZH (2008) An autophagic mechanism is involved in apoptotic death of rat striatal neurons induced by the non-N-methyl-D-aspartate receptor agonist kainic acid. Autophagy 4:1–13

Wang YQ, Wang L, Zhang MY, Wang T, Bao HJ, Liu WL, Dai DK, Zhang L, Chang P, Dong WW, Chen XP, Tao LY (2012 Sep) Necrostatin-1 suppresses autophagy and apoptosis in mice traumatic brain injury model. Neurochem Res 37(9):1849–1858

Werner C, Engelhard K (2007) Pathophysiology of traumatic brain injury. Br J Anaesth 99:4–9

Wesselborg S, Stork B (2015) Autophagy signal transduction by ATG proteins: from hierarchies to networks. Cell Mol Life Sci 72:4721–4757

Winter CD, Iannotti F, Pringle AK, Trikkas C, Clough GF, Church MK (2002) A microdialysis-method for the recovery of IL-1beta, IL-6 and nerve growth factor from human brain in vivo. J Neurosci Methods 119:45–50

Wu Q, Xia SX, Li QQ, Gao Y, Shen X, Ma L, Zhang MY, Wang T, Li YS, Wang ZF, Luo CL, Tao LY (2016) Mitochondrial division inhibitor 1 (Mdivi-1) offers neuroprotection through diminishing cell death and improving functional outcome in a mouse model of traumatic brain injury. Brain Res 1630:134–143

Xue L, Fletcher GC, Tolkovsky AM (2001) Mitochondria are selectively eliminated from eukaryotic cells after blockade of caspases during apoptosis. Curr Biol 11:361–365

Xu X, Chua KW, Chua CC, Liu CF, Hamdy RC, Chua BH (2010a) Synergistic protective effects of humanin and necrostatin-1 on hypoxia and ischemia/reperfusion injury. Brain Res 1355:189–194

Xu X, Chua CC, Zhang M, Geng D, Liu CF, Hamdy RC, Chua BH (2010b) The role of PARP activation in glutamate-induced necroptosis in HT-22 cells. Brain Res 1343:206–212

Xu X, Chua CC, Gao J, Chua KW, Wang H, Hamdy RC, Chua BH (2008) Neuroprotective effect of humanin on cerebral ischemia/reperfusion injury is mediated by a PI3K/Akt pathway. Brain Res 1227:12–18

Xu X, Chua CC, Kong J, Kostrzewa RM, Kumaraguru U, Hamdy RC, Chua BH (2007) Necrostatin-1 protects against glutamateinduced glutathione depletion and caspase-independent cell death in HT-22 cells. J Neurochem 103(5):2004–2014

Yamashima T, Kohda Y, Tsuchiya K, Ueno T, Yamashita J, Yoshioka T, Kominami E (1998) Inhibition of ischaemic hippocampal neuronal death in primates with cathepsin B inhibitor CA-074: a novel strategy for neuroprotection based on "calpain-cathepsin hypothesis". Eur J Neurosci 10:1723–1733

You Z, Savitz SI, Yang J, Degterev A, Yuan J, Cuny GD, Moskowitz MA, Whalen MJ (2008) Necrostatin-1 reduces histopathology and improves functional outcome after controlled cortical impact in mice. J Cereb Blood Flow Metab 28(9):1564–1573

Yu L, Alva A, Su H, Dutt P, Freundt E, Welsh S, Baehrecke EH, Lenardo MJ (2004) Regulation of an ATG7-beclin 1 program of autophagic cell death by caspase-8. Science 304(5676):1500–1502

Zhang H, Zhong C, Shi L, Guo Y, Fan Z (2009) Granulysin induces cathepsin B release from lysosomes of target tumor cells to attack mitochondria through processing of bid leading to necroptosis. J Immunol 182:6993–7000

Zhang M, Shan H, Wang T, Liu W, Wang Y, Wang L, Zhang L, Chang P, Dong W, Chen X, Tao L (2013a) Dynamic change of hydrogen sulfide after traumatic brain injury and its effect in mice. Neurochem Res 38(4):714–725

Zhang M, Shan H, Wang Y, Wang T, Liu W, Wang L, Zhang L, Chang P, Dong W, Chen X, Tao L (2013b) The expression changes of cystathionine-β-synthase in brain cortex after traumatic brain injury. J Mol Neurosci 51(1):57–67

Zhang M, Shan H, Chang P, Wang T, Dong W, Chen X, Tao L (2014) Hydrogen sulfide offers neuroprotection on traumatic brain injury in parallel with reduced apoptosis and autophagy in mice. PLoS One 9(1):e87241

Zhang M, Shan H, Chang P, Ma L, Chu Y, Shen X, Wu Q, Wang Z, Luo C, Wang T, Chen X, Tao L (2016) Upregulation of 3-MST relates to neuronal autophagy after traumatic brain injury in mice. Cell Mol Neurobiol. [Epub ahead of print]

Zhang N, Chen Y, Jiang R, Li E, Chen X, Xi Z, Guo Y, Liu X, Zhou Y, Che Y, Jiang X (2011) PARP and RIP 1 are required for autophagy induced by 110-deoxyverticillin A, which precedes caspase-dependent apoptosis. Autophagy 7(6):598–612

Zhang N, Wang S, Li Y, Che L, Zhao Q (2013c) A selective inhibitor of Drp1, Mdivi-1, acts against cerebral ischemia/ reperfusion injury via an anti-apoptotic pathway in rats. Neurosci Lett 535:104–109

Zhang YB, Li SX, Chen XP, Yang L, Zhang YG, Liu R, Tao LY (2008) Autophagy is activated and might protect neurons from degeneration after traumatic brain injury. Neurosci Bull 24(3):143–149

Zhao ST, Huang XT, Zhang C, Ke Y (2011) Humanin protects cortical neurons from ischemia and reperfusion injury by the increased activity of superoxide dismutase. Neurochem Res 37:153–160

Zheng YT, Shahnazari S, Brech A, Lamark T, Johansen T, Brumell JH (2009) The adaptor protein p62/SQSTM1 targets invading bacteria to the autophagy pathway. J Immunol 183:5909–5916

Zhou W, Yuan J (2014) SnapShot: Necroptosis. Cell 158:464–440

Zhu S, Zhang Y, Bai G, Li H (2011) Necrostatin-1 ameliorates symptoms in R6/2 transgenic mouse model of Huntington's disease. Cell Death Dis 2:e115

Zou H, Henzel WJ, Liu X, Lutschg A, Wang X (1997) Apaf-1, a human protein homologous to C. elegans CED-4, participates in cytochrome c-dependent activation of caspase-3. Cell 90(3):405–413

Part III
Focal Cerebral Ischemia

Chapter 6
Involvement of Apoptosis-Inducing Factor (AIF) in Neuronal Cell Death Following Cerebral Ischemia

Nikolaus Plesnila and Carsten Culmsee

Abstract Delayed neuronal death is a hallmark of infarct development and sustained functional impairment in rodent models of focal cerebral ischemia, an experimental paradigm resembling ischemic stroke in humans. The exact molecular pathophysiology of this still enigmatic event is not only of academic interest but may hold the key for novel therapeutic strategies for human stroke. There is general understanding that acute lack of perfusion leads to rapid necrotic-oncotic cell death in the core of the ischemic infarct. In contrast, conditions associated with severely reduced, but not immediately lethal reductions of cerebral blood flow in the ischemic penumbra likely result in delayed and more programmed type of neuronal cell death. Based on results first obtained from non-neuronal cells, cysteine aspartate proteases (caspases) were described as key modulators of this process. More recently, however, it became clear that also caspase-independent mechanisms play a significant role for ischemia-induced delayed neuronal cell loss. In this chapter, we review the role of one of the first described caspase-independent cell death proteins, apoptosis-inducing factor (AIF), for post-ischemic brain damage. Our conclusion is that there is compelling evidence for a causal role of AIF in neuronal cell death following experimental stroke and other neurological disorders associated with cerebral ischemia. Hence, AIF and other, more recently described subtypes of caspase-independent cell death may provide promising targets for therapeutic interventions in cerebrovascular disease.

Keywords Stroke · Focal cerebral ischemia · Apoptosis inducing factor · AIF · Programmed cell death

N. Plesnila (✉)
Institute for Stroke and Dementia Research (ISD), University of Munich Medical Center, Munich, Germany
e-mail: nikolaus.plesnila@med.uni-muenchen.de

C. Culmsee
Clinical Pharmacy—Pharmacology and Toxicology, Faculty of Pharmacy, Philipps-University of Marburg, Marburg, Germany
e-mail: culmsee@uni-marburg.de

D. G. Fujikawa (ed.), *Acute Neuronal Injury*,
https://doi.org/10.1007/978-3-319-77495-4_6

6.1 Introduction

Every year stroke is responsible for the death of 5.5 million people and thus accounts for 10% of all deaths in industrialized countries worldwide. Despite such a high incidence and mortality, therapeutic options for stroke patients are still very limited (Lo et al. 2003). Currently, the only clinical treatment options for stroke patients are recanalization of large brain supplying arteries by local or systemic administration of recombinant tissue plasminogen activator (rtPA) and/or by mechanical removal of clots/emboli with stent retrievers (Hussain et al. 2016). A major limitation of these early therapeutic approaches, however, is that both require brain CT and MRI imaging in order to exclude hemorrhagic stroke and to localize the occluded vessel. Accordingly, rtPA lysis and mechanical recanalization can only be initiated after affected patients are admitted to a specialized center. By the time diagnostic procedures have been completed the therapeutic window for both procedures, i.e. 4.5 h after the onset of ischemia, has shortened significantly or even closed. As a result less than 10% of all stroke patients are subjected to recanalization therapy (Adams et al. 2007). The remaining 90% may only hope for spontaneous reperfusion, which in most cases, however, occurs too late to prevent penumbral cell death, the main mechanism underlying infarct outgrowth and the subsequent loss of neurological function (Molina et al. 2001). Hence, novel treatment strategies are warranted which are able to prolong neuronal survival in the ischemic penumbra, i.e., under compromised cerebral blood flow conditions.

It is well accepted that even after reperfusion cell death signaling pathways triggered by the initial ischemic event remain activated and result in additional neuronal cell death under completely normal blood flow conditions. This was first demonstrated in experimental approaches where very brief ischemic episodes were induced, i.e. 30 min of middle cerebral artery occlusion in mice or rats (MCAo), which may resemble transient ischemic attacks (TIA) in patients. Under this condition neuronal cell death may occur with a delay of up to 24 h following reperfusion (Du et al. 1996; Endres et al. 1998). Subsequently, post-reperfusion cell death was also demonstrated following more severe ischemic episodes which are associated with acute infarction and, hence, resemble acute stroke in humans. In the ischemic penumbra of mice subjected to 60 min of transient MCAo neurons die with a delay of only 3–6 h (Fig. 6.1a), i.e., also post-reperfusion cell death seems to have a clinically relevant therapeutic window. Accordingly, an optimal therapeutic approach towards the treatment of stroke should include the protection of neuronal cells during the period of compromised blood flow but also the prevention of cell death after reperfusion.

6.2 Mechanisms of Delayed Cell Death Following Focal Cerebral Ischemia

The morphological hallmarks of neuronal cell death following focal cerebral ischemia are cell shrinkage and nuclear condensation, features not present in classical necrotic cell death which is associated with cell lysis and nuclear decomposition.

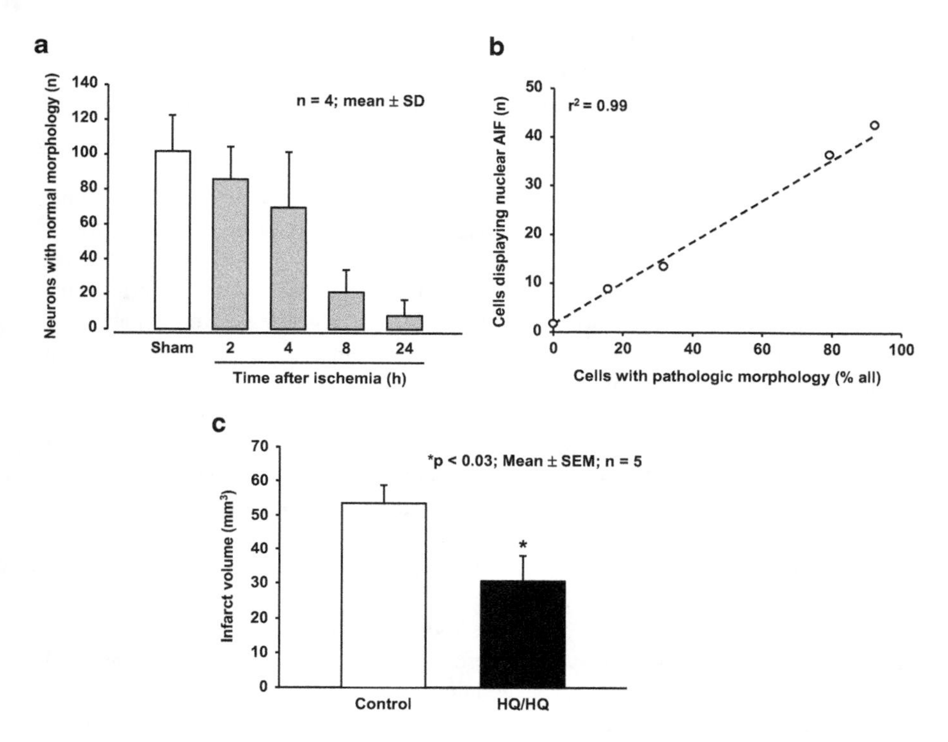

Fig. 6.1 Delayed neuronal cell death in the ischemic penumbra and correlation with nuclear AIF following transient focal cerebral ischemia in mice (**a**) Following 60 min of middle cerebral artery occlusion (MCAo) the majority of neurons (~70%) in the ischemic penumbra, i.e. the cerebral cortex, stay alive for at least 4 h. Despite sufficient blood flow 24 h after MCAo over 90% of neurons which were viable 2 h after ischemia display altered membrane and nuclear morphology indicating cell death. (**b**) Correlation of neurons displaying pathological morphology with cells showing nuclear AIF (Culmsee et al. 2005). (**c**) In Harlequin mutant mice (HQ) which have a reduced expression of AIF protein due to a proviral insertion in the *aif* gene, the infarct volume, calculated on the basis of the histomorphometric data from the individual sections, showed a 43% reduction as compared to wild type littermates (n = 5, *p < 0.03) (Culmsee et al. 2005)

Cell shrinkage and nuclear condensation following cerebral ischemia are found in brain areas affected by immediate and delayed cell death. Accordingly, the mechanisms leading to ischemic cell death seem to be very similar irrespective if affected cells are located in the infarct core where blood flow is almost absent or in the ischemic penumbra where collateral blood flow may keep cells alive for several hours (Astrup et al. 1981). For many years it remained unclear how ischemia causes the morphological findings described above. Nuclear condensation is the morphological sequel of DNA damage, which usually occurs in a highly regulated manner during various forms of programmed cell death, such as apoptosis, necroptosis, or parthanatos (Andrabi et al. 2011; Galluzzi et al. 2014; Vanden Berghe et al. 2014). More than 20 years ago Linnik et al. and Charriaut-Merlangue et al. were the first to demonstrate that nuclear condensation following cerebral ischemia was the result of DNA damage and endonuclease activation (Charriaut-Marlangue et al. 1996;

Linnik et al. 1995). These findings triggered intense search for the upstream signaling responsible for post-ischemic endonuclease activation which was finally believed to be the activation of caspase-3 (Namura et al. 1998). Namura and colleagues showed constitutive expression of inactive caspase-3 in neurons throughout the brain, most prominently in neuronal perikarya within piriform cortex and, most importantly, caspase-like enzyme activity in ischemic brain 30–60 min after reperfusion following 2 h MCAo. Active caspase-3 was detected in ischemic neurons at the time of reperfusion by immunohistochemistry. DNA laddering and TUNEL-positive cells as indicators of DNA fragmentation were detected 6–24 h after reperfusion (Namura et al. 1998). Further proof for the role of active caspase-3 for ischemic cell death was derived in the same year from experiments from the same laboratory using pan-caspase and caspase-3 specific peptide inhibitors. Post-ischemic neuronal cell death was prevented and neuronal function was improved when caspase activation was inhibited up to 6 h following reperfusion from 30 min MCAo (Endres et al. 1998). The ultimate mechanistic link between caspase-3 activation and post-ischemic DNA fragmentation was established by Gao and co-workers by showing that caspase-activated DNase (CAD), a molecule known to be cleaved and thereby activated by caspase-3, was responsible for post-ischemic DNA-fragmentation (Cao et al. 2001).

In consequence, many research groups focused on the upstream mechanisms of caspase-3 activation. Due to very low expression and activation levels of potentially involved molecules it turned out to be technically very challenging to identify respective mechanisms. Caspase-8, a molecule able to cleave caspase-3 in non-neuronal cells, was found to be activated following experimental stroke; however, caspase-8 was described to be activated in a population of neurons (lamina V) distinct from that where active caspase-3 was observed (lamina II/III) (Velier et al. 1999) and a direct link between caspase-8 and caspase-3 activation has not yet been demonstrated in models of cerebral ischemia. Further upstream factors in the cascade of caspase activation such as Fas/CD95 receptors and tumor necrosis factor-related apoptosis-inducing ligand (TRAIL), were found to be upregulated following MCAo, and lpr mice, which express dysfunctional Fas receptors, were protected from focal ischemic brain damage (Martin-Villalba et al. 1999). Despite these interesting findings it still remained unclear how caspase-3 was activated following cerebral ischemia until in 2001 it was demonstrated that the BH3-only Bcl-2 family Bid, which has a caspase-8 specific cleavage site, was truncated after experimental stroke (Plesnila et al. 2001). Cleaved/truncated Bid (tBid) translocates from the cytoplasm to the outer mitochondrial membrane where together with Bax it induces the formation of an oligomeric membrane pore (Zha et al. 2000), thereby releasing cytochrome c from mitochondria (Wei et al. 2000). After focal cerebral ischemia mitochondria of Bid-deficient mice released far less cytochrome c and cortical infarction was significantly reduced compared to wildtype littermates, thereby demonstrating the prominent role of mitochondria in post-ischemic cell death. This was supported by recent experiments on Bax- deficient mice, which showed a similar level of neuroprotection as Bid- deficient animals (D'Orsi et al. 2015). These data further imply that after focal cerebral ischemia caspase-3 may be activated through the mitochondrial pathway, i.e. by the mitochondrial release of cytochrome c (Fujimura et al. 2000) and apoptosome formation (Plesnila 2004; Plesnila et al. 2001; Yin et al. 2002). Not much later, however, this view was challenged by the fact that caspase-3 knock out mice, which became available at that time, showed much less neuroprotection than expected based on the anticipated prominent role of caspase-3 activation for ischemic neuronal cell death (Le et al. 2002). Together with the pronounced neuroprotective effect achieved by interactions with mitochondrial cell death signaling, (Cao et al. 2002; D'Orsi et al. 2015; Kilic et al. 2002; Martinou et al. 1994; Plesnila et al. 2001; Wiessner et al. 1999), i.e., mechanisms upstream of caspase-3 activation such as Bid and Bax activation, it became clear that alternative cell death pathways distinct from caspase-3 may be present downstream of mitochondria.

The hypothesis that caspase-independent neuronal cell death signaling exists downstream of mitochondria was also suggested by in vitro experiments showing that caspase inhibition provided only transient neuroprotection which was followed by a more delayed type of DNA-fragmentation-related cell death (see Rideout and Stefanis 2001 for review). It was Ruth Slack and her colleagues who identified a mitochondrial protein, apoptosis-inducing factor (AIF), to be one of the most potent molecular candidates for caspase-independent death in neurons (Cregan et al. 2002). AIF translocation from mitochondria to the nucleus was detected in damaged neurons in vitro in models of neuronal cell death relevant to the pathology of ischemic brain damage, such as glutamate neurotoxicity, DNA damage or oxygen-glucose deprivation, whereas neutralizing AIF antibodies, pharmacological inhibition of AIF release or AIF siRNA prevented neuronal cell death in these in vitro approaches (Becattini et al. 2006; Cao et al. 2002; Cregan et al. 2002; Culmsee et al. 2005).

AIF is a 67 kDa flavoprotein with significant homology to bacterial and plant oxidoreductases located in the mitochondrial intramembranous space (Susin et al. 1999). Upon release from mitochondria, AIF migrates to the nucleus where it induces large-scale (~50 kbp) DNA fragmentation and cell death in a caspase-independent manner (Daugas et al. 2000; Penninger and Kroemer 2003). Recent findings in models of glutamate neurotoxicity in cultured neurons in vitro and cerebral hypoxia/ischemia in vivo suggested that AIF translocation from mitochondria to the nucleus requires Cyclophilin A (CypA), which seems to coordinate DNA binding and chromatinolysis through complex formation with histone H2AX (Artus et al., 2010; Baritaud et al., 2010; Doti et al., 2014; Zhu et al., 2007). Finally, the Dawson laboratory identified macrophage migration inhibitory factor (MIF) as the key nuclease mediating AIF-dependent DNA degradation in paradigms of parthanatos induced by oxidative stress and DNA-damage (Wang et al., 2016). Eliminating MIF's nuclease activity exerted sustained protective effects in a model of focal cerebral ischemia, both at the level of histology and behavior.

In the brain, AIF was shown to be expressed in all so far investigated cell types, i.e. neurons and glial cells (Cao et al. 2003; Zhu et al. 2003). The expression in normal neuronal cells was confined to the mitochondria as shown by co-immunostaining with the mitochondrial marker cytochrome oxidase (Plesnila et al. 2004). Interestingly, unlike the expression pattern of many other apoptotic proteins the expression of AIF increases gradually with brain maturation and peaks in adulthood, indicating that in contrast to, e.g. caspase-3, AIF may exert its main function in adult neurons (Cao et al. 2003).

The first pathological condition where AIF was shown to play an important role for neuronal damage was cerebral hypoxia-ischemia, a model for asphyxia in newborn children. Hypoxia-ischemia in 7-day-old rats induced by ligation of the left carotid artery for 55 min together with the reduction of ambient oxygen to 7.7% in a hypoxia chamber resulted in AIF release from mitochondria and translocation to the nucleus in neurons displaying DNA fragmentation- and pyknosis (Zhu et al. 2003). Since AIF translocation was not influenced by inhibition of caspases using the pan-caspase inhibitor BAF these experiments stressed the caspase-independent manner of AIF-induced cell death. Similar findings were also observed following cardiac arrest- induced brain damage in rats, i.e., following transient global ischemia. Following 15 min of four-vessel occlusion (4-VO) AIF was found to translocate from mitochondria to the nucleus in hippocampal CA1 neurons. The temporal profile of AIF translocation coincided with the induction of large-scale DNA fragmentation (50 kbp; 24–72 h after 4-VO), a well-characterized hallmark of delayed neuronal cell death (Cao et al. 2003). In line with findings in the rodent models of transient hypoxia-ischemia in immature animals, treatment with a caspase-3 inhibitor had no effect on nuclear AIF accumulation and did not provide any long-lasting neuroprotective effects after global ischemia in adult rats (Cao et al. 2003).

At almost the same time we demonstrated the translocation of AIF from mitochondria to the nucleus following transient focal cerebral ischemia, an experimental model of ischemic stroke followed by reperfusion (Plesnila et al. 2004). Nuclear AIF was detected in single neuronal cells very early, i.e. within 1 h after 45 min of middle cerebral artery occlusion (MCAo) and peaked 24 h thereafter. The time course of AIF translocation paralleled mitochondrial cytochrome c release and apoptosis-like DNA damage as identified by hair-pin probe (HPP) staining, indicating ischemia-induced mitochondrial permeabilization and AIF-induced DNA fragmentation (Plesnila et al. 2004). Further, we showed that in the same experimental paradigm of ischemic stroke that AIF nuclear translocation was mainly found in neurons (Culmsee et al. 2005) and that the number of cells displaying pathological morphology following cerebral ischemia correlated very well ($r^2 = 0.99$) with the number of neurons showing nuclear AIF (Fig. 6.1b).

That nuclear translocation of AIF was indeed responsible for post-ischemic cell death and not only a byproduct of the morphological changes associated with neuronal cell death was first shown in 2005. Small inhibitory RNA (siRNA)-mediated downregulation of AIF expression (−80%) in HT22 hippocampal neurons and in primary cultured neurons resulted in a significant reduction of glutamate and oxygen-glucose deprivation-induced neuronal cell death, respectively (Figs. 6.2 and 6.3). Reduction of cell death was associated with a lack of nuclear AIF translocation, thereby demonstrating that AIF plays a causal role in excitotoxic and hypoxic-hypoglycaemic cell death in vitro (Culmsee et al. 2005). In the same study we demonstrated that AIF is also relevant for post-ischemic cell death in vivo. Harlequin mutant mice carry a pro-viral insertion in the AIF-gene thereby expressing only 10–20% of normal AIF protein levels (Klein et al. 2002). These mutant mice show significantly reduced post-ischemic brain damage as compared to their wild-type littermates, which express AIF at normal levels (Culmsee et al. 2005) (Fig. 6.1c).

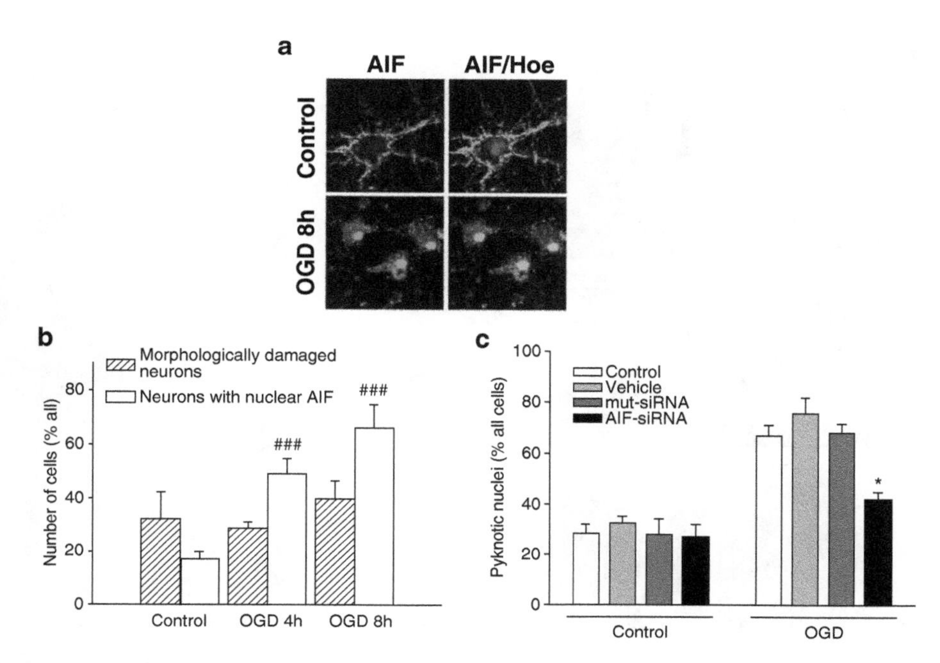

Fig. 6.2 AIF-siRNA knockdown attenuates glutamate-induced neuronal cell death in primary cultured neurons (**a**) Confocal laser scanning microscope images of AIF immunoreactivity (green) were obtained after 8 h of oxygene glucose deprivation (OGD). Co-staining with DAPI (dark blue) allowed the identification of nuclear translocation of AIF (AIF/DAPI, light blue) in damaged cells. (**b**) Number of damaged neurons and neurons displaying nuclear AIF 4 and 8 h after reoxygenation after 4 h of oxygen—glucose deprivation. AIF translocates to the nucleus before signs of morphological neuronal damage [as determined by nuclear morphology after DAPI/Hoechst staining or propidium iodide/calcein staining] become evident (n = 4; ###p < 0.001 vs. control). (**c**) Primary cultured neurons were pre-treated with vehicle (lipofectamine), non-functional mutant RNA (mut-siRNA), or AIF-siRNA for 48 h before exposure to OGD for 4 h. Cell death was quantified by counting of cells with pyknotic nuclei 24 h after re-oxygenation in medium containing glucose. In AIF siRNA- treated neurons the number of cells displaying pyknotic nuclei was reduced by ~50% (n = 4; *p < 0.01 vs. control) (Culmsee et al. 2005)

Further analysis in vitro revealed that reduced AIF expression exerts preconditioning effects at the level of mitochondria, thereby preserving mitochondrial integrity and function in conditions of glutamate toxicity (Fig. 6.3) (Oexler et al. 2012). Whether such preconditioning effects at the level of mitochondria caused by reduced AIF expression levels account for protective effects against ischemic brain damage in vivo requires further investigation.

In vitro, nuclear AIF translocation was dependent on poly(ADP-ribose) polymerase 1 (PARP1) activation, as shown by using the specific PARP1 inhibitor PJ-34 (Culmsee et al. 2005). Accordingly, these results suggest that PARP1 activation is located upstream of AIF release from mitochondria and that AIF is the major factor mediating PARP1-induced cell death, findings also supported by other laboratories using different strategies to inhibit PARP, i.e. by cilostazol or gallotannin (Lee et al.

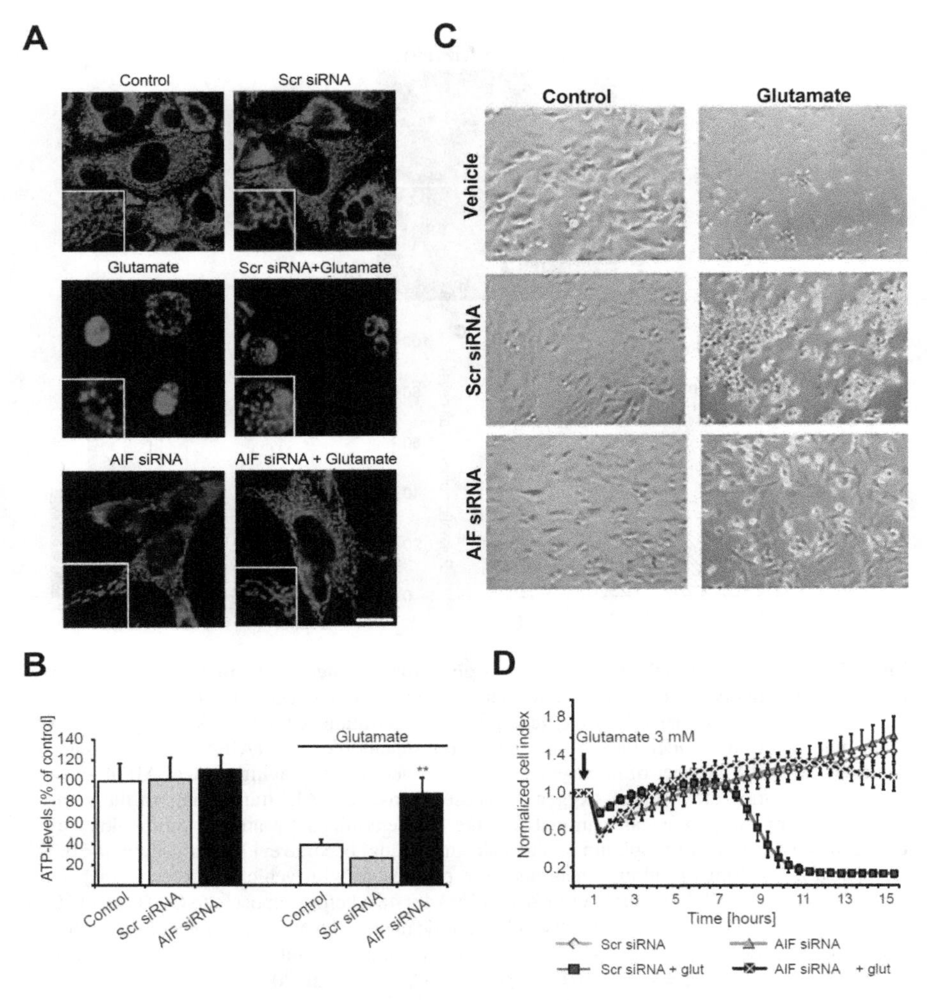

Fig. 6.3 AIF-siRNA preserves mitochondrial integrity and function, and cell viability. (**a**) Fluorescence photomicrographs show that AIF siRNA (20 nM) prevents the fission of mitochondria (stained with Mitotracker red) in glutamate-exposed (3 mM, 14 h) HT-22 cells compared to non-transfected control cells and cells transfected with scr siRNA. Scale bar 20 μm; insets show magnifications for better detection of mitochondrial morphology. (**b**) ATP levels from AIF siRNA transfected cells (20 nM) were protected from ATP depletion as determined 24 h after glutamate exposure (n = 6; **p\0.01 compared to glutamate treated control cells and scr siRNA; ANOVA, Scheffe' test). (**c**) AIF siRNA (20 nM) prevents glutamate-induced (5 mM, 12 h) cell death in neuronal HT22 cells compared to non-transfected control cells and cells transfected with scr siRNA. (**d**) xCELLigence real-time measurement: HT22 cells were treated with glutamate (glut) 72 h after transfection. AIF siRNA (20 nM) shows sustained protection over time (n = 8) (Oexler et al. 2012)

2007; Wei et al. 2007). More recently, Iduna was identified as an NMDA-receptor-induced survival protein which binds poly(ADP-ribose) polymers, thereby preventing AIF translocation to the nucleus in paradigms involving parthanatos in NMDA excitotoxicity in vitro and ischemic neuronal death in vivo (Andrabi et al. 2011). Further, activation of neuronal nitric oxide synthase (nNOS) and formation of ROS, particularly lipidperoxides, were linked to AIF-mediated neuronal cell death following experimental stroke (Li et al. 2007; Tobaben et al. 2011; Yigitkanli et al. 2017). Gene deletion of nNOS, application of a metalloporphyrin-based superoxide dismutase or inhibition of 12/15 lipidperoxidase (LOX) mimic reduced post-ischemic cell death, together with a reduction of the number of neurons displaying nuclear AIF, thereby suggesting that ROS and peroxynitrite formation may cause direct or indirect mitochondrial damage and subsequent AIF release, nuclear translocation, and large-scale DNA fragmentation (Lee et al. 2005; Li et al. 2007).

Results from our and other laboratories on the direct upstream mechanisms responsible for the release of AIF from mitochondria suggest that pro-apoptotic proteins of the bcl-2 family such as Bid interacting with regulators of mitochondrial fission such as Drp1 play an important role for this process. SiRNA-mediated knockdown and small molecule inhibitors of Bid or Drp1 fully preserved mitochondrial integrity and function, and prevented cell death, together with translocation of AIF from mitochondria to the nucleus in primary cultured neurons following oxygen-glucose deprivation and completely preserved cell and nuclear morphology following glutamate toxicity in HT22 hippocampal cells (Culmsee et al. 2005; Landshamer et al. 2008; Grohm et al. 2010, 2012). Further, the small molecular inhibitors of Drp1, MDIVI-A and MDIVI-B reduced infarct size in a model of focal cerebral ischemia in mice (Grohm et al. 2012), similar to previously reported effects of genetic Bid deletion (Plesnila et al. 2001; Yin et al. 2002)

In conclusion, the current literature suggests that AIF-mediated caspase-independent signaling pathways are of major importance for delayed neuronal cell death following experimental stroke. Caspase activation occurs during this process, however, inhibition of caspases seems to only delay and not to prevent neuronal death following focal cerebral ischemia. These findings suggest that inhibition of mitochondrial AIF release and subsequent AIF-dependent mechanisms of DNA damage may serve as novel targets for drug development aimed to mitigate cell death following stroke.

References

Adams HP Jr, del ZG, Alberts MJ, Bhatt DL, Brass L, Furlan A, Grubb RL, Higashida RT, Jauch EC, Kidwell C, Lyden PD, Morgenstern LB, Qureshi AI, Rosenwasser RH, Scott PA, Wijdicks EF (2007) Guidelines for the early management of adults with ischemic stroke: a guideline from the American Heart Association/American Stroke Association Stroke Council, Clinical Cardiology Council, Cardiovascular Radiology and Intervention Council, and the Atherosclerotic Peripheral Vascular Disease and Quality of Care Outcomes in Research Interdisciplinary Working Groups: The American Academy of Neurology affirms the value of this guideline as an educational tool for neurologists. Circulation 115:e478–e534

Andrabi SA, Kang HC, Haince JF, Lee YI, Zhang J, Chi Z, West AB, Koehler RC, Poirier GG, Dawson TM, Dawson VL (2011) Iduna protects the brain from glutamate excitotoxicity and stroke by interfering with poly(ADP-ribose) polymer-induced cell death. Nat Med 17:692–699

Artus C, Boujrad H, Bouharrour A, Brunelle MN, Hoos S, Yuste VJ, Lenormand P, Rousselle JC, Namane A, England P, Lorenzo HK, Susin SA (2010) AIF promotes chromatinolysis and caspase-independent programmed necrosis by interacting with histone H2AX. EMBO J 29:1585–1599

Astrup J, Siesjo BK, Symon L (1981) Thresholds in cerebral ischemia—the ischemic penumbra. Stroke 12:723–725

Baritaud M, Boujrad H, Lorenzo HK, Krantic S, Susin SA (2010) Histone H2AX: the missing link in AIF-mediated caspase-independent programmed necrosis. Cell Cycle 9:3166–3173

Becattini B, Culmsee C, Leone M, Zhai D, Zhang X, Crowell KJ, Rega MF, Landshamer S, Reed JC, Plesnila N, Pellecchia M (2006) Structure-activity relationships by interligand NOE-based design and synthesis of antiapoptotic compounds targeting Bid. Proc Natl Acad Sci U S A 103:12602–12606

Cao G, Pei W, Lan J, Stetler RA, Luo Y, Nagayama T, Graham SH, Yin XM, Simon RP, Chen J (2001) Caspase-activated DNase/DNA fragmentation factor 40 mediates apoptotic DNA fragmentation in transient cerebral ischemia and in neuronal cultures. J Neurosci 21:4678–4690

Cao G, Pei W, Ge H, Liang Q, Luo Y, Sharp FR, Lu A, Ran R, Graham SH, Chen J (2002) In vivo delivery of a Bcl-xL fusion protein containing the TAT protein transduction domain protects against ischemic brain injury and neuronal apoptosis. J Neurosci 22:5423–5431

Cao G, Clark RS, Pei W, Yin W, Zhang F, Sun FY, Graham SH, Chen J (2003) Translocation of apoptosis-inducing factor in vulnerable neurons after transient cerebral ischemia and in neuronal cultures after oxygen-glucose deprivation. J Cereb Blood Flow Metab 23:1137–1150

Charriaut-Marlangue C, Margaill I, Represa A, Popovici T, Plotkine M, Ben Ari Y (1996) Apoptosis and necrosis after reversible focal ischemia: an in situ DNA fragmentation analysis. J Cereb Blood Flow Metab 16:186–194

Cregan SP, Fortin A, MacLaurin JG, Callaghan SM, Cecconi F, Yu SW, Dawson TM, Dawson VL, Park DS, Kroemer G, Slack RS (2002) Apoptosis-inducing factor is involved in the regulation of caspase-independent neuronal cell death. J Cell Biol 158:507–517

Culmsee C, Zhu C, Landshamer S, Becattini B, Wagner E, Pellecchia M, Blomgren K, Plesnila N (2005) Apoptosis-inducing factor triggered by poly(ADP-ribose) polymerase and Bid mediates neuronal cell death after oxygen-glucose deprivation and focal cerebral ischemia. J Neurosci 25:10262–10272

Daugas E, Nochy D, Ravagnan L, Loeffler M, Susin SA, Zamzami N, Kroemer G (2000) Apoptosis-inducing factor (AIF): a ubiquitous mitochondrial oxidoreductase involved in apoptosis. FEBS Lett 476:118–123

D'Orsi B, Kilbride SM, Chen G, Perez AS, Bonner HP, Pfeiffer S, Plesnila N, Engel T, Henshall DC, Dussmann H, Prehn JH (2015) Bax regulates neuronal Ca2+ homeostasis. J Neurosci 35:1706–1722

Doti N, Reuther C, Scognamiglio PL, Dolga AM, Plesnila N, Ruvo M, Culmsee C (2014) Inhibition of the AIF/CypA complex protects against intrinsic death pathways induced by oxidative stress. Cell Death Dis 5:e993

Du C, Hu R, Csernansky CA, Hsu CY, Choi DW (1996) Very delayed infarction after mild focal cerebral ischemia: a role for apoptosis? J Cereb Blood Flow Metab 16:195–201

Endres M, Namura S, Shimizu-Sasamata M, Waeber C, Zhang L, Gomez-Isla T, Hyman BT, Moskowitz MA (1998) Attenuation of delayed neuronal death after mild focal ischemia in mice by inhibition of the caspase family. J Cereb Blood Flow Metab 18:238–247

Fujimura M, Morita-Fujimura Y, Noshita N, Sugawara T, Kawase M, Chan PH (2000) The cytosolic antioxidant copper/zinc-superoxide dismutase prevents the early release of mitochondrial cytochrome c in ischemic brain after transient focal cerebral ischemia in mice. J Neurosci 20:2817–2824

Galluzzi L, Kepp O, Krautwald S, Kroemer G, Linkermann A (2014) Molecular mechanisms of regulated necrosis. Semin Cell Dev Biol 35:24–32. https://doi.org/10.1016/j.semcdb.2014.02.006. Epub 2014 Feb 26

Grohm J, Plesnila N, Culmsee C (2010) Bid mediates fission, membrane permeabilization and peri-nuclear accumulation of mitochondria as a prerequisite for oxidative neuronal cell death. Brain Behav Immun 24:831–838

Grohm J, Kim SW, Mamrak U, Tobaben S, Cassidy-Stone A, Nunnari J, Plesnila N, Culmsee C (2012) Inhibition of Drp1 provides neuroprotection in vitro and in vivo. Cell Death Differ 19:1446–1458

Hussain M, Moussavi M, Korya D, Mehta S, Brar J, Chahal H, Qureshi I, Mehta T, Ahmad J, Zaidat OO, Kirmani JF (2016) Systematic review and pooled analyses of recent Neurointerventional randomized controlled trials: setting a new standard of care for acute ischemic stroke treatment after 20 years. Interv Neurol 5:39–50

Kilic E, Dietz GP, Hermann DM, Bahr M (2002) Intravenous TAT-Bcl-Xl is protective after middle cerebral artery occlusion in mice. Ann Neurol 52:617–622

Klein JA, Longo-Guess CM, Rossmann MP, Seburn KL, Hurd RE, Frankel WN, Bronson RT, Ackerman SL (2002) The harlequin mouse mutation downregulates apoptosis-inducing factor. Nature 419:367–374

Landshamer S, Hoehn M, Barth N, Duvezin-Caubet S, Schwake G, Tobaben S, Kazhdan I, Becattini B, Zahler S, Vollmar A, Pellecchia M, Plesnila N, Wagner E, Culmsee C (2008) Bid-induced release of AIF from mitochondria causes immediate neuronal cell death. Cell Death Differ 15:1553–1563

Le DA, Wu Y, Huang Z, Matsushita K, Plesnila N, Augustinack JC, Hyman BT, Yuan J, Kuida K, Flavell RA, Moskowitz MA (2002) Caspase activation and neuroprotection in caspase-3- deficient mice after in vivo cerebral ischemia and in vitro oxygen glucose deprivation. Proc Natl Acad Sci U S A 99:15188–15193

Lee BI, Chan PH, Kim GW (2005) Metalloporphyrin-based superoxide dismutase mimic attenuates the nuclear translocation of apoptosis-inducing factor and the subsequent DNA fragmentation after permanent focal cerebral ischemia in mice. Stroke 36:2712–2717

Lee JH, Park SY, Shin HK, Kim CD, Lee WS, Hong KW (2007) Poly(ADP-ribose) polymerase inhibition by cilostazol is implicated in the neuroprotective effect against focal cerebral ischemic infarct in rat. Brain Res 1152:182–190. Epub 2007

Li X, Nemoto M, Xu Z, Yu SW, Shimoji M, Andrabi SA, Haince JF, Poirier GG, Dawson TM, Dawson VL, Koehler RC (2007) Influence of duration of focal cerebral ischemia and neuronal nitric oxide synthase on translocation of apoptosis-inducing factor to the nucleus. Neuroscience 144:56–65

Linnik MD, Miller JA, Sprinkle-Cavallo J, Mason PJ, Thompson FY, Montgomery LR, Schroeder KK (1995) Apoptotic DNA fragmentation in the rat cerebral cortex induced by permanent middle cerebral artery occlusion. Brain Res Mol Brain Res 32:116–124

Lo EH, Dalkara T, Moskowitz MA (2003) Mechanisms, challenges and opportunities in stroke. Nat Rev Neurosci 4:399–415

Martinou JC, Dubois-Dauphin M, Staple JK, Rodriguez I, Frankowski H, Missotten M, Albertini P, Talabot D, Catsicas S, Pietra C (1994) Overexpression of BCL-2 in transgenic mice protects neurons from naturally occurring cell death and experimental ischemia. Neuron 13:1017–1030

Martin-Villalba A, Herr I, Jeremias I, Hahne M, Brandt R, Vogel J, Schenkel J, Herdegen T, Debatin KM (1999) CD95 ligand (Fas-L/APO-1L) and tumor necrosis factor-related apoptosis-inducing ligand mediate ischemia-induced apoptosis in neurons. J Neurosci 19:3809–3817

Molina CA, Montaner J, Abilleira S, Ibarra B, Romero F, Arenillas JF, Alvarez-Sabin J (2001) Timing of spontaneous recanalization and risk of hemorrhagic transformation in acute cardio-embolic stroke. Stroke 32:1079–1084

Namura S, Zhu J, Fink K, Endres M, Srinivasan A, Tomaselli KJ, Yuan J, Moskowitz MA (1998) Activation and cleavage of caspase-3 in apoptosis induced by experimental cerebral ischemia. J Neurosci 18:3659–3668

Oexler EM, Dolga A, Culmsee C (2012) AIF depletion provides neuroprotection through a preconditioning effect. Apoptosis 17:1027–1038

Penninger JM, Kroemer G (2003) Mitochondria, AIF and caspases-rivaling for cell death execution. Nat Cell Biol 5:97–99

Plesnila N (2004) Role of mitochondrial proteins for neuronal cell death after focal cerebral ischemia. Acta Neurochir Suppl 89:15–19

Plesnila N, Zinkel S, Le DA, Amin-Hanjani S, Wu Y, Qiu J, Chiarugi A, Thomas SS, Kohane DS, Korsmeyer SJ, Moskowitz MA (2001) BID mediates neuronal cell death after oxygen/ glucose deprivation and focal cerebral ischemia. Proc Natl Acad Sci U S A 98:15318–15323

Plesnila N, Zhu C, Culmsee C, Groger M, Moskowitz MA, Blomgren K (2004) Nuclear translocation of apoptosis-inducing factor after focal cerebral ischemia. J Cereb Blood Flow Metab 24:458–466

Rideout HJ, Stefanis L (2001) Caspase inhibition: a potential therapeutic strategy in neurological diseases. Histol Histopathol 16:895–908

Susin SA, Lorenzo HK, Zamzami N, Marzo I, Snow BE, Brothers GM, Mangion J, Jacotot E, Costantini P, Loeffler M, Larochette N, Goodlett DR, Aebersold R, Siderovski DP, Penninger JM, Kroemer G (1999) Molecular characterization of mitochondrial apoptosis-inducing factor. Nature 397:441–446

Tobaben S, Grohm J, Seiler A, Conrad M, Plesnila N, Culmsee C (2011) Bid-mediated mitochondrial damage is a key mechanism in glutamate-induced oxidative stress and AIF-dependent cell death in immortalized HT-22 hippocampal neurons. Cell Death Differ 18:282–292

Vanden Berghe T, Linkermann A, Jouan-Lanhouet S, Walczak H, Vandenabeele P (2014) Regulated necrosis: the expanding network of non-apoptotic cell death pathways. Nat Rev Mol Cell Biol 15:135–147

Velier JJ, Ellison JA, Kikly KK, Spera PA, Barone FC, Feuerstein GZ (1999) Caspase-8 and caspase-3 are expressed by different populations of cortical neurons undergoing delayed cell death after focal stroke in the rat. J Neurosci 19:5932–5941

Wang Y, An R, Umanah GK, Park H, Nambiar K, Eacker SM, Kim B, Bao L, Harraz MM, Chang C, Chen R, Wang JE, Kam TI, Jeong JS, Xie Z, Neifert S, Qian J, Andrabi SA, Blackshaw S, Zhu H, Song H, Ming GL, Dawson VL, Dawson TM (2016) A nuclease that mediates cell death induced by DNA damage and poly(ADP-ribose) polymerase-1. Science 354(6308)

Wei MC, Lindsten T, Mootha VK, Weiler S, Gross A, Ashiya M, Thompson CB, Korsmeyer SJ (2000) tBID, a membrane-targeted death ligand, oligomerizes BAK to release cytochrome c. Genes Dev 14:2060–2071

Wei G, Wang D, Lu H, Parmentier S, Wang Q, Panter SS, Frey WH, Ying W (2007) Intranasal administration of a PARG inhibitor profoundly decreases ischemic brain injury. Front Biosci 12:4986–4996

Wiessner C, Allegrini PR, Rupalla K, Sauer D, Oltersdorf T, McGregor AL, Bischoff S, Bottiger BW, van der PH (1999) Neuron-specific transgene expression of Bcl-XL but not Bcl-2 genes reduced lesion size after permanent middle cerebral artery occlusion in mice. Neurosci Lett 268:119–122

Yigitkanli K, Zheng Y, Pekcec A, Lo EH, van Leyen K (2017) Increased 12/15-Lipoxygenase leads to widespread brain injury following global cerebral ischemia. Transl Stroke Res 8:194–202

Yin XM, Luo Y, Cao G, Bai L, Pei W, Kuharsky DK, Chen J (2002) Bid-mediated mitochondrial pathway is critical to ischemic neuronal apoptosis and focal cerebral ischemia. J Biol Chem 277:42074–42081

Zha J, Weiler S, Oh KJ, Wei MC, Korsmeyer SJ (2000) Posttranslational N-myristoylation of BID as a molecular switch for targeting mitochondria and apoptosis. Science 290:1761–1765

Zhu C, Qiu L, Wang X, Hallin U, Cande C, Kroemer G, Hagberg H, Blomgren K (2003) Involvement of apoptosis-inducing factor in neuronal death after hypoxia-ischemia in the neonatal rat brain. J Neurochem 86:306–317

Zhu C, Wang X, Huang Z, Qiu L, Xu F, Vahsen N, Nilsson M, Eriksson PS, Hagberg H, Culmsee C, Plesnila N, Kroemer G, Blomgren K (2007) Apoptosis-inducing factor is a major contributor to neuronal loss induced by neonatal cerebral hypoxia-ischemia. Cell Death Differ 14:775–784

Part IV
Transient Global Cerebral Ischemia

Chapter 7
Apoptosis-Inducing Factor Translocation to Nuclei After Transient Global Ischemia

Yang Sun, Tuo Yang, Jessica Zhang, Armando P. Signore, Guodong Cao, Jun Chen, and Feng Zhang

Abstract As a common human disorder, global ischemia causes long-term cognitive dysfunction. Selective death of hippocampal CA1 neurons underlies the cognitive impairment. After global ischemia, CA1 neuronal death occurs in a delayed manner, suggesting a type of programmed cell death. Apoptosis-inducing factor (AIF) is a mitochondrial protein with an important role in energy metabolism under physiological conditions. Following ischemia, AIF leaves mitochondria, translocates into nuclei, and induces DNA cleavage and chromatin condensation, therefore playing critical roles in inducing caspase-independent programmed cell death. In this chapter, we summarize the roles of AIF in CA1 nearonal death following global ischemia, highlighting recent progress.

Keywords AIF · Apoptosis · Stroke · Neuropretection

7.1 Introduction

In humans, global cerebral ischemia occurs following cardiac arrest and resuscitation, shock, or hypoxia, which produces neuronal cell death in the brain and reduced cognitive function if the patients recover. The loss of mitochondrial membrane integrity and the subsequent release of apoptogenic factors are critical in mediating the intrinsic, or mitochondrial, apoptotic pathway (Goto et al. 2002; Cao et al. 2003; Fujimura et al. 1999; Sugawara et al. 1999). Both caspase-dependent and -independent pro-death pathways can be initiated by the intrinsic pathway (Graham and Chen 2001; Zhang et al. 2004). The key signaling molecule to initiate the caspase-independent route is apoptosis-inducing factor (AIF), which is released by the mitochondria. AIF is a mitochondrial-specific flavoprotein that normally resides in the intermembrane space (Krantic et al. 2007). Following global ischemia, AIF is

Y. Sun · T. Yang · J. Zhang · A. P. Signore · G. Cao · J. Chen (✉) · F. Zhang (✉)
Department of Neurology, Pittsburgh Institute of Brain Disorders and Recovery, University of Pittsburgh, Pittsburgh, PA, USA
e-mail: zhanfx2@upmc.edu; chenj2@upmc.edu

D. G. Fujikawa (ed.), *Acute Neuronal Injury*,
https://doi.org/10.1007/978-3-319-77495-4_7

truncated by calpain, allowing it to translocate from compromised mitochondria to the nucleus, where it degrades the nuclear genome (Cao et al. 2003, 2007).

Global ischemia can be induced in rodents using several models. The essential feature of these models is a delayed (48–72 h) but selective loss of neurons in the CA1 region of the hippocampus (Zhang and Chen 2008; Zhang et al. 2006, 2004). Neuronal death in these models occurs at least in part via an apoptotic mechanism (Jin et al. 1999; Endo et al. 2006; Xu et al. 2016). The traditional view of cell death in mammalian cells largely envisioned two distinct processes with mechanisms that shared little in common: programmed cell death (PCD) and unprogrammed cell death. PCD includes apoptosis, autophagy and paraptosis, and unprogrammed cell death mainly denotes necrosis (Krantic et al. 2007). In order to identify potential therapeutic targets and to develop successful treatments, it is crucial to understand the specific contribution of the signaling pathways activated by each of the cell death mechanisms in ischemic neuronal death.

All of the death processes mentioned above are involved in ischemic neuronal death (Xue et al. 2016; Zhao et al. 2016; Wei et al. 2015), whereas only apoptosis and necrosis are mediated by AIF (Cao et al. 2003, 2007; Xu et al. 2016). Apoptosis is a controlled, energy-dependent, and well-orchestrated degradation of cellular structures. It was first described in development, where apoptosis removes extraneous tissue to mold body and organ structures. Apoptosis is exemplified by the condensation of the nucleus and the active dismantling of cellular components. The second broadly defined process is necrosis, usually referred to as an uncontrolled or nonregulated death of cells because of sudden and accidental irreversible damage. The distinction between these modes of death has recently been blurred, and cell death is now described as a continuum of programmed cell death pathways that show characteristics from each type of cell death (Boujrad et al. 2007; Bredesen et al. 2006; Golstein and Kroemer 2007; Xu et al. 2016).

Following cerebral ischemia and reperfusion, features of both apoptosis and necrosis appear (Muller et al. 2007; Pagnussat et al. 2007; Zhang et al. 2004). The pathological process of an individual neuron depends on the type, intensity, and duration of the cell death stimuli (Pagnussat et al. 2007). The harmful release of AIF depends in large part upon the health status of the mitochondria. Therefore, the greater the mitochondrial injury, the more likely AIF will be released and actively involved in inducing cell death. This chapter now focuses on AIF activation and the mechanisms in the model of global cerebral ischemia.

7.1.1 *Structures and Functions of AIF*

The *AIF* gene is located at X chromosome, and AIF protein is synthesized as a ~67 kDa precursor, with N-terminal prodomain containing two mitochondrial localization sequences. After AIF precursor is imported into mitochondria, it is processed to a ~62 kDa mature protein (Susin et al. 1999). The mature form of AIF has three structural domains: a FAD-binding domain, a NADH-binding domain, and a C-terminal domain (Mate et al. 2002). Under normal conditions, AIF is confined to the internal mitochondrial membrane, with its N-terminal region exposed to the matrix and the C-terminal to the inter-membrane space (Otera et al. 2005). Using multiple biochemical and immunogold electron microscopic analyses of mouse brain mitochondria, a recent study showed that about 30% of AIF loosely associates with the outer mitochondrial membrane (Yu et al. 2009). It is worth noting that the primary structure of AIF's C-terminal region shares about 30% identity with several bacterial NADH-oxidoreductases, suggesting its function (Miramar et al. 2001).

The discovery of AIF's normal functions was revealed by a mutation in the mouse. The AIF-deficient mouse, known as Harlequin, harbors a viral insertion in AIF, which diminished AIF expression to 20% or less in mutant Hq/Hq mice compared to wild type mice (Klein et al. 2002). Neuronal degeneration is a hallmark of the Hq/Hq mouse. Study of Hq/Hq mice determined several physiological functions of AIF, including an important NAD oxidase activity that uses nicotinamide adenine dinucleotide (NAD+) as a cofactor (Susin et al. 1999). A deficit in AIF expression causes mitochondrial complex I dysfunction and impaired oxidative phosphorylation, evidenced by increased dependence on glycolytic glucose metabolism and progressive multifocal neuropathology (El Ghouzzi et al. 2007; Vahsen et al. 2004). Loss of central neurons was due to reduced neuronal survival during brain development and increased oxidative radical activity (Cheung et al. 2006). Increased sensitivity to peroxides occurs in Hq/Hq neurons, and neurons can aberrantly re-enter the cell cycle (Klein et al. 2002). The loss of cells in Harlequin mice is specific to the brain and retina and does not appear to occur in the heart or liver, despite little AIF expression in these tissues as well (Vahsen et al. 2004). In addition, AIF plays an important role in maintaining the integrity of mitochondrial structure via preserving the complex I and III subunits, most likely duo to post-translational mechanisms (Vahsen et al. 2004).

Given that caspase-independent cell death requires AIF activation, several studies have reported neuroprotective effects of AIF inhibition by either neutralizing intracellular AIF or genetically reducing the expression of AIF (Cao et al. 2007; Culmsee et al. 2005; Yu et al. 2002; Xu et al. 2016). In transient global ischemia, total AIF expression levels per se are not significantly altered (Cao et al. 2003; Xu et al. 2016). Instead, AIF is cleaved at position G102/L103 (in mouse) by activated calpains and/or cathepsins, resulting in a mature isoform via its N-terminal truncation (Otera et al. 2005; Cao et al. 2007). Due to a nuclear localization sequence at AIF's C-terminal domain, this isoform is then translocated into the nucleus, resulting in large-scale DNA fragmentation and chromatin condensation (Dalla Via et al. 2014). The discharge of AIF from mitochondria is also dependent on the pro-apoptotic Bcl-2 family members, Bax and Bid (Cregan et al. 2002; Culmsee et al. 2005; Van Loo et al. 2002). Through direct interaction with genomic DNA along with the activity of endonuclease G (EndoG), AIF causes chromatin condensation (Cande et al. 2002).

7.2 AIF Translocation Mechanism and Therapeutic Targets

7.2.1 The Time Course of AIF Translocation

AIF-mediated cell death is an energy-dependent process. After the period of greatest energy depletion during ischemia, neuronal death occurs during the recovery of energy after reperfusion (Pagnussat et al. 2007). The time course for AIF nuclear translocation after experimental stroke varies with the severity of injury. AIF translocation into the nucleus does not occur until after 6 h of reperfusion following short (30 min) MCAO in mice, but was seen following as little as 20 min of reperfusion after longer (1 h) MCAO. When 2 h of MCAO was used instead, AIF translocation was again delayed until 6 h reperfusion (Li et al. 2007; Plesnila et al. 2004). Interestingly, AIF is a larger protein than cytochrome c, but its translocation precedes cytochrome *c* release. The small pool (30%) of AIF on the outer mitochondrial membrane may play an important part in this phenomenon. It was discovered that the outer mitochondrial membrane accounts for the rapid release of a small pool of AIF, as 20% of the uncleaved AIF rapidly translocated to the nucleus and caused death following NMDA treatment (Yu et al. 2009). However, more studies are needed to determine the different mechanisms involved in the release of the two AIF pools.

7.2.2 Mechanism of AIF Release

7.2.2.1 Poly (ADP-ribose) Polymerase-1 (PARP-1) and AIF

The release of AIF after global ischemia occurs upon a variety of stimuli, and several cascades contribute to AIF neurotoxicity, which include the inappropriate activation of DNA reparative enzymes. Our study showed that DNA single-strand breaks is a form of DNA damage induced early in neurons following cerebral ischemia (Chen et al. 1997; Stetler et al. 2010). PARP-1 is an abundant and very active chromatin-associated enzyme involved in DNA repair, as well as histone and other nuclear protein modifications. The enzyme relies on consumption of NAD+ to form ADP-ribose polymers (for reviews see Ame et al. 2004; Ha 2004; Ame et al. 2004; Ha 2004). Enzyme activity of PARP-1 is in fact activated by DNA strand breaks, and thus PARP-1 functions as a sensor of DNA damage (Demurcia and Demurcia 1994). Reperfusion following cerebral ischemia induces the generation of oxidative stimuli such as reactive oxygen species, which can lead to DNA damage, with activation of PARP-1 (Demurcia and Demurcia 1994; Eliasson et al. 1997). PARP-1 is a crucial part of both apoptosis and necrosis because its pathological over-activation can lead to cell death, generally by depleting cellular NAD+ and ATP (Szabo and Dawson 1998; Ha and Snyder 1999; Herceg and Wang 1999; Shall and de Murcia 2000; Yu et al. 2002).

Therefore, inhibition of PARP-1 over-activation should ameliorate cell death under these toxic conditions. Indeed, PARP-1 knockout animals demonstrate resistance to stroke (Eliasson et al. 1997; Endres et al. 1997; Goto et al. 2002). It is possible to prevent the death of even highly sensitive hippocampal CA1 neurons after transient ischemia by administering the PARP inhibitor PJ34 as late as 8 h after ischemia (Hamby et al. 2007). 3-AB, a PARP-1 inhibitor, protects neurons against necrosis, which is dependent on the duration of the ischemic-reperfusion episode (Strosznajder and Walski 2004). Our previous work showed that when adequate cellular NAD+ levels were maintained (Nagayama et al. 2000) or supplied (Wang et al. 2008), inhibition of PARP-1 diminished neuronal survival in the transient global ischemia model or neuronal cultures (Nagayama et al. 2000; Wang et al. 2008). Since caspase-3-mediated cleavage of PARP-1 blocks DNA repair and concomitantly prevents a depletion of cellular NAD+ pool by PARP activity (Boulares et al. 1999; Herceg and Wang 1999), the exact role that PARP-1 plays in cell death remains contentious, and may depend on which pathway is preferentially activated in individual neurons.

Evidence indicates that the release of AIF from mitochondria is also dependent on PARP-1-initiated nuclear signals (Cipriani et al. 2005; Yu et al. 2002). Reperfusion accelerates the appearance of nuclear AIF after 1 h of transient focal brain ischemia compared with permanent ischemia, consistent with the possibility that early oxidant stress triggers the signaling pathways that stimulate AIF translocation (Li et al. 2007). In concert with it, inhibition of PARP-1 reduces nuclear AIF translocation (Culmsee et al. 2005). The PAR polymer, generated when PARP-1 is over-activated, is now known as a key signaling molecule in the PARP1 mediated cell death (Andrabi et al. 2006; Komjati et al. 2005). They reach a toxic level when PARP-1 becomes over-activated and translocates to the cytosol, inducing AIF nuclear translocation (Yu et al. 2006). AIF is a PAR polymer-binding protein and a physical interaction between PAR and AIF is required in inducing the release of AIF from the mitochondria (Gagne et al. 2008; Wang et al. 2011).

PARP-1 might also contribute to the gender differences found in cerebral ischemia that are not directly attributable to the neuroprotective effects of the female hormones. In the immature male brain, neurons display greater caspase-independent translocation of AIF after hypoxic ischemia, whereas female-derived neurons exhibit stronger activation of caspase-3 (Zhu et al. 2006). In addition, while male PARP-1 knockout mice were protected from ischemia, female brain showed exacerbated histological injury after MCAO (McCullough et al. 2005).

The importance of PARP-1 is highlighted by recent reports, which showed that the activation of PARP-1 is necessary for calpain-mediated AIF cleavage (Cao et al. 2007; Vosler et al. 2009). In addition to PARP-1, a recent study showed that PARP-2 also contributed to nuclear translocation of AIF after transient focal cerebral ischemia in male mice (Li et al. 2010). The activation of AIF by PARP signaling may occur with a significant number of variables including length of insult, moment-to-moment levels of cell energy levels, cell-specific expression, and gender. The final activation due to cellular damage can thus be highly variable and may spare some neurons for reasons not readily apparent.

7.2.2.2 Direct Activation of AIF: Truncation by Calpain

A number of cysteine proteases, including the caspases, cathepsins, and calpains, are activated in neurons after ischemic injury (Graham and Chen 2001; Windelborn and Lipton 2008; Vosler et al. 2009). In AIF-mediated neuronal death, two families of cysteine proteases play a key role: calpains and cathepsins. Calpain I, also known as u-calpain, requires micromolar order of calcium to be activated, while calpain II or m-calpain requires millimolar calcium, as measured *in vitro*. Nevertheless, calcium is not involved in the cathepsin-mediated control of AIF activation. The cellular distribution of mRNA for calpain I and calpastatin, the endogenous calpain inhibitor, are relatively uniform throughout the mouse brain. In contrast, calpain II gene expression is selectively higher in specific neuronal populations, including pyramidal neurons of the hippocampus (Li et al. 1996), which are the most sensitive population of neurons to global ischemia.

While calpains are normal complements of cellular enzymatic activities, their inappropriate-activation or over-activation will lead to pathophysiological activity and contribute to cell death following cerebral ischemia. A major target of calpains is AIF. Calpain I cleaves AIF in a caspase-independent cell death manner in liver mitochondria (Polster et al. 2005; Cao et al. 2007) and the PC12 neuronal cell line (Liou et al. 2005). Our previous work shows that calpain I induces AIF release in the nervous system after both oxygen-glucose deprivation (OGD) *in vitro* and transient global ischemia *in vivo* (Fig. 7.1) (Cao et al. 2007). In fact, N-terminal truncation by calpain I was found to be required for AIF activation. When neurons express a mutant form of AIF that could not be cleaved by calpain, AIF was not released from mitochondria and was not found in the nucleus after OGD or global ischemia (Cao et al. 2007). This important finding shows that calpain I is a direct activator of AIF release, and is specifically involved in caspase-independent cell death.

Calpain activity was previously thought to be limited to the cytosol; however, there is ample evidence showing that a mitochondrial calpain exists and that it cleaves AIF, thus allowing AIF to leave the mitochondria (Garcia et al. 2005). A new form of calpains has been discovered and named calpain-10 (Arrington et al. 2006). Calpain-10 is located at the mitochondrial outer membrane, intermembrane space, inner membrane, and matrix region, and is an important mediator of mitochondrial dysfunction via the cleavage of Complex I subunits and activation of the mitochondrial-permeable transition pore (Arrington et al. 2006). The mitochondrial calpain most closely resembles m-calpain, as anti-m-calpain antibodies can also stain the mitochondrial calpain, and the calcium dependency of m-calpain and mitochondria calpain are similar (Ozaki et al. 2007). Thus, the discovery of calpain activity in the mitochondria itself shows that there is a direct link between mitochondrial dysfunction, elevated calcium levels, and protease activity.

Calpains also cleave a number of other substrates crucial in the cell death process, illustrating how intertwined the different cell death pathways are in neurons. For example, calpain I can cleave Bid (Chae et al. 2007); and calpain II can trigger the ischemia-induced lysosomal release of cathepsins in brain (Windelborn and

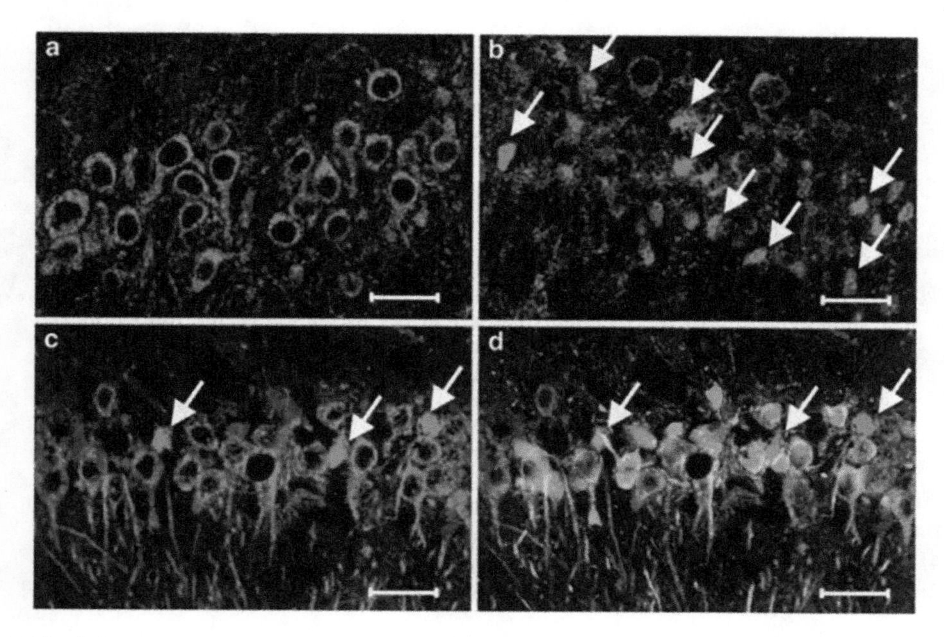

Fig. 7.1 AIF translocation in vivo following global ischemia is prevented by overexpression of calpastatin. Representative immunofluorescence of AIF (red) from non-ischemic CA1 (**a**) or 72 h after global ischemia (**b–d**). AAV–Cps (**c**, **d**) or the empty vector (**b**) was infused 14 days before ischemia, and brain sections were double-label immunostained for AIF (red) and hemagglutinin (HA) (green, **d**). Note that majority of CA1 neurons lost normal localization of AIF after ischemia (**b**, arrows), but AIF translocation was rare in Cps-overexpressed CA1 (**c**, **d**, arrows). Scale bars, 50 μm. From Cao et al. (2007)

Lipton 2008), which brings about the truncation of procaspase-3 into its active form (Blomgren et al. 2001; Mcginnis et al. 1999). This latter pathway is particularly interesting, as caspase-3 in turn reduces the activity of calpastatin, the endogenous calpain inhibitor, to form a positive feedback loop, which can result in the further activation of calpains and subsequent release of AIF (Kato et al. 2000; Porn-Ares et al. 1998; Wang et al. 1998).

7.2.2.3 Release of AIF from Mitochondria: Formation of the Mitochondrial Outer Membrane Pore

In addition to AIF and EndoG, mitochondria can release a variety of death-promoting molecules, including cytochrome c, Smac/Diablo, and Omi/HtrA2. The ability of mitochondria to release these molecules depends largely on the formation of large and nonselective pores or channels through the two layers of membrane systems in mitochondria. The inner membrane forms the mitochondrial-permeable transition pore, which is a calcium-dependent process and uses the proteins cyclophilin (Cyp) D, voltage-dependent anion channel and adenylate nucleotide translocase. The Bcl-2 family proteins, including Bid, Bax, and Bak and very likely others as yet unidentified proteins, form the outer membrane permeabilization pore, termed the mitochondrial apoptosis channel (for complete review, see Belizario et al. 2007). Many of these proteins need to undergo proteolytic cleavage before they can form any type of channel or pore, and the cleavage is usually induced by many stressors, including ischemia. For example, Bid can be truncated by caspase-8 and translocates to the outer mitochondrial membrane in ischemic brain (Gross et al. 1999). The understanding of the mechanisms behind the formation of the mitochondrial death channels is still incomplete and under debate. In general, Bid is activated and leads to BAK and BAX oligomerization, mitochondrial outer membrane permeabilization, and AIF release following stressors (for complete review, see Chipuk et al. 2010). The prevention of unwanted mitochondrial channels during normal physiological conditions is critical for maintaining cellular function and health. Regulation occurs via several anti-apoptogenic members of the Bcl-2 protein family, including Bcl-2 and Bcl-xL (Breckenridge and Xue 2004). In fact, AIF translocation and cleavage is inhibited by Bcl-2 and Bcl-xL overexpression (Cao et al. 2003; Otera et al. 2005).

7.2.3 Regulation of AIF Activity in the Cytoplasm

Once released from mitochondria to the cytosol, AIF then translocates into the nucleus. This process is regulated positively by Cyp A and negatively by heat shop protein 70 (Hsp70).

7.2.3.1 Cyp A Induces AIF Nucleus Translocation

Cyclophilins were first identified as the intracellular receptors for the immunosuppressant drug cyclosporin A (Handschumacher et al. 1984). Previous studies have shown that cyclophilins are involved in degradation of the genome during apoptosis (Montague et al. 1997; Cande et al. 2004; Zhu et al. 2007). For instance, CypA can form a complex with AIF to act as a co-factor for AIF nuclear translocation and AIF-dependent chromatinolysis following cerebral ischemia (Cande et al. 2004; Zhu et al. 2007). Elimination of Cyp A confers neuroprotection in vivo, suggesting that the lethal translocation of AIF to the nucleus requires interaction with Cyp A (Zhu et al. 2007).

7.2.3.2 Hsp70 Inhibits AIF Nucleus Translocation

Previous studies have shown that Hsp70 over-expression protects cells from death induced by various insults that cause either necrosis or apoptosis, including hypoxia and ischemia/reperfusion, by inhibiting multiple cell death pathways (Giffard and Yenari 2004). One of the mechanisms by which Hsp70 may be neuroprotective is the sequestration or neutralization of AIF by Hsp70, as evidenced by Hsp70 overexpression (Gurbuxani et al. 2003; Ravagnan et al. 2001; Wang et al. 2015). This process is associated with increased cytosolic retention of AIF when bound to Hsp70, limiting the entry of activated or cytotoxic AIF into the nucleus (Gurbuxani et al. 2003; Kroemer 2001).

7.2.3.3 Ubiquitination of AIF Via XIAP

X-linked inhibitor of apoptosis (XIAP) is an inhibitor of caspases and apoptosis (Suzuki et al. 2001). XIAP is also involved in the signal transduction and regulation of ubiquitin-ligase activity in the cellular system (Reffey et al. 2001; Yamaguchi et al. 1999; Yang et al. 2000). Previous work shows that XIAP may participate in the ubiquitinization of AIF (Wilkinson et al. 2008), which leads to AIF proteosomal degradation (Wilkinson et al. 2008). Further study shows that lysine 255 of AIF is critical to bind DNA and degrade chromatin, and the target of XIAP is this lysine residue (Lewis et al. 2011).

7.2.4 AIF-Induced DNA Fragmentation

Although AIF is involved in the breakdown of neuronal DNA, AIF itself is devoid of any nuclease activity (Susin et al. 2000, 1999). AIF translocates to the nucleus where it directly interacts with DNA by virtue of positive charges, which are clustered on the surface of AIF. DNA binding is therefore required for the death-promoting action of AIF (Ye et al. 2002). The binding of AIF to DNA induces chromatin condensation by interacting directly with DNA and possibly displacing chromatin-associated proteins. AIF could then disrupt normal chromatin structure, leading to the appearance of nuclear condensation. The remodeling of chromatin upon AIF modulation may increase the susceptibility of DNA to nucleases (Ye et al. 2002). The binding site within AIF is the same for distinct nucleic acid species, with no clear sequence specificity (Vahsen et al. 2006).

EndoG is another death-promoting factor released from mitochondria along with AIF (Susin et al. 1999). In the cerebral cortex, 4 h after ischemia, endoG level is significantly increased in the nucleus, correspondingly with decreased mitochondrial endoG content. EndoG may also interact with AIF in the nucleus after cerebral ischemia (Lee et al. 2005). In these ways, AIF truncation and release lead to the disruption of neuronal DNA via both direct (DNA binding) and indirect (EndoG) mechanisms.

7.2.4.1 Cyclophilins

Cyp A has been stated as a co-factor for AIF nuclear translocation and AIF-dependent chromatinolysis. Cyp D, on the other hand, is thought to be one of the components that forms the inner permeability transition pore and is thus involved in the release of death-promoting factors from the mitochondria (see above). A clearer understanding of how Cyp D participates in the formation of the mitochondrial permeability transition pore will provide important answers to the role these proteins play in caspase independent cell death.

7.2.4.2 Histone H2AX

Histone H2AX is another key factor in AIF-mediated apoptosis. As a member of the histone H2A family, H2AX participates in forming the histone nucleosome core. Previous studies have shown that the function of H2AX is primarily associated with DNA damage repair. On exposure of cells to inducers of double-strand breaks DNA damage, H2AX is phosphorylated at Ser139 in the nucleosomes surrounding the break point (Thiriet and Hayes 2005). Phosphorylated H2AX (γH2AX) renders damaged DNA sites accessible to repairing factors (Pilch et al. 2003). On the other hand, H2AX is crucial for AIF- mediated neuronal death. After DNA alkylating agent treatment, H2AX Ser139 phosphorylation is required for AIF mediated cell death (Artus et al. 2010; Baritaud et al. 2012).

7.3 Conclusion

Current knowledge about caspase-independent, AIF-induced neuronal death is incomplete. Though a large part of programmed cell death pathways involve the activation of caspases, inhibition of caspases alone as a therapeutic strategy is not sufficient to rescue damaged neurons. The caspase-independent pathway mediated by AIF provokes a compound network of signaling cascades that in and of themselves, can account for some of the specific cell death seen in the hippocampus following global ischemia. In vivo, the caspase-dependent and caspase-independent death pathways are, however, highly interconnected and often not easily distinguished from each other. The over-activation of calpains, the specific activator of AIF, maybe a promising target for drug intervention using calpain-specific inhibitors.

References

Ame JC, Spenlehauer C, de Murcia G (2004) The PARP superfamily. BioEssays 26:882–893

Andrabi SA, Kim NS, Yu SW, Wang H, Koh DW, Sasaki M, Klaus JA, Otsuka T, Zhang Z, Koehler RC, Hurn PD, Poirier GG, Dawson VL, Dawson TM (2006) Poly(ADP-ribose) (PAR) polymer is a death signal. Proc Natl Acad Sci U S A 103:18308–18313

Arrington DD, Van Vleet TR, Schnellmann RG (2006) Calpain 10: a mitochondrial calpain and its role in calcium-induced mitochondrial dysfunction. Am J Physiol Cell Physiol 291:C1159–C1171

Artus C, Boujrad H, Bouharrour A, Brunelle MN, Hoos S, Yuste VJ, Lenormand P, Rousselle JC, Namane A, England P, Lorenzo HK, Susin SA (2010) AIF promotes chromatinolysis and caspase-independent programmed necrosis by interacting with histone H2AX. EMBO J 29:1585–1599

Baritaud M, Cabon L, Delavallee L, Galan-Malo P, Gilles ME, Brunelle-Navas MN, Susin SA (2012) AIF-mediated caspase-independent necroptosis requires ATM and DNA-PK-induced histone H2AX Ser139 phosphorylation. Cell Death Dis 3:e390

Belizario JE, Alves J, Occhiucci JM, Garay-Malpartida M, Sesso A (2007) A mechanistic view of mitochondrial death decision pores. Braz J Med Biol Res 40:1011–1024

Blomgren K, Zhu C, Wang X, Karlsson JO, Leverin AL, Bahr BA, Mallard C, Hagberg H (2001) Synergistic activation of caspase-3 by m-calpain after neonatal hypoxia-ischemia: a mechanism of "pathological apoptosis"? J Biol Chem 276:10191–10198

Boujrad H, Gubkina O, Robert N, Krantic S, Susin SA (2007) AIF-mediated programmed necrosis: a highly regulated way to die. Cell Cycle 6:2612–2619

Boulares AH, Yakovlev AG, Ivanova V, Stoica BA, Wang G, Iyer S, Smulson M (1999) Role of poly(ADP-ribose) polymerase (PARP) cleavage in apoptosis. Caspase 3-resistant PARP mutant increases rates of apoptosis in transfected cells. J Biol Chem 274:22932–22940

Breckenridge DG and Xue D (2004) Regulation of mitochondrial membrane permeabilization by BCL-2 family proteins and caspases. Curr Opin Cell Biol 16:647–652

Bredesen DE, Rao RV, Mehlen P (2006) Cell death in the nervous system. Nature 443:796–802

Cande C, Cohen I, Daugas E, Ravagnan L, Larochette N, Zamzami N, Kroemer G (2002) Apoptosis-inducing factor (AIF): a novel caspase-independent death effector released from mitochondria. Biochimie 84:215–222

Cande C, Vahsen N, Kouranti I, Schmitt E, Daugas E, Spahr C, Luban J, Kroemer RT, Giordanetto F, Garrido C, Penninger JM, Kroemer G (2004) AIF and cyclophilin A cooperate in apoptosis-associated chromatinolysis. Oncogene 23:1514–1521

Cao G, Clark RS, Pei W, Yin W, Zhang F, Sun FY, Graham SH, Chen J (2003) Translocation of apoptosis-inducing factor in vulnerable neurons after transient cerebral ischemia and in neuronal cultures after oxygen-glucose deprivation. J Cereb Blood Flow Metab 23:1137–1150

Cao G, Xing J, Xiao X, Liou AK, Gao Y, Yin XM, Clark RS, Graham SH, Chen J (2007) Critical role of calpain I in mitochondrial release of apoptosis-inducing factor in ischemic neuronal injury. J Neurosci 27:9278–9293

Chae SU, Ha KC, Piao CS, Chae SW, Chae HJ (2007) Estrogen attenuates cardiac ischemia-reperfusion injury via inhibition of calpain-mediated bid cleavage. Arch Pharm Res 30:1225–1235

Chen J, Jin K, Chen M, Pei W, Kawaguchi K, Greenberg DA, Simon RP (1997) Early detection of DNA strand breaks in the brain after transient focal ischemia: implications for the role of DNA damage in apoptosis and neuronal cell death. J Neurochem 69:232–245

Cheung EC, Joza N, Steenaart NA, Mcclellan KA, Neuspiel M, Mcnamara S, Maclaurin JG, Rippstein P, Park DS, Shore GC, Mcbride HM, Penninger JM, Slack RS (2006) Dissociating the dual roles of apoptosis-inducing factor in maintaining mitochondrial structure and apoptosis. EMBO J 25:4061–4073

Chipuk JE, Moldoveanu T, Llambi F, Parsons MJ, Green DR (2010) The BCL-2 family reunion. Mol Cell 37:299–310

Cipriani G, Rapizzi E, Vannacci A, Rizzuto R, Moroni F, Chiarugi A (2005) Nuclear poly(ADP-ribose) polymerase-1 rapidly triggers mitochondrial dysfunction. J Biol Chem 280:17227–17234

Cregan SP, Fortin A, Maclaurin JG, Callaghan SM, Cecconi F, Yu SW, Dawson TM, Dawson VL, Park DS, Kroemer G, Slack RS (2002) Apoptosis-inducing factor is involved in the regulation of caspase-independent neuronal cell death. J Cell Biol 158:507–517

Culmsee C, Zhu CL, Landshamer S, Becattini B, Wagner E, Pellecchia M, Blomgren K, Plesnila N (2005) Apoptosis-inducing factor triggered by poly(ADP-ribose) polymerase and bid mediates neuronal cell death after oxygen-glucose deprivation and focal cerebral ischemia (vol 25, pg 10262, 2005). J Neurosci 25

Dalla Via L, Garcia-Argaez AN, Martinez-Vazquez M, Grancara S, Martinis P, Toninello A (2014) Mitochondrial permeability transition as target of anticancer drugs. Curr Pharm Des 20:223–244

David G Breckenridge, Ding Xue, (2004) Regulation of mitochondrial membrane permeabilization by BCL-2 family proteins and caspases. Current Opinion in Cell Biology 16 (6):647–652

Demurcia G, Demurcia JM (1994) Poly(ADP-Ribose) polymerase—a molecular nick-sensor. Trends Biochem Sci 19:172–176

El Ghouzzi V, Csaba Z, Olivier P, Lelouvier B, Schwendimann L, Dournaud P, Verney C, Rustin P, Gressens P (2007) Apoptosis-inducing factor deficiency induces early mitochondrial degeneration in brain followed by progressive multifocal neuropathology. J Neuropathol Exp Neurol 66:838–847

Eliasson MJL, Sampei K, Mandir AS, Hurn PD, Traystman RJ, Bao J, Pieper A, Wang ZQ, Dawson TM, Snyder SH, Dawson VL (1997) Poly(ADP-ribose) polymerase gene disruption renders mice resistant to cerebral ischemia. Nat Med 3:1089–1095

Endo H, Kamada H, Nito C, Nishi T, Chan PH (2006) Mitochondrial translocation of p53 mediates release of cytochrome c and hippocampal CA1 neuronal death after transient global cerebral ischemia in rats. J Neurosci 26:7974–7983

Endres M, Wang ZQ, Namura S, Waeber C, Moskowitz MA (1997) Ischemic brain injury is mediated by the activation of poly(ADP-ribose)polymerase. J Cereb Blood Flow Metab 17:1143–1151

Fujimura M, Morita-Fujimura Y, Kawase M, Copin JC, Calagui B, Epstein CJ, Chan PH (1999) Manganese superoxide dismutase mediates the early release of mitochondrial cytochrome C and subsequent DNA fragmentation after permanent focal cerebral ischemia in mice. J Neurosci 19:3414–3422

Gagne JP, Isabelle M, Lo KS, Bourassa S, Hendzel MJ, Dawson VL, Dawson TM, Poirier GG (2008) Proteome-wide identification of poly(ADP-ribose) binding proteins and poly(ADP-ribose)-associated protein complexes. Nucleic Acids Res 36:6959–6976

Garcia M, Bondada V, Geddes JW (2005) Mitochondrial localization of mu-calpain. Biochem Biophys Res Commun 338:1241–1247

Giffard RG, Yenari MA (2004) Many mechanisms for Hsp70 protection from cerebral ischemia. J Neurosurg Anesthesiol 16:53–61

Golstein P, Kroemer G (2007) Cell death by necrosis: towards a molecular definition. Trends Biochem Sci 32:37–43

Goto S, Xue R, Sugo N, Sawada M, Blizzard KK, Poitras MF, Johns DC, Dawson TM, Dawson VL, Crain BJ, Traystman RJ, Mori S, Hurn PD (2002) Poly(ADP-ribose) polymerase impairs early and long-term experimental stroke recovery. Stroke 33:1101–1106

Graham SH, Chen J (2001) Programmed cell death in cerebral ischemia. J Cereb Blood Flow Metab 21:99–109

Gross A, Yin XM, Wang K, Wei MC, Jockel J, Millman C, Erdjument-Bromage H, Tempst P, Korsmeyer SJ (1999) Caspase cleaved BID targets mitochondria and is required for cytochrome c release, while BCL-X-L prevents this release but not tumor necrosis factor-R1/Fas death. J Biol Chem 274:1156–1163

Gurbuxani S, Schmitt E, Cande C, Parcellier A, Hammann A, Daugas E, Kouranti I, Spahr C, Pance A, Kroemer G, Garrido C (2003) Heat shock protein 70 binding inhibits the nuclear import of apoptosis-inducing factor. Oncogene 22:6669–6678

Ha HC (2004) Defective transcription factor activation for proinflammatory gene expression in poly(ADP-ribose) polymerase 1-deficient glia. Proc Natl Acad Sci U S A 101:5087–5092

Ha HC, Snyder SH (1999) Poly(ADP-ribose) polymerase is a mediator of necrotic cell death by ATP depletion. Proc Natl Acad Sci U S A 96:13978–13982

Hamby AM, Suh SW, Kauppinen TM, Swanson RA (2007) Use of a poly(ADP-ribose) polymerase inhibitor to suppress inflammation and neuronal death after cerebral ischemia-reperfusion. Stroke 38:632–636

Handschumacher RE, Harding MW, Rice J, Drugge RJ (1984) Cyclophilin—a specific cytosolic binding-protein for cyclosporin-A. Science 226:544–547

Herceg Z, Wang ZQ (1999) Failure of poly(ADP-ribose) polymerase cleavage by caspases leads to induction of necrosis and enhanced apoptosis. Mol Cell Biol 19:5124–5133

Jin KL, Chen J, Nagayama T, Chen MZ, Sinclair J, Graham SH, Simon RP (1999) In situ detection of neuronal DNA strand breaks using the Klenow fragment of DNA polymerase I reveals different mechanisms of neuron death after global cerebral ischemia. J Neurochem 72:1204–1214

Kato M, Nonaka T, Maki M, Kikuchi H, Imajoh-Ohmi S (2000) Caspases cleave the amino-terminal calpain inhibitory unit of calpastatin during apoptosis in human Jurkat T cells. J Biochem 127:297–305

Klein JA, Longo-Guess CM, Rossmann MP, Seburn KL, Hurd RE, Frankel WN, Bronson RT, Ackerman SL (2002) The harlequin mouse mutation down-regulates apoptosis-inducing factor. Nature 419:367–374

Komjati K, Besson VC, Szabo C (2005) Poly (ADP-ribose) polymerase inhibitors as potential therapeutic agents in stroke and neurotrauma. Curr Drug Targets CNS Neurol Disord 4:179–194

Krantic S, Mechawar N, Reix S, Quirion R (2007) Apoptosis-inducing factor: a matter of neuron life and death. Prog Neurobiol 81:179–196

Kroemer G (2001) Heat shock protein 70 neutralizes apoptosis-inducing factor. ScientificWorld J 1:590–592

Lee BI, Lee DJ, Cho KJ, Kim GW (2005) Early nuclear translocation of endonuclease G and subsequent DNA fragmentation after transient focal cerebral ischemia in mice. Neurosci Lett 386:23–27

Lewis EM, Wilkinson AS, Davis NY, Horita DA, Wilkinson JC (2011) Nondegradative ubiquitination of apoptosis inducing factor (AIF) by X-linked inhibitor of apoptosis at a residue critical for AIF-mediated chromatin degradation. Biochemistry 50:11084–11096

Li JH, Grynspan F, Berman S, Nixon R, Bursztajn S (1996) Regional differences in gene expression for calcium activated neutral proteases (calpains) and their endogenous inhibitor calpastatin in mouse brain and spinal cord. J Neurobiol 30:177–191

Li X, Klaus JA, Zhang J, Xu Z, Kibler KK, Andrabi SA, Rao K, Yang Z-J, Dawson TM, Dawson VL, Koehler RC (2010) Contributions of poly(ADP-ribose) polymerase-1 and -2 to nuclear translocation of apoptosis-inducing factor and injury from focal cerebral ischemia. J Neurochem 113:1012–1022

Li X, Nemoto M, Xu Z, Yu SW, Shimoji M, Andrabi SA, Haince JF, Poirier GG, Dawson TM, Dawson VL, Koehler RC (2007) Influence of duration of focal cerebral ischemia and neuronal nitric oxide synthase on translocation of apoptosis-inducing factor to the nucleus. Neuroscience 144:56–65

Liou AKF, Zhou ZG, Pei W, Lim TM, Yin XM, Chen J (2005) BimEL up-regulation potentiates AIF translocation and cell death in response to MPTP. FASEB J 19:1350–1352

Mate MJ, Ortiz-Lombardia M, Boitel B, Haouz A, Tello D, Susin SA, Penninger J, Kroemer G, Alzari PM (2002) The crystal structure of the mouse apoptosis-inducing factor AIF. Nat Struct Biol 9(6):442

McCullough LD, Zeng Z, Blizzard KK, Debchoudhury I, Hurn PD (2005) Ischemic nitric oxide and poly (ADP-ribose) polymerase-1 in cerebral ischemia: male toxicity, female protection. J Cereb Blood Flow Metab 25:502–512

Mcginnis KM, Gnegy ME, Park YH, Mukerjee N, Wang KK (1999) Procaspase-3 and poly(ADP) ribose polymerase (PARP) are calpain substrates. Biochem Biophys Res Commun 263:94–99

Miramar MD, Costantini P, Ravagnan L, Saraiva LM, Haouzi D, Brothers G, Penninger JM, Peleato ML, Kroemer G, Susin SA (2001) NADH oxidase activity of mitochondrial apoptosis-inducing factor. J Biol Chem 276:16391–16398

Montague JW, Hughes FM Jr, Cidlowski JA (1997) Native recombinant cyclophilins A, B, and C degrade DNA independently of peptidylprolyl cis-trans-isomerase activity potential roles of cyclophilins in apoptosis. J Biol Chem 272:6677–6684

Muller GJ, Lassmann H, Johansen FF (2007) Anti-apoptotic signaling and failure of apoptosis in the ischemic rat hippocampus. Neurobiol Dis 25:582–593

Nagayama T, Simon RP, Chen DX, Henshall DC, Pei W, Stetler RA, Chen J (2000) Activation of poly(ADP-Ribose) polymerase in the rat hippocampus may contribute to cellular recovery following sublethal transient global ischemia. J Neurochem 74:1636–1645

Otera H, Ohsakaya S, Nagaura Z, Ishihara N, Mihara K (2005) Export of mitochondrial AIF in response to proapoptotic stimuli depends on processing at the intermembrane space. EMBO J 24:1375–1386

Ozaki T, Tomita H, Tamai M, Ishiguro SI (2007) Characteristics of mitochondrial calpains. J Biochem 142:365–376

Pagnussat AD, Faccioni-Heuser MC, Netto CA, Achaval M (2007) An ultrastructural study of cell death in the CA1 pyramidal field of the hippocapmus in rats submitted to transient global ischemia followed by reperfusion. J Anat 211:589–599

Pilch DR, Sedelnikova OA, Redon C, Celeste A, Nussenzweig A, Bonner WM (2003) Characteristics of gamma-H2AX foci at DNA double-strand breaks sites. Biochem Cell Biol 81:123–129

Plesnila N, Zhu CL, Culmsee C, Groger M, Moskowitz MA, Blomgren K (2004) Nuclear translocation of apoptosis-inducing factor after focal cerebral ischemia. J Cereb Blood Flow Metab 24:458–466

Polster BM, Basanez G, Etxebarria A, Hardwick JM, Nicholls DG (2005) Calpain I induces cleavage and release of apoptosis-inducing factor from isolated mitochondria. J Biol Chem 280:6447–6454

Porn-Ares MI, Samali A, Orrenius S (1998) Cleavage of the calpain inhibitor, calpastatin, during apoptosis. Cell Death Differ 5:1028–1033

Ravagnan L, Gurbuxani S, Susin SA, Maisse C, Daugas E, Zamzami N, Mak T, Jaattela M, Penninger JM, Garrido C, Kroemer G (2001) Heat-shock protein 70 antagonizes apoptosis-inducing factor. Nat Cell Biol 3:839–843

Reffey SB, Wurthner JU, Parks WT, Roberts AB, Duckett CS (2001) X-linked inhibitor of apoptosis protein functions as a cofactor in transforming growth factor-beta signaling. J Biol Chem 276:26542–26549

Shall S, de Murcia G (2000) Poly(ADP-ribose) polymerase-1: what have we learned from the deficient mouse model? Mutat Res 460:1–15

Stetler RA, Gao YQ, Zukin RS, Vosler PS, Zhang LL, Zhang F, Cao GD, Bennett MVL, Chen J (2010) Apurinic/apyrimidinic endonuclease APE1 is required for PACAP-induced neuroprotection against global cerebral ischemia. Proc Natl Acad Sci U S A 107:3204–3209

Strosznajder RP, Walski M (2004) Effects 3-aminobenzamide on ultrastructure of hippocampal CA1 layer after global ischemia in gerbils. J Physiol Pharmacol 55(Suppl 3):127–133

Sugawara T, Fujimura M, Morita-Fujimura Y, Kawase M, Chan PH (1999) Mitochondrial release of cytochrome c corresponds to the selective vulnerability of hippocampal CA1 neurons in rats after transient global cerebral ischemia. J Neurosci 19:Rc39

Susin SA, Daugas E, Ravagnan L, Samejima K, Zamzami N, Loeffler M, Costantini P, Ferri KF, Irinopoulou T, Prevost MC, Brothers G, Mak TW, Penninger J, Earnshaw WC, Kroemer G (2000) Two distinct pathways leading to nuclear apoptosis. J Exp Med 192:571–580

Susin SA, Lorenzo HK, Zamzami N, Marzo I, Snow BE, Brothers GM, Mangion J, Jacotot E, Costantini P, Loeffler M, Larochette N, Goodlett DR, Aebersold R, Siderovski DP, Penninger JM, Kroemer G (1999) Molecular characterization of mitochondrial apoptosis-inducing factor. Nature 397:441–446

Suzuki Y, Nakabayashi Y, Nakata K, Reed JC, Takahashi R (2001) X-linked inhibitor of apoptosis protein (XIAP) inhibits caspase-3 and -7 in distinct modes. J Biol Chem 276:27058–27063

Szabo C, Dawson VL (1998) Role of poly(ADP-ribose) synthetase in inflammation and ischaemia-reperfusion. Trends Pharmacol Sci 19:287–298

Thiriet C, Hayes JJ (2005) Chromatin in need of a fix: phosphorylation of H2AX connects chromatin to DNA repair. Mol Cell 18:617–622

Vahsen N, Cande C, Briere JJ, Benit P, Joza N, Larochette N, Mastroberardino PG, Pequignot MO, Casares N, Lazar V, Feraud O, Debili N, Wissing S, Engelhardt S, Madeo F, Piacentini M, Penninger JM, Schagger H, Rustin P, Kroemer G (2004) AIF deficiency compromises oxidative phosphorylation. EMBO J 23:4679–4689

Vahsen N, Cande C, Dupaigne P, Giordanetto F, Kroemer RT, Herker E, Scholz S, Modjtahedi N, Madeo F, Le Cam E, Kroemer G (2006) Physical interaction of apoptosis-inducing factor with DNA and RNA. Oncogene 25:1763–1774

Van Loo G, Saelens X, Van Gurp M, Macfarlane M, Martin SJ, Vandenabeele P (2002) The role of mitochondrial factors in apoptosis: a Russian roulette with more than one bullet. Cell Death Differ 9:1031–1042

Vosler PS, Sun D, Wang S, Gao Y, Kintner DB, Signore AP, Cao G, Chen J (2009) Calcium dysregulation induces apoptosis-inducing factor release: cross-talk between PARP-1- and calpain-signaling pathways. Exp Neurol 218:213–220

Wang F, Dai AY, Tao K, Xiao Q, Huang ZL, Gao M, Li H, Wang X, Cao WX, Feng WL (2015) Heat shock protein-70 neutralizes apoptosis inducing factor in Bcr/Abl expressing cells. Cell Signal 27:1949–1955

Wang KK, Posmantur R, Nadimpalli R, Nath R, Mohan P, Nixon RA, Talanian RV, Keegan M, Herzog L, Allen H (1998) Caspase-mediated fragmentation of calpain inhibitor protein calpastatin during apoptosis. Arch Biochem Biophys 356:187–196

Wang S, Xing Z, Vosler PS, Yin H, Li W, Zhang F, Signore AP, Stetler RA, Gao Y, Chen J (2008) Cellular NAD replenishment confers marked neuroprotection against ischemic cell death: role of enhanced DNA repair. Stroke 39:2587–2595

Wang Y, Kim NS, Haince JF, Kang HC, David KK, Andrabi SA, Poirier GG, Dawson VL, Dawson TM (2011) Poly(ADP-ribose) (PAR) binding to apoptosis-inducing factor is critical for PAR polymerase-1-dependent cell death (parthanatos). Sci Signal 4:ra20

Wei T, Kang Q, Ma B, Gao S, Li X, Liu Y (2015) Activation of autophagy and paraptosis in retinal ganglion cells after retinal ischemia and reperfusion injury in rats. Exp Ther Med 9:476–482

Wilkinson JC, Wilkinson AS, Galban S, Csomos RA, Duckett CS (2008) Apoptosis-inducing factor is a target for ubiquitination through interaction with XIAP. Mol Cell Biol 28:237–247

Windelborn JA, Lipton P (2008) Lysosomal release of cathepsins causes ischemic damage in the rat hippocampal slice and depends on NMDA-mediated calcium influx, arachidonic acid metabolism, and free radical production. J Neurochem 106:56–69

Xu Y, Wang J, Song X, Qu L, Wei R, He F, Wang K, Luo B (2016) RIP3 induces ischemic neuronal DNA degradation and programmed necrosis in rat via AIF. Sci Rep 6:29362

Xue LX, Xu ZH, Wang JQ, Cui Y, Liu HY, Liang WZ, Ji QY, He JT, Shao YK, Mang J, Xu ZX (2016) Activin A/Smads signaling pathway negatively regulates Oxygen Glucose Deprivation-induced autophagy via suppression of JNK and p38 MAPK pathways in neuronal PC12 cells. Biochem Biophys Res Commun 480(3):355–361

Yamaguchi K, Nagai S, Ninomiya-Tsuji J, Nishita M, Tamai K, Irie K, Ueno N, Nishida E, Shibuya H, Matsumoto K (1999) XIAP, a cellular member of the inhibitor of apoptosis protein family, links the receptors to TAB1-TAK1 in the BMP signaling pathway. EMBO J 18:179–187

Yang Y, Fang SY, Jensen JP, Weissman AM, Ashwell JD (2000) Ubiquitin protein ligase activity of IAPs and their degradation in proteasomes in response to apoptotic stimuli. Science 288:874–877

Ye H, Cande C, Stephanou NC, Jiang SL, Gurbuxani S, Larochette N, Daugas E, Garrido C, Kroemer G, Wu H (2002) DNA binding is required for the apoptogenic action of apoptosis inducing factor. Nat Struct Biol 9:680–684

Yu SW, Andrabi SA, Wang H, Kim NS, Poirier GG, Dawson TM, Dawson VL (2006) Apoptosis-inducing factor mediates poly(ADP-ribose) (PAR) polymer-induced cell death. Proc Natl Acad Sci U S A 103:18314–18319

Yu SW, Wang HM, Poitras MF, Coombs C, Bowers WJ, Federoff HJ, Poirier GG, Dawson TM, Dawson VL (2002) Mediation of poly(ADP-ribose) polymerase-1-dependent cell death by apoptosis-inducing factor. Science 297:259–263

Yu SW, Wang Y, Frydenlund DS, Ottersen OP, Dawson VL, Dawson TM (2009) Outer mitochondrial membrane localization of apoptosis-inducing factor: mechanistic implications for release. ASN Neuro 1:275–281

Zhang F, Chen J (2008) Leptin protects hippocampal CA1 neurons against ischemic injury. J Neurochem 107:578–587

Zhang F, Signore AP, Zhou Z, Wang S, Cao G, Chen J (2006) Erythropoietin protects CA1 neurons against global cerebral ischemia in rat: potential signaling mechanisms. J Neurosci Res 83:1241–1251

Zhang F, Yin W, Chen J (2004) Apoptosis in cerebral ischemia: executional and regulatory signaling mechanisms. Neurol Res 26:835–845

Zhao Y, Huang G, Chen S, Gou Y, Dong Z, Zhang X (2016) Homocysteine aggravates cortical neural cell injury through neuronal autophagy overactivation following rat cerebral ischemia-reperfusion. Int J Mol Sci 17(8):1196

Zhu C, Wang X, Deinum J, Huang Z, Gao J, Modjtahedi N, Neagu MR, Nilsson M, Eriksson PS, Hagberg H, Luban J, Kroemer G, Blomgren K (2007) Cyclophilin A participates in the nuclear translocation of apoptosis-inducing factor in neurons after cerebral hypoxia-ischemia. J Exp Med 204(8):1741

Zhu CL, Xu FL, Wang XY, Shibata M, Uchiyama Y, Blomgren K, Hagberg H (2006) Different apoptotic mechanisms are activated in male and female brains after neonatal hypoxia-ischaemia. J Neurochem 96:1016–1027

Chapter 8
Necroptosis in Cerebral Ischemia

Marta M. Vieira and Ana Luísa Carvalho

Abstract Necroptosis is a form of regulated necrotic cell death, which is mediated by receptor-interacting protein 1 kinase (RIPK1) and RIPK3, and the downstream effector mixed lineage kinase domain-like (MLKL). Necroptotic signals induced by death receptors, such as TNF receptor 1, Toll-like receptors or interferon receptors lead to the formation of the necrosome, and result in cell death with morphological features of necrosis. The execution of necroptosis involves oligomerization of MLKL upon phosphorylation by RIPK3, its translocation to the plasma membrane and consequent plasma membrane permeabilization. Necroptosis participates in physiological functions but is also involved in cell death associated with several pathophysiological conditions. Here, we discuss key features of necroptosis, and evidence implicating necroptotic cell death in brain ischemia.

Keywords Necroptosis · Brain ischemia · Receptor interacting protein kinase 1/3 · Mixed lineage kinase domain-like (MLKL) · Necrosome

8.1 General Considerations

Cell death has been classically divided into two major types, programmed (or regulated) cell death (PCD) and unregulated cell death. The archetypal form of PCD is apoptosis, whereas necrosis is generally considered a type of "accidental" cell death. The concept of PCD arises from the fact that upon specific stimuli (e.g. heat, radiation, hypoxia) cells activate a cascade of events that leads to a highly regulated cellular demise process. Apoptosis induces several cyto-architectural alterations,

---

M. M. Vieira
Receptor Biology Section, National Institute of Neurological Disorders and Stroke, National Institutes of Health, Bethesda, MD, USA

A. L. Carvalho (✉)
Department of Life Sciences, CNC—Center for Neuroscience and Cell Biology, University of Coimbra, Coimbra, Portugal
e-mail: alc@cnc.uc.pt

D. G. Fujikawa (ed.), *Acute Neuronal Injury*,
https://doi.org/10.1007/978-3-319-77495-4_8

including membrane blebbing, with formation of apoptotic bodies, chromatin condensation, DNA laddering and fragmentation of the nucleus, organelle fragmentation and release of protein content from the mitochondria to the cytoplasm. After cellular deconstruction, the debris are removed by phagocytes, avoiding activation of inflammatory responses and damage to the neighboring cells (Taylor et al. 2008). There are two major pathways of apoptosis, the extrinsic (or death receptor—DR—pathway) and the intrinsic (or mitochondrial pathway) one, which are thought to cross-talk and influence one another (Igney and Krammer 2002). In contrast, necrosis induces plasma membrane damage, swollen mitochondria and slow mitochondrial membrane depolarization, subsequent to nuclear and cytoplasmic morphological changes. Because the cellular debris are not so promptly removed as during the final stage of apoptosis, necrosis is thought to induce an inflammatory component and to have a propagating effect, causing damage to neighboring cells that may eventually die. In addition to these events, neuronal necrosis involves programmed, or regulated, biochemical cascades triggered by excessive calcium entry through NMDA-receptor-operated cation channels (Fujikawa 2015).

Interestingly, during the past two decades a novel concept of programmed necrosis has emerged. This notion arose from the observation that, in conditions of apoptosis inhibition, certain cells still initiate a process of regulated cell death, with necrotic phenotype. Indeed, this type of PCD, termed necroptosis, displays necrotic features, such as compromise of the cellular membrane, reactive oxygen species (ROS) production and mitochondrial damage, in the absence of typical apoptotic features, such as cytochrome c release or caspase-3-dependent cleavage of poly-ADP ribose polymerase (PARP), chromatin condensation or oligonucleosomal DNA degradation (He et al. 2009; Holler et al. 2000; Vandenabeele et al. 2010). Whether or not there are interactions between necroptosis and excitotoxic programmed necrosis has not been determined.

Mounting evidence has demonstrated that ischemic insults induce a combination of different processes of neuronal cell death, namely necroptosis and excitotoxic programmed necrosis (Meloni et al. 2011; Vieira et al. 2014; Xu et al. 2016; Vosler et al. 2009). Although we still lack a complete understanding of these mechanisms in neurons, the diversity of cell death processes activated following ischemia seems to imply a certain redundancy, possibly to ensure the demise of damaged neurons. Such hypothesis is further corroborated by the fact that necroptosis actually occurs in conditions of apoptosis inhibition (Degterev et al. 2005; Qu et al. 2016; Vieira et al. 2014), suggesting that this may be a backup pathway activated under unfavorable conditions.

In this chapter, we first describe necroptosis, with a focus on the mechanisms that trigger and mediate this form of regulated necrosis. We then explore the evidence supporting the implication of necroptosis in brain ischemia, and how this pathway could be therapeutically targeted.

8.2 Death Receptor Signaling

Death receptors (DRs) are ubiquitous cell membrane proteins whose signaling underlies complex cell fate events. The decision between life and death is initiated upon death ligand (DL) stimulation (classically, TNFα), with the recruitment of death domain (DD)-containing proteins to the vicinity of the TNFR1—complex I [for a review of the diverse DRs and their DLs see (Han et al. 2011)]. This complex comprises proteins such as RIPK1, TNFR-associated factor 2 (TRAF2) and TNFR-associated via death domain (TRADD) (Micheau and Tschopp 2003), among others. Cellular inhibitor of apoptosis protein 1 (cIAP1) may also be found in this complex, which leads to survival signaling mediated by nuclear factor kappa B (NF-κB) transcriptional activity, by acting as an inhibitor of DL-induced apoptosis (Gaither et al. 2007; Geserick et al. 2009; Wang et al. 1998; Yang and Du 2004). Increased NF-κB-mediated gene transcription contributes to cell survival and inflammation. However, depending on the molecular context of the cell, and upon impaired NF-κB signaling, DRs may also activate two distinct mechanisms of programmed cell death (Fig. 8.1). This outcome is achieved via the assembly of a

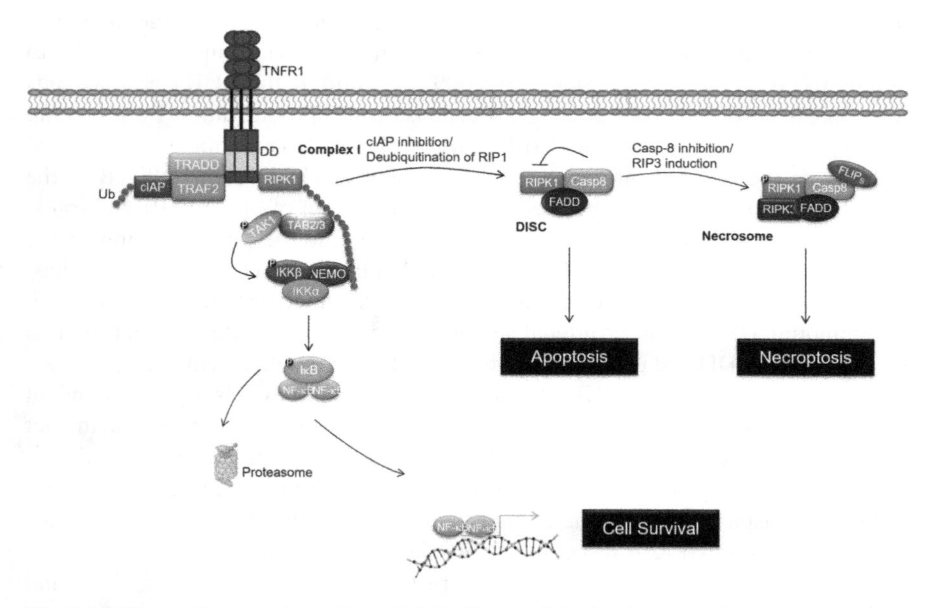

Fig. 8.1 DR signaling complexes. Upon DL binding, the DR recruits a complex of DD-containing proteins to its vicinity (complex I). In this complex, RIPK1 is ubiquitinated by cIAPs, initiating signaling to NF-κB activation. This contributes to cell survival via transcription of survival genes. Upon deubiquitination of RIPK1, the complex dissociates from the receptor and recruits FADD and procaspase-8, forming a complex called DISC. Caspase-8 auto-activates, cleaves RIPK1 inhibiting necroptosis, and initiates apoptosis. In conditions of low caspase-8 activity, RIPK3 is recruited to interact with RIPK1, assembling the necrosome, and together they activate necroptosis

second high molecular weight complex, complex II, also named death inducing signaling complex (DISC). This complex comprises RIPK1, TRADD, TRAF2, Fas-associated protein with death domain (FADD) and procaspase-8 and is dissociated from TNFR1, which is internalized. Assembly of the DISC generally leads to induction of apoptosis, through processing of caspases-8, −2 and −3 (Micheau and Tschopp 2003). In certain conditions, depending on the protein environment in this complex, additional proteins are recruited to assemble a third signaling complex—complex IIb or necrosome (Declercq et al. 2009). The necrosome then activates necroptosis via a distinct mechanism of PCD (Declercq et al. 2009; Weinlich et al. 2011).

8.3 RIPK1: A Multi-Talented DR Effector

Receptor interacting protein kinase 1 (RIPK1) is a 60 kDa ubiquitous protein that is recruited to complex I, following TNFα binding to TNFRs. The ubiquitination state of this protein seems to be determinant for the decision between life and death. This post-translational modification of RIPK1 is highly regulated by ubiquitinases and deubiquitinases (DUBs) in complex I. Indeed, cIAP1 and cIAP2 are able to ubiquitinate RIPK1 and upon inhibition of these enzymes the recruitment of RIPK1 to complex II is increased (Geserick et al. 2009). In addition, the DUB Cezanne leads to suppression of NF-κB signaling in response to TNFα activation by removing Lys63 polyubiquitin chains from RIPK1, in complex I. This contributes to increased stability of the inhibitor of kappa B (IκB) complex, which retains NF-κB in the cytoplasmic compartment (Enesa et al. 2008). Inhibition of both cIAP1 and transforming growth factor-β activated kinase-1 (TAK1) induces the formation of the necrosome and subsequent ROS production (Vanlangenakker et al. 2011). Thus, when ubiquitinated with Lys63 chains, RIPK1 seems to contribute to cell survival, by promoting NF-κB transcriptional activity. Once deubiquitinated, RIPK1 recruits proteins like FADD and procaspase8, thereby allowing DISC assembly.

The association between RIPK1 and caspase-8 in complex II is highly dependent on FADD availability (Micheau and Tschopp 2003). This protein can have an impact on the phenotype of cell death. Different domains of FADD have the ability to induce caspase-dependent apoptosis (death effector domain—DED) or necrosis (DD), in a caspase-independent manner, due to their ability to selectively interact with caspase-8 or RIPK-1, respectively (Vanden Berghe et al. 2004).

Autoactivation of caspase-8 is regulated by the cFLIP isoform ($FLIP_L$, $FLIP_S$ and $FLIP_R$) that is present in DISC (Kataoka et al. 1998; Scaffidi et al. 1999). The expression of $FLIP_L$, which is regulated by NF-κB activity, exerts a negative regulatory effect on complex II, through inhibition of caspase-8 (Micheau and Tschopp 2003). The $FLIP_L$ protein possesses caspase-like domains, but has no catalytic activity (Tschopp et al. 1998), thus it interacts with procaspase-8, allowing its allosteric activation and partial processing. This event inhibits induction of apoptosis due to caspase-8 limited activity, but allows this enzyme to process local substrates, like

RIPK1 (Kavuri et al. 2011; Krueger et al. 2001; Micheau et al. 2002), therefore facilitating concomitant inhibition of apoptosis and necroptosis (Oberst et al. 2011). When complex I displays limited activity, then the expression of $FLIP_L$ is lower and complex II has the ability to initiate apoptotic signaling (Micheau and Tschopp 2003). The $FLIP_S$ isoform, on the other hand, inhibits caspase-8 auto-activation (Golks et al. 2005; Krueger et al. 2001), thereby limiting the cleavage of RIPK1, which is then capable of initiating necroptosis signaling.

Necroptosis induction is achieved via the recruitment of a RIPK1-related protein—RIPK3—which has been shown to be specifically recruited to the necrosome (Cho et al. 2009; He et al. 2009; Zhang et al. 2009), implying that the presence of RIPK3 in complex II is specifically associated with necroptotic signaling. Although it is known that TNFα receptors are present in neuronal plasma membranes (Olmos and Llado 2014; Jara et al. 2007), whether TNFα activates the TNFα receptors in neurons, which then activate RIPK1 is not known. Thus, the factor or factors responsible for upregulation of RIPK3 and RIPK1 following oxygen-glucose deprivation of hippocampal neuronal cultures and upregulation of RIPK3 following transient global cerebral ischemia must be determined (Vieira et al. 2014).

8.4 Necroptosis Molecular Complexes

Recruitment of RIPK3 to complex II is thought to be the molecular prompt that shifts the cell death program to necroptosis. Accordingly, the pronecrotic complex II, called necrosome (Fig. 8.2), is distinct from the apoptotic DISC, in that the interaction between RIPK1 and RIPK3 appears to occur specifically under necroptotic-inducing conditions (Cho et al. 2009). Although the complete picture of the events downstream of the necrosome is still not clear, two molecular mechanisms seem to have a profound impact on necroptosis induction: RIPK1/RIPK3 interaction and their kinase activities (see Box 8.1 for a brief timeline of necroptosis discovery).

RIPK1 and RIPK3 were shown to associate via their RIP homotypic interaction motif (RHIM) domains (Sun et al. 2002), forming an amyloid complex upon induction of necroptosis (Li et al. 2012). Notably, viral proteins that disrupt this interaction have been shown to suppress necroptosis induction (Upton et al. 2010), and mutations in the RHIM domain of RIPK3 abolished its ability to induce necroptosis (He et al. 2009), thereby demonstrating the crucial role of RIPK1-RIPK3 interaction in this process of cell death. Additionally, RIPK3 was shown to autophosphorylate and phosphorylate RIPK1. Phosphorylation of the latter seems to act as a cue to shut-off NF-κB signaling, which contributes to cell death (Sun et al. 2002), and it may contribute to the stabilization of the association of RIPK1 within complex II (Cho et al. 2009). The intermediate domain (ID) of RIPK1 is important for induction of necroptosis, as deletion of this domain in the protein shifts the type of cell death from necroptosis to apoptosis, due to increased recruitment of RIPK1 and caspase-8 to FADD (Duprez et al. 2012). The RIPK3 kinase activity is required for

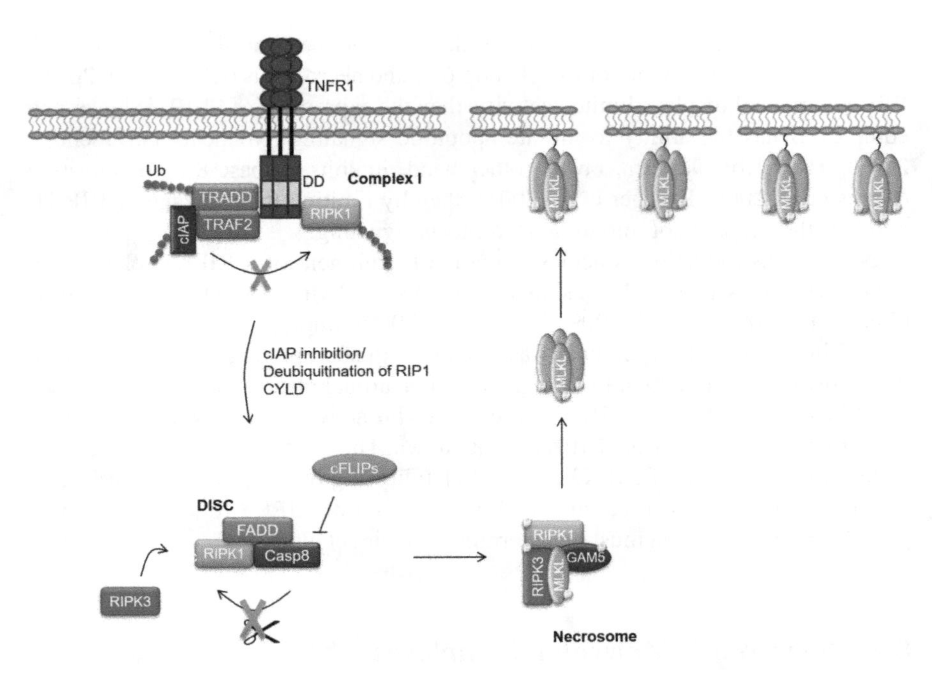

Fig. 8.2 Assembly of the necrosome. When caspase-8 is downregulated, RIPK1 is capable of recruiting RIPK3. RIPK1 and RIPK3 interact through their RHIM domains and autoactivate via phosphorylation. The necrosome then phosphorylates its substrates, including PGAM5 and MLKL. Once phosphorylated, MLKL oligomerizes and is recruited to the cellular membrane, where it interacts with phosphatidylinositol phosphates. This binding promotes formation of membrane pores, which is one of the hallmarks of necroptotic neuronal death

nuclear localization of the protein and for the induction of caspase-independent cell death (Feng et al. 2007).

Despite the prominent role of RIPK1 in scaffolding these complexes, RIPK1-independent (but RIPK3-dependent) necroptosis has been shown to occur upon knockdown of both RIPK1 and caspase-8, evidencing the negative regulation exerted by caspase-8 and suggesting that RIPK3 may be sufficient to induce necroptosis in certain cellular contexts (Vanlangenakker et al. 2011).

The cellular signaling network underlying the necroptotic process involves *de novo* protein synthesis (Yu et al. 2004), mitochondrial dysfunction, ROS accumulation and ATP depletion (Cho et al. 2009; He et al. 2009; Irrinki et al. 2011). These phenomena are promoted by the association between RIPK3 and a group of metabolism-related enzymes: phosphorylase-glycogen-liver (PYGL), glutamate-ammonia ligase (GLUL) and glutamate dehydrogenase 1 (GLUD1). The protein PYGL is the rate-limiting enzyme in glycogen degradation, while GLUL and GLUD1 are determinant enzymes for oxidative phosphorylation (Zhang et al. 2009). This association seems to contribute to mitochondrial damage (Kim et al. 2011).

Box 8.1: Timeline for the discovery of necroptosis

The identification of a form of caspase-independent cell death in response to death ligands (DL), such as tumor necrosis factor α (TNFα), was first reported in 1998 in L929 cells stimulated with this cytokine and treated with an array of caspase inhibitors. This cellular model is sensitized to a form of necrotic cell death, involving reactive oxygen species (ROS) formation, upon apoptosis inhibition (Vercammen et al. 1998). In 2000, a similar caspase-independent T-cell death mechanism downstream of Fas, TNF and TNF-related apoptosis-inducing ligand (TRAIL) ligands was identified (Holler et al. 2000). The type of cell death induced in these conditions was considered to have a necrotic phenotype and to induce late mitochondrial damage. In this same report, cells deficient in either FADD or RIPK1 were shown to be resistant to this form of caspase-independent cell death. Notably, the requirement for RIPK1 kinase activity to necrosis induction was first identified in the same study (Holler et al. 2000) and the ROS accumulation subsequent to necroptosis induction was confirmed a few years later (Lin et al. 2004).

The term necroptosis was first coined in 2005 by Degterev and colleagues (Degterev et al. 2005). They used this term to describe a type of non-apoptotic cell death that is mediated by Fas/TNF receptor activation, already observed in diverse cellular systems at the time. This cell death program was described to occur in the presence of the apoptotic inhibitor zVAD.fmk and as exhibiting necrotic morphology, such as plasma membrane integrity loss and mitochondrial dysfunction (Degterev et al. 2005). An important milestone in the understanding of necroptosis was achieved in this study, with the identification of the specific necroptosis inhibitor, Necrostatin-1 (Nec-1). This inhibitor was able to rescue cell viability following TNF/zVAD insults, but with no effect against purely apoptotic insults, like Fas ligand/cycloheximide treatment. The mechanism of Nec-1 inhibition was demonstrated to occur via RIPK1 kinase inhibition, confirming the requirement of kinase activity of this protein for necroptosis induction (Degterev et al. 2005).

Another important milestone was reached in 2009, with the identification of receptor-interacting protein kinase-3 (RIPK3) as a key mediator of necroptotic signaling by three independent groups using distinct approaches (Cho et al. 2009; He et al. 2009; Zhang et al. 2009). Cho et al. (2009) performed a siRNA screen in FADD-deficient Jurkat cells, which are known to die by necroptosis in response to TNFα. They identified both RIPK1 and RIPK3 as crucial mediators of this type of cell death. Interestingly, they also found that RIPK3 knockdown had no effect on TNF- and FasL-induced apoptosis or nuclear factor kappa B (NF-kB) activation, emphasizing the specific role of RIPK3 on the necroptotic mechanism. This protein was shown to be specifically recruited to complex II and, unlike RIPK1, was not present in complex I (Cho et al. 2009). He et al. (2009) performed a genome-wide siRNA screen and transfected HT-29 cells with different siRNA pools to assess which of

(continued)

Box 8.1 (continued)

these were protective against necroptotic stimulation. Besides RIPK1, which had been previously identified, they found RIPK3 to be important for this process. Importantly, this protein was specifically correlated with necroptosis since it did not interfere with induction of apoptosis. In fact, the ability of different cell lines to undergo necroptosis seemed to be correlated with RIPK3 expression. Accordingly, cells that expressed RIPK3 had the ability to induce necroptosis (for example Jurkat, U937, L929 and MEFs) while cells with no observable expression of RIPK3 were resistant to necroptotic stimulation, including HEK293T, MCF-7 and U2OS. Interestingly, when RIPK3 was expressed in "resistant" cells, like T98G cells, they were rendered susceptible to necroptosis (He et al. 2009). Finally, Zhang et al. (2009) studied the mechanism of necroptosis in two lines of NIH3T3 cells, N and A cells, which are, respectively, susceptible and resistant to necroptosis. A microarray analysis was performed to detect changes in genes that might be responsible for this differential vulnerability. RIPK3 was found to be responsible for induction of necroptosis in N cells, while A cells lack expression of this protein. Specifically, these authors demonstrated that knockdown of RIPK3 abolishes the component of necroptotic death in N cells, whereas expression of RIPK3 in A cells renders them susceptible to TNF-induced necroptosis (Zhang et al. 2009).

Despite the identification of a few necroptotic substrates downstream of the protein complex containing both RIPK1 and RIPK3, the complete understanding of the molecular signaling events underlying this type of cell death still remains elusive. However, it is increasingly clear that MLKL plays a central role as an executor of necroptosis. Phosphorylation of MLKL by RIPK3 leads to its homo-oligomerization and translocation to the plasma membrane (Cai et al. 2014; Wang et al. 2014a; Chen et al. 2014), where it forms membrane disrupting pores (Wang et al. 2014a; Ros et al. 2017) that likely mediate final plasma membrane disruption in necroptosis.

Despite the known role of RIPK1/3 kinase activities in necroptosis, few substrates have been identified so far. Mixed lineage kinase domain-like (MLKL) was shown to be a molecular target of the necrosome (Sun et al. 2012; Zhao et al. 2012). Its role in necroptosis is emphasized by complete abrogation of necroptosis upon MLKL knock down or in the presence of necrosulfonamide, a MLKL inhibitor. Phosphorylated MLKL forms homo-oligomers and translocates to the plasma membrane (Cai et al. 2014; Wang et al. 2014a; Chen et al. 2014). Phosphorylation of MLKL promotes its binding to phosphatidylinositol phosphates (PIPs), allowing plasma membrane recruitment (Dondelinger et al. 2014). Additional higher affinity PIP binding sites in MLKL are exposed upon binding to the membrane, resulting in binding stabilization (Quarato et al. 2016). At the plasma membrane, oligomers of MLKL have been shown to form membrane disrupting pores (Wang et al. 2014a), and to cause Ca^{2+} influx mediated by the transient receptor potential melastatin related 7 (TRPM7) (Cai et al. 2014). In addition, a recent study demonstrated that osmotic forces mediate necroptotic plasma membrane rupture, which involves the formation of 4 nm diameter membrane pores (Ros et al. 2017). Accordingly, osmoprotectants reduce cell death in an in vivo renal model of ischemia-reperfusion injury that features necroptosis (Ros et al. 2017). The MLKL-dependent formation of membrane pores may mediate final plasma membrane disruption in necroptosis (Fig. 8.2).

The mitochondrial protein Ser/Thr-protein phosphatase (PGAM5) is also phosphorylated by RIPK3 and associates with the RIPK1/3-MLKL complex. The short isoform of PGAM5 (PGAM5S) is proposed to be an effector downstream of the necrosome since its RIPK3-mediated phosphorylation is abrogated by necrosulfonamide. PGAM5S then dephosphorylates the mitochondrial protein dynamin related protein 1 (Drp-1), resulting in its activation, which contributes to mitochondrial fragmentation (Wang et al. 2012).

A genome-wide siRNA screening allowed the identification of 432 genes related to necroptosis signaling (Hitomi et al. 2008). Interestingly, some of the genes identified in this study have enriched expression both in the immune and the nervous systems, both of which have been shown to undergo paradigms of necroptotic cell death. Additionally, a subset of genes that regulate both apoptosis and necroptosis, that included TNFR1a and CYLD, was also reported.

While a complex cascade of protein interactions from TNFRs to necrosome has been demonstrated to be required for necroptotic cell death, spontaneous formation of a protein module similar to complex II has also been described (Feoktistova et al. 2011; Tenev et al. 2011). Due to its independence from DR signaling, this high molecular weight complex comprising RIPK1, FADD and caspase-8, among other proteins, was called ripoptosome. The formation of this complex occurs following genotoxic stress (Tenev et al. 2011) or TLR3 stimulation (Feoktistova et al. 2011), both in conditions of cIAP depletion, suggesting that these proteins inhibit the assembly of the ripoptosome. This complex seems to have the ability to induce either apoptosis or necroptosis, depending on the cellular context (Tenev et al. 2011). Notably, in in vitro ischemia this complex does not seem to contribute to neuronal death, since treatment with a TNFα neutralizing antibody blocks induction of necroptosis in this cellular model (Vieira et al. 2014), suggesting that necroptosis in this context is TNFR-dependent and thereby induced by the necrosome, not by ripoptosome formation.

Overall, one of the most striking evidence from these diverse studies is the complementarity between apoptosis and necroptosis. Such notion is corroborated by occlusion of lethality observed in caspase-$8^{-/-}$/RIPK$3^{-/-}$ double knock out mice (Kaiser et al. 2011; Oberst et al. 2011). These results also emphasize the role of caspase-8 as a negative regulator of necroptosis. This regulation is achieved by the formation of catalytically active dimers of caspase-8 and $FLIP_L$, which inhibit the assembly of the necrosome, without inducing apoptosis (Micheau et al. 2002; Oberst et al. 2011).

8.5 Necroptosis in Disease

Deregulation of necroptosis contributes to the pathogenesis of various inflammatory, ischemic and neurodegenerative diseases characterized by excessive cell death and inflammation [for a recent review see (Galluzzi et al. 2017)]. Given the proinflammatory nature of necroptosis, it is also relevant in cancer conditions. However, it is still unclear how necroptosis influences tumorigenesis and metastasis, since while in some cancers there is loss of RIPK3 expression, associated with reduced necroptosis, in others inflammation associated with necroptosis in the tumor environment promotes cancer metastasis, but also cancer immunity [reviewed in Najafov et al. (2017)].

Necroptosis plays a role in the pathogenesis of several acute and chronic neurological diseases. Inhibition of necroptosis is neuroprotective in adult stroke and neonatal hypoxia/ischemia (see next section), and necroptosis was shown to occur in retinal neurons following ischemic damage. Necrostatin-1 was effective in reducing specifically this component of cell death, without affecting apoptotic neuronal death, and in improving the functional outcome in the ischemia challenged retina (Rosenbaum et al. 2010). Also, in a mouse model of retinal detachment, both inhibition of RIPK1 as well as RIPK3 ablation were shown to be neuroprotective, demonstrating the activation of necroptotic mechanisms upon apoptosis inhibition (Trichonas et al. 2010).

Certain neuronal death paradigms have been shown to induce necroptosis. Indeed, following treatment with 24(S)-hydroxycholesterol (24S-OHC), which is cytotoxic, both human neuroblastoma SH-SY5Y cells and cortical neurons were shown to die by necroptosis (Yamanaka et al. 2011). Additionally, menadione, a compound that induces superoxide production, with concomitant mitochondrial dysfunction and ATP depletion, was shown to induce caspase-independent necroptotic-like death in HT-22 cells, a cell line of immortalized mouse hippocampal neurons (Fukui et al. 2012). A recent study found that the environmental neurotoxicant polychlorinated biphenyl (PCB)-95 increases RIPK1, RIPK3 and MLKL expression in cortical neurons, while decreasing caspase-8 levels (Guida et al. 2017). Necrostatin-1 or siRNA-mediated knockdown of RIPK1, RIPK3 or MLKL expression significantly reduced neuronal death induced by PCB-95. In addition, this study showed that PCB-95 increases the levels of RE1-silencing transcription factor (REST), which represses caspase-8 and cAMP Responsive Element Binding Protein (CREB) expression. Since CREB binds to the promoter regions of the RIPK1, RIPK3 and MLKL genes, its down-regulation increases the levels of these proteins (Guida et al. 2017).

Necrostatin-1 was shown to be effective in reducing the damage associated with injury in a mouse model of traumatic brain injury (You et al. 2008), and in models of spinal cord injury (Wang et al. 2014b). Interestingly, necroptosis was proposed to have a neuroprotective effect by being activated in inflamed microglia, which results in less damage to neurons. Upon inflammatory stimulation, inhibition of caspase-8 activity is beneficial to neurons by promoting necroptosis in microglia, whereas

Nec-1 rescues neurodegeneration. This implies that caspase inhibitors could promote neuronal survival against certain insults by promoting microglial demise (Fricker et al. 2013).

Necroptosis has been recently implicated in amyotrophic lateral sclerosis (ALS) (Ito et al. 2016; Re et al. 2014). In this disease, motor neuron axonal degeneration occurs via RIPK1 and RIPK3-mediated mechanisms, suggesting that inhibiting RIPK1 may be protective and useful in the treatment of ALS. Accordingly, a RIPK1 inhibitor is in Phase I human clinical trials for the treatment of ALS (by Denali Therapeutics). In cortical lesions of human multiple sclerosis (MS) pathological samples, activation of RIPK1, RIPK3 and MLKL has been detected (Ofengeim et al. 2015). This study suggests that TNF α, implicated in MS, induces oligodendrocyte degeneration mediated by necroptosis, and shows that RIPK1 inhibition protects oligodendrocytes from cell death in animal models of MS.

8.6 Necroptosis Induction in Cerebral Ischemia

Cerebral ischemia is a leading cause of disability and death (Flynn et al. 2008). This pathology lacks effective treatments, a direct consequence of a limited time window for intervention before major neurodegeneration ensues, as well as of the lack of specificity of some of the treatments developed so far. The interruption in the blood supply to the brain causes a deprivation of oxygen and glucose, leading to a failure in ATP production, which is pernicious due to the brain high metabolic demand (Moskowitz et al. 2010). This results in neuronal depolarization and excessive glutamate release, accompanied by dysfunction of the glutamate reuptake mechanisms in glia cells and neurons, as a result of the dissipation of the Na^+ gradient, due to the depletion of ATP. Thus, an excessive glutamate accumulation at the synapse causes the overactivation of glutamate receptors leading to an intracellular Ca^{2+} overload that may trigger cytotoxicity, a phenomenon called excitotoxicity. Depending on the duration and intensity of the insult, other non-excitotoxic mechanisms are activated, contributing to Ca^{2+} overload (Besancon et al. 2008; Szydlowska and Tymianski 2010). Deregulation of Ca^{2+} homeostasis activates deleterious intracellular mechanisms, including mitochondrial dysfunction and oxidative and nitrosative stress, that induce inflammation and ultimately neuronal death (Iadecola and Anrather 2011; Moskowitz et al. 2010). In global ischemia and in the penumbra area, neuronal degeneration ensues within 24–72 h after the ischemic insult, a process called delayed neuronal death. This delay in neuronal demise onset suggests the reliance on transcriptional changes, leading to programmed cell death mechanisms.

Cerebral ischemia can be addressed experimentally, using several research models, both in vitro and in vivo. The most common in vivo models to study transient global ischemia are the four vessel occlusion (4-VO) and the 2-VO combined with hypotension, both in the rat. As for focal ischemia, the middle cerebral artery occlusion (MCAO) model is generally used. Regarding the in vitro models, the closest to *in vivo* ischemia is the oxygen and glucose deprivation (OGD) challenge, which is most commonly used either in primary cultures of hippocampal or cortical neurons or in forebrain or organotypic hippocampal slices. Due to its characteristics, the OGD challenge of hippocampal neurons is a good model for global ischemia and allows for the dissection of molecular pathways underlying cerebral ischemia. This challenge consists of placing cultured neurons or slices in a glucose-free medium inside an anaerobic chamber, thereby combining the deprivation of these two factors which mimics, in a simplified system, what happens in the brain during the interruption of the blood flow.

The extent of pathways activated following brain ischemia underlies, as mentioned, a variety of cell death mechanisms, posing the difficulty of distinguishing the different types of cell death experimentally. For this purpose, a few methods have been advanced, which allow distinguishing apoptotic from necroptotic cells based on differential staining using AnnexinV/PI (apoptotic cells are AnnexinV positive/PI negative, while necroptotic cells are AnnexinV negative/PI positive), cell and nuclear morphology, and RIPK1 kinase activity, which has been correlated specifically to necroptosis induction (Degterev et al. 2014; Miao and Degterev 2009). Distinguishing between classical necrosis and necroptosis, however, poses other problems, since morphologically these two forms of cell death are very similar, despite one being a form of "accidental" cell death and the other a type of programmed cell death (Vanden Berghe et al. 2010). In this respect, identification of Nec-1 as a specific RIPK1 inhibitor (Degterev et al. 2008) was an important advancement in necroptosis research. As mentioned above, RIPK1 activity was demonstrated to be required solely for necroptotic cell death (Vanden Berghe et al. 2010), thus underscoring the importance of Nec-1 as a valuable tool in necroptosis studies. More recently, other necrostatins have been identified, with varying degrees of specificity to RIPK1 (Jagtap et al. 2007; Wang et al. 2007; Zheng et al. 2008).

The first evidence of necroptosis following a cerebral ischemic insult was reported by Degterev et al., in 2005. In this work, the authors described Nec-1 as a specific necroptosis inhibitor and observed significant neuroprotection by Nec-1 against a focal ischemic insult. Indeed, they reported a significant reduction in infarct size afforded by Nec-1 treatment, as well as improvement of neurological scores (Degterev et al. 2005). In recent years the neuroprotective effect afforded by Nec-1 treatment has also been observed following OGD (Vieira et al. 2014), as well as other in vivo ischemic paradigms (Degterev et al. 2005; Yin et al. 2015), even at 30d post-insult (Xu et al. 2016). These data argue for a component of necroptotic neuronal death following ischemic insults, both in vitro and in vivo, in addition to the apoptotic neuronal death previously described. However, a recent study found that brain injury following MCAO model of stroke or hypoxia-induced cerebral edema was not ameliorated in knock out mice for RIPK3 (Newton et al. 2016), despite the improvement observed by these mice in models of cardiac and kidney ischemia/reperfusion (reviewed in Galluzzi et al. 2017).

Several features of necroptosis were observed in the in vitro model of ischemia, OGD, including a DR-dependent component of neuronal death that is not inhibited by caspase inhibitors like zVAD.fmk. This component was sensitive, however, to Nec-1 treatment and displayed morphological features of necrosis, such as cytoplasmic membrane rupture, as well as nuclear membrane compromise, allowing for positive PI staining (Vieira et al. 2014). Furthermore, following OGD, RIPK3 mRNA is upregulated, with concomitant downregulation of caspase-8 mRNA (Vieira et al. 2014), suggesting a cellular mechanism permissive to necroptosis induction, given the negative regulatory role of caspase-8 on necroptosis activation. At the protein level, RIPK3 expression is specifically upregulated following global ischemic insults, both in vitro and in vivo (Vieira et al. 2014), and RIPK3 translocates to the nucleus (Xu et al. 2016; Yin et al. 2015), where it interacts with AIF (Xu et al. 2016). The other necrosome component, RIPK1 protein, is also upregulated at least following in vitro insults (Qu et al. 2016; Vieira et al. 2014). Notably, overexpression of either component of the necrosome has a detrimental effect following OGD, whereas knockdown of RIPK3 or inhibition of RIPK1 are neuroprotective, reducing the extent of necroptotic neuronal death (Vieira et al. 2014).

Although the molecular mechanism downstream of the necrosome are still not completely resolved, in ischemic contexts, they involve MLKL. Indeed, the interaction between RIPK1-RIPK3 and MLKL increases following an OGD challenge, and downregulation of MLKL by siRNA exerts a neuroprotective effect against in vivo insults (Qu et al. 2016). PGAM5, another necrosome substrate, was initially described as a downstream target of RIPK1/3 contributing to necroptosis, but some recent evidence points to a protective role for PGAM5, by promoting mitophagy. Absence of this protein is proposed to have a deleterious effect by leading to an increase in the number of defective mitochondria and subsequent ROS production, which contributes to the necroptotic damage following heart and brain ischemia (Lu et al. 2016). The role of this protein is thus still not completely unveiled.

The MAPK ERK pathway has been suggested as a putative downstream signaling effector of necroptosis, since ERK inhibitors blocked glutamate-induced necroptosis in HT-22 cells, and Nec-1 reduced glutamate-induced ERK phosphorylation, thereby decreasing its activation, but not that of JNK or p38 MAPKs (Zhang et al. 2013). Furthermore, HDAC6, a histone deacetylase, was also found to contribute to necroptotic damage, since it is upregulated following in vitro ischemia, and inhibition of its activity decreases the extent of OGD-induced necroptosis, possibly via ROS production inhibition (Yuan et al. 2015).

Necroptotic cell death is implicated in brain ischemia, but also in ischemic injury in other organs, such as the kidney and the heart, with common features, namely the upregulation of the necrosome components RIPK1 and RIPK3, as well as the protective effect exerted by Nec-1, in the kidney (Linkermann et al. 2012), in cardiac tissue (Luedde et al. 2014; Tuuminen et al. 2016) and in retinal cells (Ding et al. 2015; Rosenbaum et al. 2010). These effects have also been observed in lung epithelial cells following ischemic insults associated with renal transplants, due to crosstalk between these organs (Zang et al. 2013; Zhao et al. 2015). Additionally, similar to what was observed in neurons, overexpression of the necrosome components is detrimental, whereas knockdown of RIPK3 is protective against ischemic insults in myocytes (Luedde et al. 2014) and retinal cells (Ding et al. 2015).

One of the challenges in stroke/global cerebral ischemia treatment is the fact that diverse modes of cell death seem to be activated concomitantly. Indeed, neurons may die via necrosis, apoptosis, necroptosis or autophagy (Mehta et al. 2007; Meloni et al. 2011; Vieira et al. 2014). Thus, the best approach to ensure maximal neuronal survival may be the use of combinatorial strategies targeting the diverse modes of neuronal death that are induced by brain ischemia. As new data is reported, a more complete picture of the necroptosis process, as well as a better knowledge of the research models of cerebral ischemia, may shed light into the possibilities afforded by inhibition of an array of cell death programs in the treatment and management of the injury induced following ischemic paradigms.

References

Besancon E, Guo S, Lok J, Tymianski M, Lo EH (2008) Beyond Nmda and Ampa glutamate receptors: emerging mechanisms for ionic imbalance and cell death in stroke. Trends Pharmacol Sci 29:268–275

Cai Z, Jitkaew S, Zhao J, Chiang HC, Choksi S, Liu J, Ward Y, Wu LG, Liu ZG (2014) Plasma membrane translocation of trimerized MLKL protein is required for TNF-induced necroptosis. Nat Cell Biol 16:55–65

Chen X, Li W, Ren J, Huang D, He WT, Song Y, Yang C, Li W, Zheng X, Chen P, Han J (2014) Translocation of mixed lineage kinase domain-like protein to plasma membrane leads to necrotic cell death. Cell Res 24:105–121

Cho YS, Challa S, Moquin D, Genga R, Ray TD, Guildford M, Chan FK (2009) Phosphorylation-driven assembly of the RIP1-RIP3 complex regulates programmed necrosis and virus-induced inflammation. Cell 137:1112–1123

Declercq W, Vanden Berghe T, Vandenabeele P (2009) RIP kinases at the crossroads of cell death and survival. Cell 138:229–232

Degterev A, Huang Z, Boyce M, Li Y, Jagtap P, Mizushima N, Cuny GD, Mitchison TJ, Moskowitz MA, Yuan J (2005) Chemical inhibitor of nonapoptotic cell death with therapeutic potential for ischemic brain injury. Nat Chem Biol 1:112–119

Degterev A, Hitomi J, Germscheid M, Ch'en IL, Korkina O, Teng X, Abbott D, Cuny GD, Yuan C, Wagner G, Hedrick SM, Gerber SA, Lugovskoy A, Yuan J (2008) Identification of RIP1 kinase as a specific cellular target of necrostatins. Nat Chem Biol 4:313–321

Degterev A, Zhou W, Maki JL, Yuan J (2014) Assays for necroptosis and activity of RIP kinases. Methods Enzymol 545:1–33

Ding W, Shang L, Huang JF, Li N, Chen D, Xue LX, Xiong K (2015) Receptor interacting protein 3-induced RGC-5 cell necroptosis following oxygen glucose deprivation. BMC Neurosci 16:49

Dondelinger Y, Declercq W, Montessuit S, Roelandt R, Goncalves A, Bruggeman I, Hulpiau P, Weber K, Sehon CA, Marquis RW, Bertin J, Gough PJ, Savvides S, Martinou JC, Bertrand MJ, Vandenabeele P (2014) MLKL compromises plasma membrane integrity by binding to phosphatidylinositol phosphates. Cell Rep 7:971–981

Duprez L, Bertrand MJ, Vanden Berghe T, Dondelinger Y, Festjens N, Vandenabeele P (2012) Intermediate domain of receptor-interacting protein kinase 1 (RIPK1) determines switch between necroptosis and RIPK1 kinase-dependent apoptosis. J Biol Chem 287:14863–14872

Enesa K, Zakkar M, Chaudhury H, Luong Le A, Rawlinson L, Mason JC, Haskard DO, Dean JL, Evans PC (2008) NF-kappaB suppression by the deubiquitinating enzyme Cezanne: a novel negative feedback loop in pro-inflammatory signaling. J Biol Chem 283:7036–7045

Feng S, Yang Y, Mei Y, Ma L, Zhu DE, Hoti N, Castanares M, Wu M (2007) Cleavage of RIP3 inactivates its caspase-independent apoptosis pathway by removal of kinase domain. Cell Signal 19:2056–2067

Feoktistova M, Geserick P, Kellert B, Dimitrova DP, Langlais C, Hupe M, Cain K, Macfarlane M, Hacker G, Leverkus M (2011) cIAPs block Ripoptosome formation, a RIP1/caspase-8 containing intracellular cell death complex differentially regulated by cFLIP isoforms. Mol Cell 43:449–463

Flynn RW, Macwalter RS, Doney AS (2008) The cost of cerebral ischaemia. Neuropharmacology 55:250–256

Fricker M, Vilalta A, Tolkovsky AM, Brown GC (2013) Caspase inhibitors protect neurons by enabling selective necroptosis of inflamed microglia. J Biol Chem 288:9145–9152

Fujikawa DG (2015) The role of excitotoxic programmed necrosis in acute brain injury. Comput Struct Biotechnol J 13:212–221

Fukui M, Choi HJ, Zhu BT (2012) Rapid generation of mitochondrial superoxide induces mitochondrion-dependent but caspase-independent cell death in hippocampal neuronal cells that morphologically resembles necroptosis. Toxicol Appl Pharmacol 262:156–166

Gaither A, Porter D, Yao Y, Borawski J, Yang G, Donovan J, Sage D, Slisz J, Tran M, Straub C, Ramsey T, Iourgenko V, Huang A, Chen Y, Schlegel R, Labow M, Fawell S, Sellers WR, Zawel L (2007) A Smac mimetic rescue screen reveals roles for inhibitor of apoptosis proteins in tumor necrosis factor-alpha signaling. Cancer Res 67:11493–11498

Galluzzi L, Kepp O, Chan FK, Kroemer G (2017) Necroptosis: mechanisms and relevance to disease. Annu Rev Pathol 12:103–130

Geserick P, Hupe M, Moulin M, Wong WW, Feoktistova M, Kellert B, Gollnick H, Silke J, Leverkus M (2009) Cellular IAPs inhibit a cryptic CD95-induced cell death by limiting RIP1 kinase recruitment. J Cell Biol 187:1037–1054

Golks A, Brenner D, Fritsch C, Krammer PH, Lavrik IN (2005) c-FLIPR, a new regulator of death receptor-induced apoptosis. J Biol Chem 280:14507–14513

Guida N, Laudati G, Serani A, Mascolo L, Molinaro P, Montuori P, Di Renzo G, Canzoniero LMT, Formisano L (2017) The neurotoxicant PCB-95 by increasing the neuronal transcriptional repressor REST down-regulates caspase-8 and increases Ripk1, Ripk3 and MLKL expression determining necroptotic neuronal death. Biochem Pharmacol 142:229–241

Han J, Zhong CQ, Zhang DW (2011) Programmed necrosis: backup to and competitor with apoptosis in the immune system. Nat Immunol 12:1143–1149

He S, Wang L, Miao L, Wang T, Du F, Zhao L, Wang X (2009) Receptor interacting protein kinase-3 determines cellular necrotic response to TNF-alpha. Cell 137:1100–1111

Hitomi J, Christofferson DE, Ng A, Yao J, Degterev A, Xavier RJ, Yuan J (2008) Identification of a molecular signaling network that regulates a cellular necrotic cell death pathway. Cell 135:1311–1323

Holler N, Zaru R, Micheau O, Thome M, Attinger A, Valitutti S, Bodmer JL, Schneider P, Seed B, Tschopp J (2000) Fas triggers an alternative, caspase-8-independent cell death pathway using the kinase RIP as effector molecule. Nat Immunol 1:489–495

Iadecola C, Anrather J (2011) The immunology of stroke: from mechanisms to translation. Nat Med 17:796–808

Igney FH, Krammer PH (2002) Death and anti-death: tumour resistance to apoptosis. Nat Rev Cancer 2:277–288

Irrinki KM, Mallilankaraman K, Thapa RJ, Chandramoorthy HC, Smith FJ, Jog NR, Gandhirajan RK, Kelsen SG, Houser SR, May MJ, Balachandran S, Madesh M (2011) Requirement of FADD, NEMO, and BAX/BAK for aberrant mitochondrial function in tumor necrosis factor alpha-induced necrosis. Mol Cell Biol 31:3745–3758

Ito Y, Ofengeim D, Najafov A, Das S, Saberi S, Li Y, Hitomi J, Zhu H, Chen H, Mayo L, Geng J, Amin P, Dewitt JP, Mookhtiar AK, Florez M, Ouchida AT, Fan JB, Pasparakis M, Kelliher MA, Ravits J, Yuan J (2016) RIPK1 mediates axonal degeneration by promoting inflammation and necroptosis in ALS. Science 353:603–608

Jagtap PG, Degterev A, Choi S, Keys H, Yuan J, Cuny GD (2007) Structure-activity relationship study of tricyclic necroptosis inhibitors. J Med Chem 50:1886–1895

Jara JH, Singh BB, Floden AM, Combs CK (2007) Tumor necrosis factor alpha stimulates NMDA receptor activity in mouse cortical neurons resulting in ERK-dependent death. J Neurochem 100:1407–1420

Kaiser WJ, Upton JW, Long AB, Livingston-Rosanoff D, Daley-Bauer LP, Hakem R, Caspary T, Mocarski ES (2011) RIP3 mediates the embryonic lethality of caspase-8-deficient mice. Nature 471:368–372

Kataoka T, Schroter M, Hahne M, Schneider P, Irmler M, Thome M, Froelich CJ, Tschopp J (1998) FLIP prevents apoptosis induced by death receptors but not by perforin/granzyme B, chemotherapeutic drugs, and gamma irradiation. J Immunol 161:3936–3942

Kavuri SM, Geserick P, Berg D, Dimitrova DP, Feoktistova M, Siegmund D, Gollnick H, Neumann M, Wajant H, Leverkus M (2011) Cellular FLICE-inhibitory protein (cFLIP) isoforms block CD95- and TRAIL death receptor-induced gene induction irrespective of processing of caspase-8 or cFLIP in the death-inducing signaling complex. J Biol Chem 286:16631–16646

Kim JY, Kim YJ, Lee S, Park JH (2011) BNip3 is a mediator of TNF-induced necrotic cell death. Apoptosis 16:114–126

Krueger A, Schmitz I, Baumann S, Krammer PH, Kirchhoff S (2001) Cellular FLICE-inhibitory protein splice variants inhibit different steps of caspase-8 activation at the CD95 death-inducing signaling complex. J Biol Chem 276:20633–20640

Li J, Mcquade T, Siemer AB, Napetschnig J, Moriwaki K, Hsiao YS, Damko E, Moquin D, Walz T, Mcdermott A, Chan FK, Wu H (2012) The RIP1/RIP3 necrosome forms a functional amyloid signaling complex required for programmed necrosis. Cell 150:339–350

Lin Y, Choksi S, Shen HM, Yang QF, Hur GM, Kim YS, Tran JH, Nedospasov SA, Liu ZG (2004) Tumor necrosis factor-induced nonapoptotic cell death requires receptor-interacting protein-mediated cellular reactive oxygen species accumulation. J Biol Chem 279:10822–10828

Linkermann A, Brasen JH, Himmerkus N, Liu S, Huber TB, Kunzendorf U, Krautwald S (2012) Rip1 (receptor-interacting protein kinase 1) mediates necroptosis and contributes to renal ischemia/reperfusion injury. Kidney Int 81:751–761

Lu W, Sun J, Yoon JS, Zhang Y, Zheng L, Murphy E, Mattson MP, Lenardo MJ (2016) Mitochondrial protein PGAM5 regulates mitophagic protection against cell necroptosis. PLoS One 11:e0147792

Luedde M, Lutz M, Carter N, Sosna J, Jacoby C, Vucur M, Gautheron J, Roderburg C, Borg N, Reisinger F, Hippe HJ, Linkermann A, Wolf MJ, Rose-John S, Lullmann-Rauch R, Adam D, Flogel U, Heikenwalder M, Luedde T, Frey N (2014) RIP3, a kinase promoting necroptotic cell death, mediates adverse remodelling after myocardial infarction. Cardiovasc Res 103:206–216

Mehta SL, Manhas N, Raghubir R (2007) Molecular targets in cerebral ischemia for developing novel therapeutics. Brain Res Rev 54:34–66

Meloni BP, Meade AJ, Kitikomolsuk D, Knuckey NW (2011) Characterisation of neuronal cell death in acute and delayed in vitro ischemia (oxygen-glucose deprivation) models. J Neurosci Methods 195:67–74

Miao B, Degterev A (2009) Methods to analyze cellular necroptosis. Methods Mol Biol 559:79–93

Micheau O, Tschopp J (2003) Induction of TNF receptor I-mediated apoptosis via two sequential signaling complexes. Cell 114:181–190

Micheau O, Thome M, Schneider P, Holler N, Tschopp J, Nicholson DW, Briand C, Grutter MG (2002) The long form of FLIP is an activator of caspase-8 at the Fas death-inducing signaling complex. J Biol Chem 277:45162–45171

Moskowitz MA, Lo EH, Iadecola C (2010) The science of stroke: mechanisms in search of treatments. Neuron 67:181–198

Najafov A, Chen H, Yuan J (2017) Necroptosis and cancer. Trends Cancer 3:294–301

Newton K, Dugger DL, Maltzman A, Greve JM, Hedehus M, Martin-Mcnulty B, Carano RA, Cao TC, Van Bruggen N, Bernstein L, Lee WP, Wu X, Devoss J, Zhang J, Jeet S, Peng I, Mckenzie BS, Roose-Girma M, Caplazi P, Diehl L, Webster JD, Vucic D (2016) RIPK3 deficiency or catalytically inactive RIPK1 provides greater benefit than MLKL deficiency in mouse models of inflammation and tissue injury. Cell Death Differ 23:1565–1576

Oberst A, Dillon CP, Weinlich R, Mccormick LL, Fitzgerald P, Pop C, Hakem R, Salvesen GS, Green DR (2011) Catalytic activity of the caspase-8-FLIP(L) complex inhibits RIPK3-dependent necrosis. Nature 471:363–367

Ofengeim D, Ito Y, Najafov A, Zhang Y, Shan B, Dewitt JP, Ye J, Zhang X, Chang A, Vakifahmetoglu-Norberg H, Geng J, Py B, Zhou W, Amin P, Berlink Lima J, Qi C, Yu Q, Trapp B, Yuan J (2015) Activation of necroptosis in multiple sclerosis. Cell Rep 10:1836–1849
Olmos G, Llado J (2014) Tumor necrosis factor alpha: a link between neuroinflammation and excitotoxicity. Mediat Inflamm 2014:861231
Qu Y, Shi J, Tang Y, Zhao F, Li S, Meng J, Tang J, Lin X, Peng X, Mu D (2016) MLKL inhibition attenuates hypoxia-ischemia induced neuronal damage in developing brain. Exp Neurol 279:223–231
Quarato G, Guy CS, Grace CR, Llambi F, Nourse A, Rodriguez DA, Wakefield R, Frase S, Moldoveanu T, Green DR (2016) Sequential engagement of distinct MLKL phosphatidylinositol-binding sites executes necroptosis. Mol Cell 61:589–601
Re DB, Le Verche V, Yu C, Amoroso MW, Politi KA, Phani S, Ikiz B, Hoffmann L, Koolen M, Nagata T, Papadimitriou D, Nagy P, Mitsumoto H, Kariya S, Wichterle H, Henderson CE, Przedborski S (2014) Necroptosis drives motor neuron death in models of both sporadic and familial ALS. Neuron 81:1001–1008
Ros U, Pena-Blanco A, Hanggi K, Kunzendorf U, Krautwald S, Wong WW, Garcia-Saez AJ (2017) Necroptosis execution is mediated by plasma membrane nanopores independent of calcium. Cell Rep 19:175–187
Rosenbaum DM, Degterev A, David J, Rosenbaum PS, Roth S, Grotta JC, Cuny GD, Yuan J, Savitz SI (2010) Necroptosis, a novel form of caspase-independent cell death, contributes to neuronal damage in a retinal ischemia-reperfusion injury model. J Neurosci Res 88:1569–1576
Scaffidi C, Schmitz I, Krammer PH, Peter ME (1999) The role of c-FLIP in modulation of CD95-induced apoptosis. J Biol Chem 274:1541–1548
Sun X, Yin J, Starovasnik MA, Fairbrother WJ, Dixit VM (2002) Identification of a novel homotypic interaction motif required for the phosphorylation of receptor-interacting protein (RIP) by RIP3. J Biol Chem 277(11):9505
Sun L, Wang H, Wang Z, He S, Chen S, Liao D, Wang L, Yan J, Liu W, Lei X, Wang X (2012) Mixed lineage kinase domain-like protein mediates necrosis signaling downstream of RIP3 kinase. Cell 148:213–227
Szydlowska K, Tymianski M (2010) Calcium, ischemia and excitotoxicity. Cell Calcium 47:122–129
Taylor RC, Cullen SP, Martin SJ (2008) Apoptosis: controlled demolition at the cellular level. Nat Rev Mol Cell Biol 9:231–241
Tenev T, Bianchi K, Darding M, Broemer M, Langlais C, Wallberg F, Zachariou A, Lopez J, Macfarlane M, Cain K, Meier P (2011) The Ripoptosome, a signaling platform that assembles in response to genotoxic stress and loss of IAPs. Mol Cell 43:432–448
Trichonas G, Murakami Y, Thanos A, Morizane Y, Kayama M, Debouck CM, Hisatomi T, Miller JW, Vavvas DG (2010) Receptor interacting protein kinases mediate retinal detachment-induced photoreceptor necrosis and compensate for inhibition of apoptosis. Proc Natl Acad Sci U S A 107:21695–21700
Tschopp J, Irmler M, Thome M (1998) Inhibition of fas death signals by FLIPs. Curr Opin Immunol 10:552–558
Tuuminen R, Holmstrom E, Raissadati A, Saharinen P, Rouvinen E, Krebs R, Lemstrom KB (2016) Simvastatin pretreatment reduces caspase-9 and RIPK1 protein activity in rat cardiac allograft ischemia-reperfusion. Transpl Immunol 37:40–45
Upton JW, Kaiser WJ, Mocarski ES (2010) Virus inhibition of RIP3-dependent necrosis. Cell Host Microbe 7:302–313
Vanden Berghe T, Van Loo G, Saelens X, Van Gurp M, Brouckaert G, Kalai M, Declercq W, Vandenabeele P (2004) Differential signaling to apoptotic and necrotic cell death by Fas-associated death domain protein FADD. J Biol Chem 279:7925–7933
Vanden Berghe T, Vanlangenakker N, Parthoens E, Deckers W, Devos M, Festjens N, Guerin CJ, Brunk UT, Declercq W, Vandenabeele P (2010) Necroptosis, necrosis and secondary necrosis converge on similar cellular disintegration features. Cell Death Differ 17:922–930

Vandenabeele P, Galluzzi L, Vanden Berghe T, Kroemer G (2010) Molecular mechanisms of necroptosis: an ordered cellular explosion. Nat Rev Mol Cell Biol 11:700–714

Vanlangenakker N, Vanden Berghe T, Bogaert P, Laukens B, Zobel K, Deshayes K, Vucic D, Fulda S, Vandenabeele P, Bertrand MJ (2011) cIAP1 and TAK1 protect cells from TNF-induced necrosis by preventing RIP1/RIP3-dependent reactive oxygen species production. Cell Death Differ 18:656–665

Vercammen D, Beyaert R, Denecker G, Goossens V, Van Loo G, Declercq W, Grooten J, Fiers W, Vandenabeele P (1998) Inhibition of caspases increases the sensitivity of L929 cells to necrosis mediated by tumor necrosis factor. J Exp Med 187:1477–1485

Vieira M, Fernandes J, Carreto L, Anuncibay-Soto B, Santos M, Han J, Fernandez-Lopez A, Duarte CB, Carvalho AL, Santos AE (2014) Ischemic insults induce necroptotic cell death in hippocampal neurons through the up-regulation of endogenous RIP3. Neurobiol Dis 68:26–36

Vosler PS, Sun D, Wang S, Gao Y, Kintner DB, Signore AP, Cao G, Chen J (2009) Calcium dysregulation induces apoptosis-inducing factor release: cross-talk between PARP-1- and calpain-signaling pathways. Exp Neurol 218:213–220

Wang CY, Mayo MW, Korneluk RG, Goeddel DV, Baldwin AS Jr (1998) NF-kappaB antiapoptosis: induction of TRAF1 and TRAF2 and c-IAP1 and c-IAP2 to suppress caspase-8 activation. Science 281:1680–1683

Wang K, Li J, Degterev A, Hsu E, Yuan J, Yuan C (2007) Structure-activity relationship analysis of a novel necroptosis inhibitor, necrostatin-5. Bioorg Med Chem Lett 17:1455–1465

Wang Z, Jiang H, Chen S, Du F, Wang X (2012) The mitochondrial phosphatase PGAM5 functions at the convergence point of multiple necrotic death pathways. Cell 148:228–243

Wang H, Sun L, Su L, Rizo J, Liu L, Wang LF, Wang FS, Wang X (2014a) Mixed lineage kinase domain-like protein MLKL causes necrotic membrane disruption upon phosphorylation by RIP3. Mol Cell 54:133–146

Wang Y, Wang H, Tao Y, Zhang S, Wang J, Feng X (2014b) Necroptosis inhibitor necrostatin-1 promotes cell protection and physiological function in traumatic spinal cord injury. Neuroscience 266:91–101

Weinlich R, Dillon CP, Green DR (2011) Ripped to death. Trends Cell Biol 21:630–637

Xu Y, Wang J, Song X, Qu L, Wei R, He F, Wang K, Luo B (2016) RIP3 induces ischemic neuronal DNA degradation and programmed necrosis in rat via AIF. Sci Rep 6:29362

Yamanaka K, Saito Y, Yamamori T, Urano Y, Noguchi N (2011) 24(S)-hydroxycholesterol induces neuronal cell death through necroptosis, a form of programmed necrosis. J Biol Chem 286:24666–24673

Yang QH, Du C (2004) Smac/DIABLO selectively reduces the levels of c-IAP1 and c-IAP2 but not that of XIAP and livin in HeLa cells. J Biol Chem 279:16963–16970

Yin B, Xu Y, Wei RL, He F, Luo BY, Wang JY (2015) Inhibition of receptor-interacting protein 3 upregulation and nuclear translocation involved in Necrostatin-1 protection against hippocampal neuronal programmed necrosis induced by ischemia/reperfusion injury. Brain Res 1609:63–71

You Z, Savitz SI, Yang J, Degterev A, Yuan J, Cuny GD, Moskowitz MA, Whalen MJ (2008) Necrostatin-1 reduces histopathology and improves functional outcome after controlled cortical impact in mice. J Cereb Blood Flow Metab 28:1564–1573

Yu L, Alva A, Su H, Dutt P, Freundt E, Welsh S, Baehrecke EH, Lenardo MJ (2004) Regulation of an ATG7-beclin 1 program of autophagic cell death by caspase-8. Science 304:1500–1502

Yuan L, Wang Z, Liu L, Jian X (2015) Inhibiting histone deacetylase 6 partly protects cultured rat cortical neurons from oxygenglucose deprivationinduced necroptosis. Mol Med Rep 12:2661–2667

Zang D, Shao Y, Li X (2013) Ultrastructural pathology of rat lung injury induced by ischemic acute kidney injury. Ultrastruct Pathol 37:433–439

Zhang DW, Shao J, Lin J, Zhang N, Lu BJ, Lin SC, Dong MQ, Han J (2009) RIP3, an energy metabolism regulator that switches TNF-induced cell death from apoptosis to necrosis. Science 325:332–336

Zhang M, Li J, Geng R, Ge W, Zhou Y, Zhang C, Cheng Y, Geng D (2013) The inhibition of ERK activation mediates the protection of necrostatin-1 on glutamate toxicity in HT-22 cells. Neurotox Res 24:64–70

Zhao J, Jitkaew S, Cai Z, Choksi S, Li Q, Luo J, Liu ZG (2012) Mixed lineage kinase domain-like is a key receptor interacting protein 3 downstream component of TNF-induced necrosis. Proc Natl Acad Sci U S A 109:5322–5327

Zhao H, Ning J, Lemaire A, Koumpa FS, Sun JJ, Fung A, Gu J, Yi B, Lu K, Ma D (2015) Necroptosis and parthanatos are involved in remote lung injury after receiving ischemic renal allografts in rats. Kidney Int 87:738–748

Zheng W, Degterev A, Hsu E, Yuan J, Yuan C (2008) Structure-activity relationship study of a novel necroptosis inhibitor, necrostatin-7. Bioorg Med Chem Lett 18:4932–4935

Chapter 9
Histological and Elemental Changes in Ischemic Stroke

M. Jake Pushie, Vedashree R. Meher, Nicole J. Sylvain, Huishu Hou, Annalise T. Kudryk, Michael E. Kelly, and Roland N. Auer

Abstract Stroke is a leading cause of serious long-term disability in adults and a leading cause of death in developed nations. Following an ischemic stroke the metabolic profile of the affected tissue is significantly altered, with the infarct representing the most severely affected tissue, and the surrounding penumbra, or peri-infarct zone (PIZ), containing a gradient of metabolic states progressing from severely impacted toward an otherwise healthy profile. The penumbra contains potentially salvageable tissue and is the focus in many stroke treatments. In this chapter, we employ the photothrombotic stroke model (a widely used animal model for studying focal ischemia) to study the histopathological and bioelemental changes that occur post-stroke. Synchrotron-based X-ray fluorescence imaging allows simultaneous measurement of multiple elements in situ within biological tissues, as their naturally-occurring concentrations. Images of elemental distributions are compared to conventional histopathological changes in the infarct and penumbra. Understanding the bioelemental changes associated with the post-stroke brain provides opportunities to expand our understanding of the underlying cellular and tissue changes associated with ischemic stroke and can ultimately be used to guide development of future treatment methods targeting the penumbra.

Keywords Ischemic stroke · Penumbra · Photothrombotic stroke · Excitotoxicity · X-ray fluorescence imaging · Elemental mapping

---

M. Jake Pushie · N. J. Sylvain · H. Hou · M. E. Kelly
Division of Neurosurgery, Department of Surgery, College of Medicine, University of Saskatchewan, Saskatoon, SK, Canada

V. R. Meher
Department of Health Sciences, College of Medicine, University of Saskatchewan, Saskatoon, SK, Canada

A. T. Kudryk
College of Medicine, University of Saskatchewan, Saskatoon, SK, Canada

R. N. Auer (✉)
Department of Pathology and Laboratory Medicine, College of Medicine, University of Saskatchewan, Saskatoon, SK, Canada

Department of Pathology, Royal University Hospital, Saskatoon, SK, Canada

D. G. Fujikawa (ed.), *Acute Neuronal Injury*,
https://doi.org/10.1007/978-3-319-77495-4_9

Stroke is a leading worldwide cause of serious long-term disability and death in developed countries (Thrift et al. 2017), and now developing countries (Feigin et al. 2009). Ischemic stroke incidence is roughly 6.6 million Americans, and the prevalence is 2.6%. Roughly 87% of strokes are ischemic (Mozaffarian et al. 2016), and once blood supply to the brain is reduced below an ischemic threshold (Heiss 1992), biochemical cascades, irreparable tissue injury and cell death can occur within only 2–4 min of global ischemia. In focal ischemia, the tolerable duration of occlusion is longer (Kaplan et al. 1991), but the region of tissue most severely impacted is the ischemic core. Surrounding the ischemic core (Hossmann 1994) is the peri-infarct zone, or penumbra, which contains tissue that is partially perfused by collateral arteries, and is therefore salvageable if blood flow is fully restored (Bandera et al. 2006; Latchaw et al. 2003). Due to the rapid loss of tissue viability in the ischemic core, the ischemic penumbra contains the only tissue that can potentially be rescued through intervention post stroke.

The ability to identify the ischemic penumbra and quantify dynamic changes in its size are critically important for many areas of stroke research, as therapeutic interventions and drug development rely on accurately identifying and quantifying penumbral salvageable tissue. A penumbra can be defined histologically by identifying regions of selective neuronal necrosis along the border of pan-necrotic tissue with normal brain. While this identifies a border rim of variable extent, the spatial extent of penumbra that can be salvaged can only be determined by shrinkage of an infarct using an effective treatment, retrospectively leaving the penumbra, perforce, as the tissue rim that was salvaged around an infarct that is reduced in size. Thus, new methods for identifying and defining the extent of the penumbra would aid studies that target the penumbra, and allow quantitative determination of the size and the neuroanatomical extent of the penumbra—critical metrics beneficial for determining treatment efficacy.

In this chapter we discuss the histopathology and elemental changes within tissue following induction of a focal ischemic stroke, and explore the tissue-level changes using the photothrombotic mouse model of stroke. Imaging multiple elemental distributions in tissue provides a means of measuring its metabolic status. For nervous tissue, elemental changes associated with altered membrane potential and excitotoxicity are particularly useful for studying ischemic stroke. Understanding cellular features that contribute to tissue vulnerability will help us better identify therapeutic options for stroke management and recovery.

9.1 The Photothrombotic Model

Our team employs the photothrombotic stroke model in mice (Labat-gest and Tomasi 2013), using isoflurane anesthesia, and head immobilization in a stereotactic frame. After skull is exposure via a midline scalp incision, and the area over the primary somatosensory cortex is identified (−1 to −2.5 mm midline to lateral, and +1 to −0.5 mm antero-posterior to bregma, right hemisphere) (Carmichael 2005;

Winship and Murphy 2008; Hackett et al. 2016). A photoactive dye, Rose Bengal, is injected intraperitoneally, which enters the circulation in under 5 min. A sterile blocking mask is applied over the area of interest and 5 min after the dye is injected, the area is photoexcited using a laser (532 nm) located 3 cm above the skull, for 20-min (Labat-gest and Tomasi 2013; Carmichael 2005; Winship and Murphy 2008).

Photoexcitation of the dye induces localized production of singlet O_2, endothelial damage, and platelet activation. This induces formation of a thrombus, resulting in a focal cortical stroke lesion which is highly reproducible. Blood flow to the affected region of the cortex ceases, mimicking an ischemic stroke, although the model is thought to contain little secondary oxidative damage due to the lack of reperfusion-associated production of free radicals (Carmichael 2005; Kim et al. 2000). Nevertheless, the model is popular due to its reproducibility and reliability (Bandera et al. 2006; Carmichael 2005; Watson et al. 1985), compared to other potentially more variable stroke models, such as the MCAO model (Fluri et al. 2015; Liu and McCullough 2011).

Following induction of the photothrombotic stroke, Marcaine anesthetic is applied to the incision area for pain relief. Twenty-four h after the laser illumination is ceased, the animals are sacrificed by decapitation and the heads are immediately frozen in liquid N_2 in order to preserve elemental distributions (Hackett et al. 2015) and to avoid artifacts introduced by conventional tissue processing methods (Hackett et al. 2011, 2012).

9.2 Tissue Preparation for X-Ray Fluorescence Imaging

Preservation of elemental distributions and concentration levels *in situ* is of paramount importance for elemental mapping (by "elemental" we are referring to the labile cations and anions, such as K^+ and Cl^-, as well as atoms bound within molecules, such as P, S, and Fe). Conventional animal euthanization and tissue processing methods, including perfusion, post-mortem interval, application of tissue fixative, and cryoprotection, can each alter the biodistribution of elements in tissue—particularly mobile ions and chelatable trace metals (Hackett et al. 2015, 2011). For this reason, careful consideration must be given to tissue preparation, and ideally as little sample handling as possible is preferred. We have developed a protocol which optimizes elemental preservation of brain tissue. This method, however, is at the expense of techniques which are commonly employed to improve the appearance of the tissue. Furthermore, due to the necrotic changes that take place in the stroke lesion, tears and cracks are not uncommon, as the underlying tissue is largely without structural integrity. Due to the absence of tissue fixative or cryoprotection, we also observe micro-cracks and bubbles throughout the tissue. These artifacts are typically small enough that they do not affect elemental imaging results and have no impact on the quantities of elements present.

Brains are chiseled out from the skull while still frozen, and are maintained at −20 °C during cryosectioning of 30 μm-thick sections, cut using Teflon-coated blades to minimize transfer of metals from the blade to the tissue. Cut sections are mounted on metal-free plastic coverslips for X-ray fluorescence imaging (XFI). Careful selection of coverslips is necessary for X-ray fluorescence experiments as most plastic or glass coverslips and microscope slides contain an abundance of unwanted elements related to their associated manufacturing processes, which will overwhelm detection of relatively weaker fluorescence signals from trace elements of interest in samples being analyzed. For example, the thermanox coverslips we use contain traces of cobalt, but since cobalt is not one of the elements of interest for our analysis, we use these particular coverslips.

9.3 Introduction to Synchrotron-Based X-Ray Fluorescence Imaging

With the current state of technology, XFI for biological specimens can only be performed at a synchrotron with beamlines that support such imaging capabilities. These are multi-user facilities where access is typically granted through a competitive proposal application system.

The experimental facilities for XFI can typically achieve beam spot sizes of a few microns, either through use of apertures, which reduce flux, or through the use of focusing optics. The sample is mounted on a mechanical stage in the path of the incident X-ray beam, which is typically tuned to an energy sufficient to excite a range of elements of interest. The energy of the X-ray fluorescence that is emitted is dependent on the fluorescing element. Acquisition of the full X-ray fluorescence spectrum, using an energy dispersive detector, provides the means to collect the fluorescence from all elements within the area of sample illuminated by the incident beam. By raster scanning the sample in the beam, a map of the fluorescence spectrum at each position (pixel) can be obtained. It is also important to point out that XFI data must be collected prior to any other staining techniques used for light microscopy. For a more complete review on the topic of XFI see reference (Pushie et al. 2014).

9.4 Histopathology

Normal brain tissue consists of cell bodies embedded in neuropil, the fine, bubbly, reticulated, and relatively structureless tissue between cell bodies as observed by light microscopy. Using electron microscopy, neuropil is resolved to axons, dendrites and glial processes. The axons and dendrites are neuronal processes, whereas

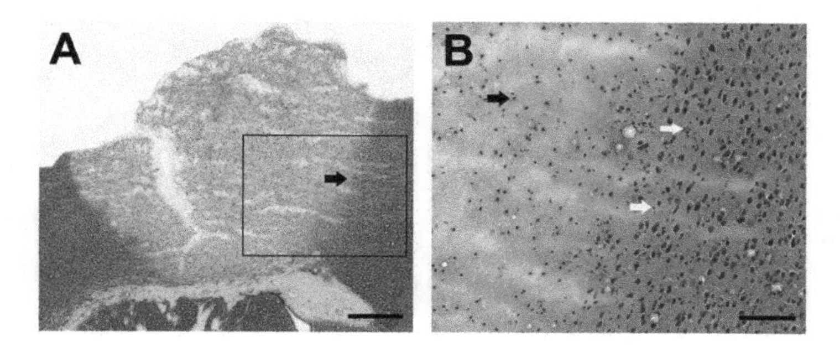

Fig. 9.1 Comparative hematoxylin and eosin (H&E) stained images showing pan-necrosis and selective neuronal necrosis of ischemic tissue. The presence of cracks and other artifacts is due to the lack of tissue preservatives which are undesirable for XFI analysis of the tissue. (**a**) Low magnification view of ischemic infarct boundary in the right somatosensory cortex, including the histological penumbra (black arrow), scale bar = 500 μm. (**b**) Higher magnification view showing pan-necrosis of neurons along the border of the infarct, which is indicated by pink-stained cytoplasm surrounding shrunken nuclei (black arrow) and selective neuronal loss in the penumbra indicated by scarce neuronal death where dead neurons are still surrounded by viable tissue (white arrows), scale bar = 100 μm

the glia, mostly astrocytes, fill the gaps between neurites, leaving little extracellular space (estimated to be 20% for diffusion of substances) (Nicholson et al. 2011). The cell bodies themselves are either neuronal or glial cells, with glia being further divided into macroglia (oligodendrocytes and astrocytes) and microglia. Microglia patrol the nervous system, subserving immune functions. Both astrocytes and oligodendrocytes, each have two forms. Astrocytes in the white matter tend to be fibrillated and are exposed to interstitial fluid, which contains metabolic waste products. It is thus not surprising that astrocytes within the white matter are more densely fibrillated (with glial fibrillary acidic protein) than those in the gray matter (termed protoplasmic astrocytes). Oligodendrocytes similarly have two localizations—one in the white matter, maintaining myelin, and the other in the cerebral cortex, where they surround and maintain neurons.

There are two forms of ischemic brain damage: selective neuronal necrosis (SNN) and pan-necrosis. SNN kills neurons but not glia or neuropil and is mediated by excitotoxicity of glutamate (Benveniste 1991). Pan-necrosis involves death of the entire tissue, affecting cellular elements indiscriminately (*i.e.* Fig. 9.1a). In the context of cerebral ischemia, pan-necrosis is termed infarction—where infarction is considered a tissue-level damage. Additionally, necrosis is often sharply demarcated, whereby neurons inside the infarct undergo pyknotic changes (black arrows in Fig. 9.1b). This sharp demarcation necessarily cuts cross axons, dendrites, and glial processes in the neuropil, thereby demonstrating this is a tissue-level mechanism as opposed to a cell-level mechanism. Another characteristic hallmark of necrosis includes mitochondrial flocculent densities, shown in the electron micrograph image in Fig. 9.2.

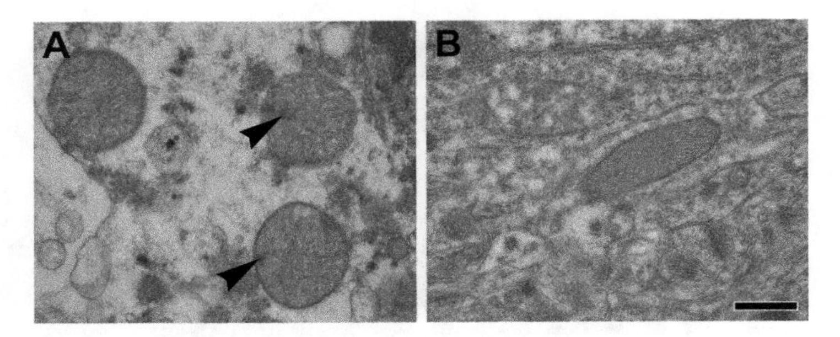

Fig. 9.2 Electron micrograph showing the morphology of necrotic changes in brain tissue following ischemia. The appearance of necrosis is marked by mitochondrial flocculent densities (indicated by the black arrows) inside the infarct (**a**) as opposed to normal non-necrotic tissue from the contralateral side, with normal mitochondria (**b**). Scale bar = 0.5 μm

9.5 Ischemic Cell Death Is Induced by Excitotoxicity

The brain is a highly perfused organ and neurons require a continuous supply of nutrients for proper functioning. Interruption of blood flow for just 2–4 min has been shown to be sufficient for neurons to die (Smith et al. 1984), whereas it can take from 20 to 40 min for other cell types, such as cardiomyocytes and kidney cells, to die as a result of ischemia (Lee et al. 2000). Excitotoxicity, the pathological process where over-excitation of receptors occurs by excitatory neurotransmitters with deleterious consequences, can also activate intracellular and extracellular pathways involved in cell death. Ischemia, hypoglycemia, epilepsy (Auer and Siesjö 1988), and domoic acid poisoning (Debonnel et al. 1989) are well-known examples of excitotoxicity-induced cell death.

There are also intracellular signalling pathways involved in information processing which are triggered by excitotoxicity, which cause cell death under ischemic conditions through induction of free radical production, membrane failure, and an inflammatory response (Lee et al. 2000).

Other important considerations during ischemia-induced cell death are acidosis, disruption of the blood brain barrier (BBB), and infiltration of leukocytes (Woodruff et al. 2011). Arterial occlusion leads to a reduction in tissue pO_2 and increase in pCO_2. Under hypoxic conditions, the tissues undergo an anaerobic glycolysis from the glucose supplied by the residual blood flow (Kalimo et al. 1981). A build-up of lactic acid triggers a drop in local tissue pH within the stroke lesion.

Due to bioenergetic failure under ischemic conditions, there is an increased influx of Na^+ and increased efflux of K^+ as well as glutamate (Paschen 1996), which results in anoxic depolarization (Hansen and Olsen 1980). Failure of glutamate uptake systems causes over- excitation of glutamate ionotropic receptors [*N*-methyl-D-aspartate (NMDA), α-amino-3-hydroxy-5-methyl-4-isoxazole propionic acid (AMPA) and kainate], resulting in excessive intracellular influx of Ca^{2+}. High intracellular Ca^{2+} leads to mitochondrial failure, activates catabolic enzymes and production of reactive oxygen species (Kalogeris et al. 2012).

9.6 Elemental Imaging

Element-based imaging remains a growing research area applicable to a wide range of fields across the basic sciences and health research fields (Hackett et al. 2016; Caine et al. 2016). There are numerous experimental methods for mapping elemental distributions in tissue, including magnetically-susceptible nuclei for MRI (*e.g.* ^{23}Na), mass spectrometry-based imaging—specifically laser ablation inductively coupled plasma mass spectrometry (LA-ICP-MS) (Egger et al. 2014), whereas matrix-assisted laser desorption and ionization (MALDI) imaging is applicable to molecules as opposed to individual elements, *per se* (Aichler and Walch 2015)—and fluorescent chelators designed to coordinate specific elements (Lindahl and Moore 2016; Karim and Petering 2016). While it is beyond the scope of this chapter to review these techniques, they are compared in a recent review (Pushie et al. 2014).The mass spectrometric techniques are destructive techniques which alter the specimen under analysis and require some level of pre-treatment. Fluorescence probes are often regarded as specific for a particular element; however their binding is, at best, selective for a particular element but may also bind chemically similar targets. Such probe-based approaches also rely on labile or exchangeable target ions, and the probe molecules themselves may have variable penetration into particular tissues or subcellular structures, or may mobilize an element into a region preferable to the chemical properties of the chelator.

The type of synchrotron-based XFI experiments described herein are particularly amenable to mapping elemental distributions within biological tissues at their naturally-occurring trace levels, and is not dependent on sample preparation method *per se*, but is best applied with minimal sample pre-treatment or manipulation in order to preserve the biodistribution in its original state. However, there are technical challenges with obtaining data for light elements below P and S, making observation of Na^+ beyond the ability of the technique to observe within biological specimens.

9.7 Results of X-Ray Fluorescence Imaging

XFI detects the intrinsic fluorescence produced by elements when photons of sufficient energy are absorbed, thereby promoting a ground state electron (*i.e.* a 1 s electron) into an excited state orbital. For XFI experiments synchrotrons are used to produce X-rays, as they are intensely bright sources of photons which can be tuned over a wide range to a desired energy (Egger et al. 2014). The X-ray beam spot can be tailored to the desired application and a range of elements can be imaged simultaneously using an energy dispersive detector combined with spectral fitting *in silico*. This affords the means to record contributions from many elements to the total fluorescence spectrum and generate element-specific maps by selectively displaying the fitted contribution of a specific element to the collected data.

Not to be confused with computed tomography-type imaging using a synchrotron light source (Lin et al. 2013; Zhang et al. 2014), synchrotron-based XFI is still a relatively new technique and there are only a small number of researchers at present investigating topics related to stroke (Caine et al. 2016).

9.8 Contrasting Elemental Images with the Elementary Dogma of Ischemic Stroke

With the benefit of new imaging modalities comes new insights and an opportunity to refine current dogma. The ischemic core of dead or dying tissue with low cerebral blood flow and high extracellular K^+ is regarded as a chief hallmark of ischemic cell damage. Although K^+ is no longer within the cell, XFI data, even as early as 1 h post-stroke, reveals that the infarct region is largely devoid of essential K^+. The infarct is readily identified histopathologically as the demarcated region with reduced hematoxylin staining, and appearance of eosinophilic neurons with shrunken cell body (Figs. 9.2 and 9.3) (Zille et al. 2012). XFI reveals these components are lost from the tissue almost entirely, indicating that the consequence of initial depolarization and eventual cell membrane degradation is not only loss of essential chemical messengers and redistribution of ions into the extracellular space. As the detected X-ray fluorescence is based on inherent atomic physics-based properties, the technique is sensitive to the elements themselves, and not to their chemical form, localization, or bioavailability. Due to concomitant disruption of the BBB during the early stages of ischemia the extracellular space of the affected tissue is in exchange with the blood, thereby providing a much larger reservoir for cellular components, like K^+ lost through depolarization or cell membrane breakdown. These components then follow their concentration gradient, such as from the affected tissue, to the blood, where the K^+ concentration is in the range of ~5 mM, compared to ~150 mM in cells of the brain. Comparing concentrations of key ions implicated in ischemic changes (Na^+, Cl^-, K^+, and Ca^{2+}), we observe that each of these ions follows its concentration gradient between the cell environment and blood (Fig. 9.4).

XFI data provides further evidence of this concentration-dependent exchange between the ischemic tissue and blood, revealing profound changes in Ca^{2+} and Cl^- content within the infarct (Fig. 9.5). Along with K^+, Na^+ and Ca^{2+} channels, Cl^- channels are also dysregulated in ischemic stroke. These channels allow passive diffusion of Cl^- into cells and decrease the membrane potential, resulting in tonic inhibition of excitatory neurotransmission, and lost Cl^- channel function is associated with neuropathologies such as epilepsy (Puljak and Kilic 2006). Ca^{2+} is a critically important chemical messenger within the brain, and is employed sparingly for functions such as activation of signaling pathways (Brini et al. 2014) and regulating gene expression (West et al. 2001). Ca^{2+}, while regarded as vital for neuronal health and function, is maintained at a very low concentration within the brain, with

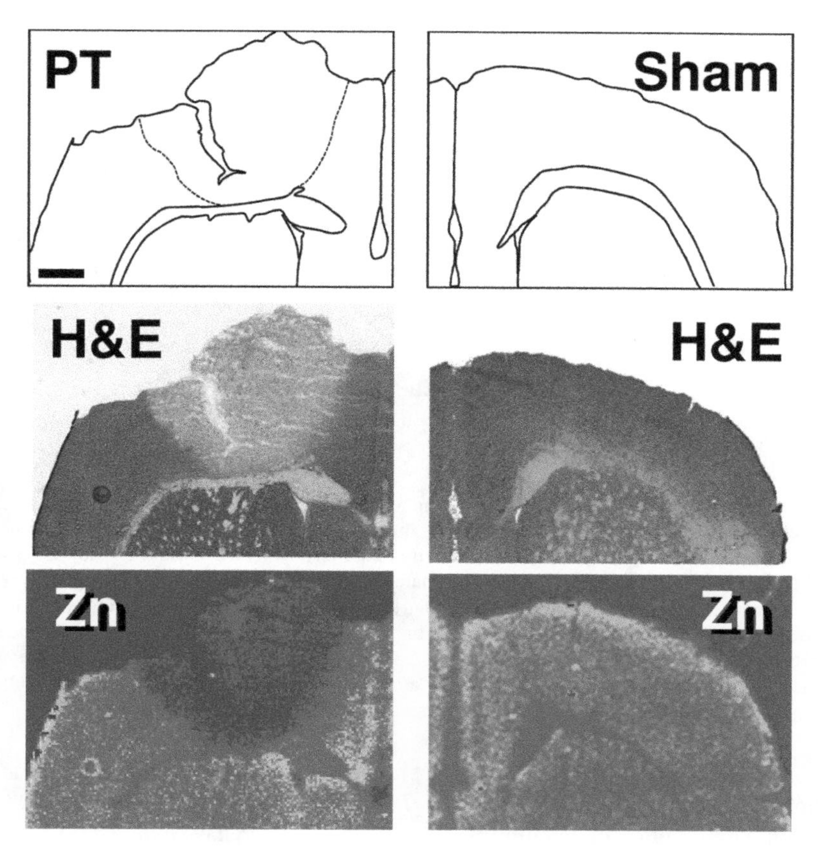

Fig. 9.3 Comparison of synchrotron X-ray fluorescence maps of Zn^{2+}, which best shows neuroanatomical detail, and conventional H&E staining of the same tissue. Zn^{2+} is present in all cells, and its concentration is dependent on cell type and tissue localization. Following ischemia, low molecular weight Zn^{2+} complexes are lost through depolarization, while the reduced pH of the infarct and competition by other divalent cations likely drives loss of Zn^{2+} from protein binding sites. The mouse photothrombotic model is compared to a sham mouse model. The stroke lesion demonstrates marked differences from normal tissue. The authors note that the cracking of the stroke lesion tissue is an artifact of the extent of necrosis and the tissue preparation method, which does not employ fixative or cryoprotectant, and requires that the head be snap-frozen. This procedure can lead to tissue cracking during the freezing procedure or during cryosectioning. Scale bar = 600 μm

average Ca^{2+} concentration in the 100 nM range. Following ischemic stroke the infarct demonstrates a significant elevation in Ca^{2+} and Cl^- content. Due to the nature of the X-ray fluorescence imaging, Na^+ (and other light element) images are not obtainable for these types of samples, due in part to low X-ray fluorescence yields, poor penetration of the relatively low energy Na^+ fluorescence, and absorption losses to the air in the sample-to-detector path length, mentioned earlier.

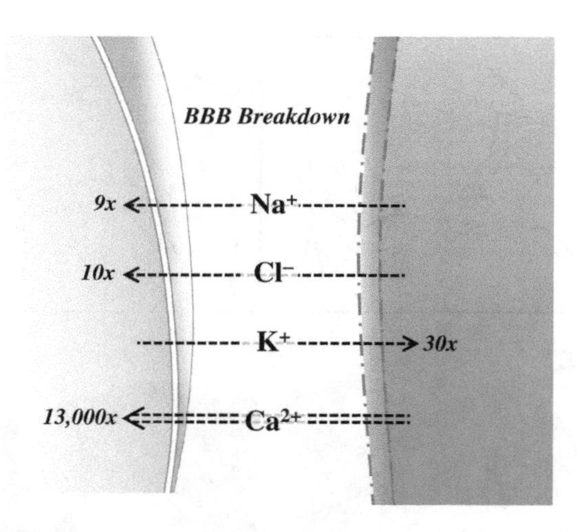

Fig. 9.4 Breakdown of the BBB allows exchange between the blood and the extracellular fluid, and between the extracellular and intracellular fluids. According to the zeroth law of thermodynamics, if two thermodynamic systems are each in equilibrium with a third, then they are also in equilibrium with each other. Therefore, as exchange between these compartments begins to reach equilibrium the small volume of the intracellular environment will more closely resemble that of blood. Comparing average concentrations for the most commonly implicated ions that undergo dysregulation in ischemic stroke (*i.e.* mobile ions) we find that the concentration differences will lead to a shift into (in the case of Na^+, Cl^- and Ca^{2+}) or out of (as in the case of K^+) the intracellular environment within the brain—an observation recapitulated in elemental maps (Fig. 9.5)

Nevertheless, the changes in Cl^- and Ca^{2+} reflect their significantly elevated concentrations in blood relative to the brain (Fig. 9.4).

With the exception of calcium-binding proteins, which have evolved high affinities for binding Ca^{2+} within specific coordination environments for signaling or as a cofactor for functionally requirements, solvation of the ions Cl^-, K^+ and Ca^{2+} is highly thermodynamically favourable. Upon breakdown of neuronal cell membrane potentials and disruption of the BBB these ions follow their chemical gradient, resulting in a redistribution of these ions from compartments of high concentration to compartments with lower concentration, which is highly entropically favourable. Other elements, however, including P and S, are associated with a wide range of biomolecules across a wide range of molecular weights. These elements are covalently bonded to multiple additional atoms, and in the case of proteins (Cys, Met, and P-Ser and P-Tyr), lipids, DNA, RNA, and other large molecules, the localization of these elements is dictated by the compartmentalization of their host molecule. P and S may also be components of small molecules, for example, taurine and GSH are sulfur-containing molecules which may also have relatively high mobility if membrane permeability of the host cell is disrupted. Taurine is a sulfur containing amino acid synthesized endogenously by metabolism of methionine and cysteine or exogenously by obtaining taurine rich food products. Under ischemic conditions, taurine provides neuroprotective effects against glutamate-induced excitotoxity, oxidative stress and inflammation (Menzie et al. 2013). Glutathione is a tripeptide

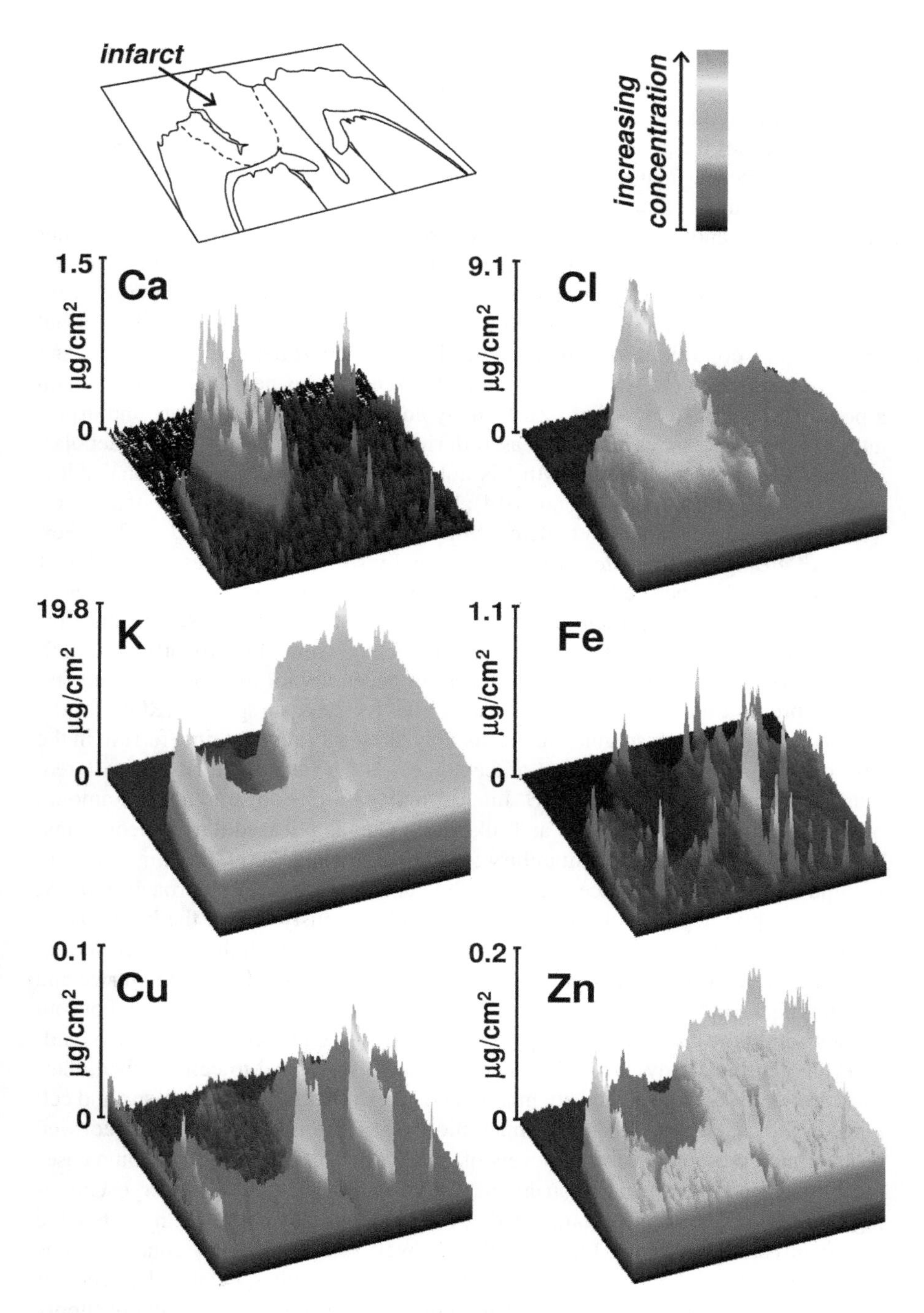

Fig. 9.5 Synchrotron XFI results for select elements rendered as 3D heat maps, demonstrating the quantitative changes that occur 24-h post- stroke in the photothrombotic model. The 2D neuroanatomical rendering is oriented with the same relative perspective as the 3D images. Normal levels for each element can be seen as the background signal within the brain section outside the stroke lesion

that exists in a reduced state (GSH) or oxidized state (GSSG). Ischemic reperfusion injury influences the generation of ROS. GSH donates either a proton or electron to the oxidant, thereby neutralizing it (Song et al. 2015).

Relative to most other organs the brain is highly enriched in trace metals, including Fe^{2+}, Cu^{2+}, and Zn^{2+}. While Zn^{2+}, by virtue of its spherically symmetric outer shell of electrons, has some similarities with the other ions discussed above, the transition metals generally possess substantially different binding properties with biomolecules and in many instances are tightly bound to host proteins under normal physiological conditions and are not typically regarded as being highly mobile and readily exchangeable. Zn^{2+} is again a notable exception, as neuronal cells maintain a pool of low molecular weight Zn^{2+}, likely complexed with glutamate and aspartate, in synaptic vesicles and is released during synaptic transmission and depolarization (Kitamura et al. 2006). Mn^{2+} is normally maintained at exceptionally low levels in tissue (Jensen and Jenson 2014); however, XFI provides sufficient sensitivity and high spatial resolution to detect Mn^{2+} traces (Robison et al. 2012). Both Fe^{2+} and Cu^{2+} are tightly regulated (Cu^{2+} exceptionally so, with estimates of less than 1 free atom of Cu^{2+} per cell) (Rae et al. 1999), and these comprise active sites in functionally-important proteins or are bound within storage proteins. Fe^{2+} and Cu^{2+} can readily partake in electron-transfer reactions, and while this property is capitalized by biological systems to catalyze reactions necessary for life, this same reactivity can promote production of deleterious reactive oxygen species (ROS) in the presence of O_2 when dysregulated. Fortunately, these storage proteins (ferritin in the case of Fe^{2+}, and MT3 in the case of Cu^{2+}), keep their metal payloads sequestered, thus preventing production of ROS. Interestingly, neurons do store a small amount of Cu^{2+} within synaptic vesicles and, like Zn^{2+}, Cu^{2+} is released during neurotransmission and depolarization (Bitanihirwe and Cunningham 2009). Within the small volume of the synaptic space the local Cu^{2+} concentration can approach 300 μM during synaptic depolarization. Zn^{2+} is found ubiquitously within the brain and is employed for a myriad of roles vital to cellular function, including proper protein folding and function, transcription, and signalling, to name a few (Bitanihirwe and Cunningham 2009). Zn^{2+} is therefore distributed relatively evenly throughout all tissues within the brain and Zn maps present the most approachable images to landmarking neuroanatomy. While Fe^{2+} and Cu^{2+} are also critical to cell survival, their roles are more specialized and are maintained at levels dictated by the tissue and cell type. In the case of Fe^{2+} the neuroanatomical distribution is largely overshadowed by the much more concentrated levels of Fe^{2+} within blood vessels. In some cases these spikes in Fe^{2+} concentration are displaced from the vessels during cryosectioning. Like Fe^{2+}, Cu^{2+} is also maintained at levels dictated by cell type; most notable are ependymal cells lining the ventricles, as well as elevated Cu^{2+} concentrations within periventricular regions, which are associated with specialized astrocytes which appear to serve a specialized role in the brain to store Cu^{2+} and likely supply neurons with essential Cu^{2+} (Pushie et al. 2011; Sullivan et al. 2017). These observations of elevated Cu^{2+} in the ependyma and periventricular regions of the rodent brain (mouse and rat models) have not been observed in similar human brain images, but may be an artifact of long post-mortem intervals, extended storage in fixative, or natural loss of ependymal cells to gliosis, which begins in early adolescence (Del Bigio 2002). Under ischemic conditions we observe a significant decrease in Fe^{2+}, Cu^{2+}, and Zn^{2+} within the infarct (Fig. 9.5). Although there is diminished concentration of these metals following an ischemic stroke, some trace amount of metal content remains. We hypothesize that the chemical form of these trace amounts of metal ions that remain, largely represent dysregulated metal ions which are no longer bound to their normal targets and in the case of Fe^{2+} and Cu^{2+} likely to contribute to ROS-mediated damage upon reperfusion of tissue due to Fenton and Haber-Weiss reactions (Birben et al. 2012; Selim and Ratan 2004; Spasojevic et al. 2010).

We hypothesize that the reason for the decreases in elemental content shown in Fig. 9.5 are multi-factorial, and are a combination of (i) cell depolarization, (ii) reduced pH due to acidosis, which lowers the affinity of most metalloproteins for their target metals, (iii) oxidative modification and degradation of host proteins and biomolecules, and (iv) breakdown of the host cell membrane, resulting is loss of cellular contents and potential mobilization of higher molecular weight metalloproteins. These hypotheses are supported at least in part by observations that the region of the infarct is under acidic conditions following an ischemic (Tombaugh and Sapolsky 1993; Siesjö et al. 1996) episode and there is a concomitant reduction in protein and lipid content and an increase in misfolded proteins following ischemic stroke (Nowak and Jacewicz 1994).

Taken together the elemental mapping clearly reveals that the current dogma of ion relocalization from within neuronal cells into the extracellular space following ischemia-induced depolarization is insufficiently descriptive. In addition, it is clear that the influx of Cl^- and Ca^{2+} (and likely Na^+) does not involve extracellular ions alone, but significantly elevated levels beyond those that normally exist in the brain. In combination with the ischemia-induced increase in BBB permeability the extracellular contents of the ischemic lesion are consequently in exchange with blood (and vice versa), resulting in dilution of extracellular ions in the larger blood volume, while highly concentrated ions found in blood also permeate the lesion.

9.9 The Concept of the Ischemic Penumbra

The ischemic penumbra corresponds to tissue that has experienced limited blood flow and may be "potentially destined for infarction but not yet been irreversibly injured" (Fisher 1997). Cells within this region are generally regarded as being under metabolic stress, but otherwise maintain their membrane potential.

While the idea of the ischemic penumbra is conceptually straight forward, there are many definitions, related to various imaging modalities and other metrics, such as blood flow, that have been proposed to identify this region. Identification of the ischemic penumbra is critically important for quantitative assessment of potential post-stroke therapies and treatments, as previously mentioned.

Elemental maps of the region of tissue surrounding the infarct, reveals that there are consistent elemental changes that occur (Fig. 9.5). For some elements there is a relatively abrupt border demarcating otherwise normal levels of the element from the dysregulated elemental levels found in the stroke lesion (*i.e.* P, S, Ca^{2+}, Fe^{2+}, Cu^{2+}, and Zn^{2+}). Even for these elements, there is a gradient that typically spans ~100 μm, which defines the border. For other elements like Cl and K this gradient may span 500 μm or more. These observations indicates that the tissue within this region is heterogeneous, with regions closer to the core of the infarct more closely mimicking the elemental levels seen in the necrotic region while regions farther away have elemental levels approaching normal levels. As the elemental levels are a measure of the metabolic state of the cells within the tissue, the elemental gradients can be collectively analyzed using soft clustering methods (Ward et al. 2013; Crawford 2015) to differentiate statistically similar regions (Fig. 9.6). The clustering shown in takes into account the elemental changes for P, S, Cl^-, K^+ and Zn^{2+}, and reveals a statistically distinct band of tissue that surrounds the ischemic core. At early post-stroke time points (*i.e.* 1 h) it is possible to identify more than one narrow concentric band surrounding the central ischemic core.

An important finding in the elemental data and their statistical analysis is that the region of tissue demarcates metabolically distinct regions of tissue. It is clear from the elemental analysis that this region of tissue surrounding the infarct has altered elemental content, which would require that the membrane potential of the underlying cells be compromised. This paints a picture of a gradation of cell states, even in the region of the penumbra, ranging from tissue that is close to the normal healthy state, to tissue closely resembling the pan-necrotic tissue of the ischemic core. It is important to highlight that this band of tissue, that resembles the stroke penumbra, contains cells which are otherwise indistinguishable from those within otherwise healthy tissue. In Fig. 9.6, the penumbral zone does not show a clear region of selective neuronal death, indicating that in this model, the histological border is a demarcation between severely compromised pan-necrosis and histologically intact tissue, with no zone of selective neuronal death in between. Consideration of this indicates that in this model, any therapeutic benefit will be seen as a shrinkage of the infarct, not elimination of a rim of selective neuronal death, as the therapeutic gain.

The statistical differentiation of tissue regions, based on elemental content, provides a quantitative measure of tissue area corresponding to the penumbra and its elemental composition. Synchrotron-based XFI, combined with the application of advanced data analysis methods, is therefore a promising new tool to analyze efficacy and underlying metabolic changes in future stroke treatments and therapies in animal models. In combination with complimentary X-ray absorption spectroscopy, which can be performed at the same experimental facilities as XFI in some cases, it is possible to also obtain information on the chemical forms of any of the elements being investigated (Pushie et al. 2014, 2011; Egger et al. 2014; Puljak and Kilic 2006; Harris et al. 2008; Lins et al. 2016; Popescu et al. 2017), thereby providing a wealth of additional chemical and metabolic information. While this is still a relatively young field of research, the future of biological XFI, in combination with conventional histopathology, looks particularly bright.

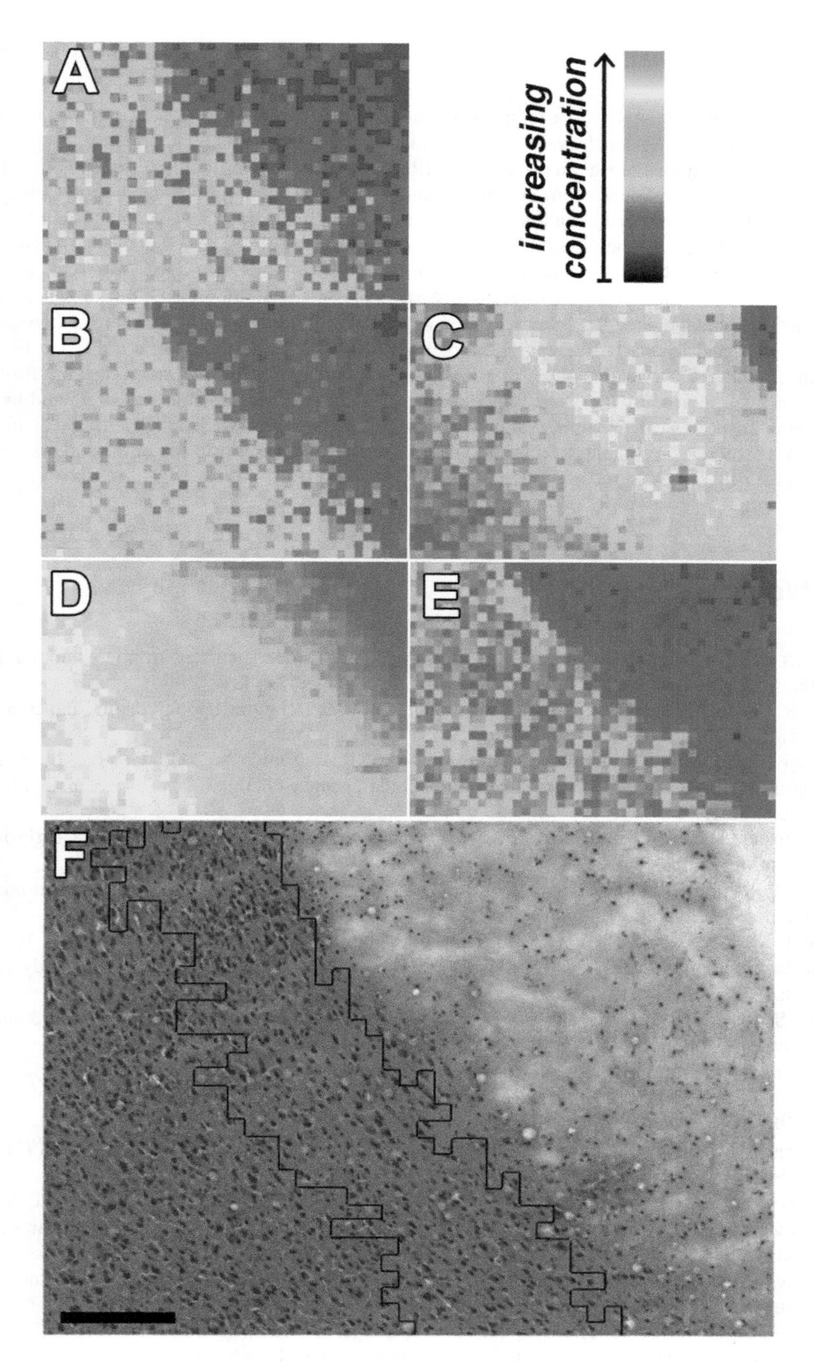

Fig. 9.6 Overlay of the border separating statistically distinct regions of tissue based on soft clustering of P (**a**), S (**b**), Cl^- (**c**), K^+ (**d**), and Zn^{2+} (**e**) distributions. These elements represent those most significantly altered across the stroke lesion and are the exemplar of elemental metabolic dysfunction between unaffected tissue and the infarct. The demarcated border overlaying the H&E image (**f**) shows (from left-to-right) unaffected tissue, the penumbra (outlined in black), and the infarct, as identified by cluster analysis. Scale bar = 200 μm; each pixel in the elemental maps = 30 μm

Acknowledgements AK was a recipient of the College of Medicine Dean's summer research award. MEBK is the Saskatchewan Clinical Stroke Research Chair and is supported by grants from the Canadian Institutes of Health research (CIHR), the Saskatchewan Health Research Foundation, the Heart and Stroke Foundation, Saskatchewan, and the University of Saskatchewan, College of Medicine. Research described in this chapter was performed in part at the Canadian Light Source, which is supported by the Natural Sciences and Engineering Research Council of Canada, the National Research Council Canada, CIHR, and Province of Saskatchewan, Western Economic Diversification Canada, and the University of Saskatchewan. In addition, the Stanford Synchrotron Radiation Lightsource (SSRL), SLAC National Accelerator Laboratory, was used for this research and is supported by the U.S. Department of Energy, Office of Science, Office of Basic Energy Sciences under Contract No. DE-AC02-76SF00515. The SSRL Structural Molecular Biology Program is supported by the DOE Office of Biological and Environmental Research, and by the National Institutes of Health, National Institute of General Medical Sciences (including P41GM103393). The contents of this publication are solely the responsibility of the authors and do not necessarily represent the official views of NIGMS or NIH.

References

Aichler M, Walch A (2015) MALDI Imaging mass spectrometry: current frontiers and perspectives in pathology research and practice. Lab Investig 95:422–431

Auer RN, Siesjö BK (1988) Biological differences between ischemia, hypoglycemia, and epilepsy. Ann Neurol 24:699–707

Bandera E, Botteri M, Minelli C, Sutton A, Abrams KR, Latronico N (2006) Cerebral blood flow threshold of ischemic penumbra and infarct core in acute ischemic stroke: a systematic review. Stroke 37:1334–1339

Benveniste H (1991) The excitotoxin hypothesis in relation to cerebral ischemia. Cerebrovasc Brain Metab Rev 3:213–245

Birben E, Sahiner UM, Sackesen C, Erzurum S, Kalayci O (2012) Oxidative stress and antioxidant defense. World Allergy Organ J 5:9–19

Bitanihirwe BK, Cunningham MG (2009) Zinc: the brain's dark horse. Synapse 63:1029–1049

Brini M, Calì T, Ottolini D, Carafoli E (2014) Neuronal calcium signaling: function and dysfunction. Cell Mol Life Sci 71:2787–2814

Caine S, Hackett MJ, Hou H, Kumar S, Maley J, Ivanishvili Z, Suen B, Szmigielski A, Jiang Z, Sylvain NJ, Nichol H, Kelly ME (2016) A novel multi-modal platform to image molecular and elemental alterations in ischemic stroke. Neurobiol Dis 91:132–142

Carmichael ST (2005) Rodent models of focal stroke: size, mechanism, and purpose. NeuroRx 2:396–409

Crawford AM (2015) Mblank computer program. Methodologies in XRF Cytometry (Thesis). University of Michigan, Michigan

Debonnel G, Beauschesne L, de Montigny C (1989) Domoic acid, the alleged "mussel toxin", might produce its neurotoxic effect through kainate receptor activation: an electrophysiological study in the rat dorsal hippocampus. Can J Physiol Pharmacol 67:29–33

Del Bigio MR (2002) Glial linings of the brain. In: Walz W (ed) The neuronal environment: brain homeostasis in health and disease. Humana Press, Totowa, NJ.

Egger AE, Theiner S, Kornauth C, Heffeter P, Berger W, Keppler BK, Hartinger CG (2014) Quantitative bioimaging by LA-ICP-MS: a methodological study on the distribution of Pt and Ru in viscera originating from cisplatin- and KP1339-treated mice. Metallomics 6:1616–1625

Feigin VL, Lawes CM, Bennett DA, Barker-Collo SL, Parag V (2009) Worldwide stroke incidence and early case fatality reported in 56 population-based studies: a systematic review. Lancet Neurol 8:355–369

Fisher M (1997) Characterizing the target of acute stroke therapy. Stroke 28:866–872
Fluri F, Schuhmann MK, Kleinschnitz C (2015) Animal models of ischemic stroke and their application in clinical research. Drug Des Devel Ther 9:3445–3454
Hackett MJ, McQuillan JA, El-Assaad F, Aitken JB, Levina A, Cohen DD, Siegele R, Carter EA, Grau GE, Hunt NH, Lay PA (2011) Chemical alterations to murine brain tissue induced by formalin fixation: implications for biospectroscopic imaging and mapping studies of disease pathogenesis. Analyst 136:2941–2952
Hackett MJ, Smith SE, Paterson PG, Nichol H, Pickering IJ, George GN (2012) X-ray absorption spectroscopy at the sulfur K-edge: a new tool to investigate the biochemical mechanisms of neurodegeneration. ACS Chem Neurosci 3:178–185
Hackett MJ, Britz CJ, Paterson PG, Nichol H, Pickering IJ, George GN (2015) In situ biospectroscopic investigation of rapid ischemic and postmortem induced biochemical alterations in the rat brain. ACS Chem Neurosci 6:226–238
Hackett MJ, Sylvain NJ, Hou H, Caine S, Alaverdashvili M, Pushie MJ, Kelly ME (2016) Concurrent glycogen and lactate imaging with FTIR spectroscopy to spatially localize metabolic parameters of the glial response following brain ischemia. Anal Chem 88:10949–10956
Hansen AJ, Olsen CE (1980) Brain extracellular space during spreading depression and ischemia. Acta Physiol Scand 108:355–365
Harris HH, Vogt S, Eastgate H, Legnini DG, Hornberger B, Cai Z, Lai B, Lay PA (2008) Migration of mercury from dental amalgam through human teeth. J Synchrotron Radiat 15:123–128
Heiss WD (1992) Experimental evidence of ischemic thresholds and functional recovery. Stroke 23:1668–1672
Hossmann K-A (1994) Viability thresholds and the penumbra of focal ischemia. Ann Neurol 36:557–565
Jensen AN, Jenson LT (2014) Manganese transport, trafficking and function in invertebrates. Manganese in health and disease. RSC Publishing, London, pp 1–33
Kalimo H, Rehncrona S, Söderfeldt H, Olsson Y, Siesjö B (1981) Brain lactic acidosis and ischemic cell damage. 2. Histopathology. J Cereb Blood Flow Metab 1:313–327
Kalogeris T, Baines CP, Krenz M, Korthuis RJ (2012) Cell biology of ischemia/reperfusion injury. Int Rev Cell Mol Biol 298:229–317
Kaplan B, Brint S, Tanabe J, Jacewicz M, Wang XJ, Pulsinelli W (1991) Temporal thresholds for neocortical infarction in rats subjected to reversible focal cerebral ischemia. Stroke 22:1032–1039
Karim MR, Petering DH (2016) Newport Green, a fluorescent sensor of weakly bound cellular Zn(2+): competition with proteome for Zn(2). Metallomics 8:201–210
Kim GW, Sugawara T, Chan PH (2000) Involvement of oxidative stress and caspase-3 in cortical infarction after photothrombotic ischemia in mice. J Cereb Blood Flow Metab 20:1690–1701
Kitamura Y, Iida Y, Abe J, Ueda M, Mifune M, Kasuya F, Ohta M, Igarashi K, Saito Y, Saji H (2006) Protective effect of zinc against ischemic neuronal injury in a middle cerebral artery occlusion model. J Pharmacol Sci 100:142–148
Labat-gest V, Tomasi S (2013) Photothrombotic ischemia: a minimally invasive and reproducible photochemical cortical lesion model for mouse stroke studies. J Vis Exp. https://doi.org/10.3791/50370
Latchaw RE, Yonas H, Hunter GJ, Yuh WT, Ueda T, Sorensen AG, Sunshine JL, Biller J, Wechsler L, Higashida R, Hademenos G, Council on Cardiovascular Radiology of the American Heart Association (2003) Guidelines and recommendations for perfusion imaging in cerebral ischemia: a scientific statement for healthcare professionals by the writing group on perfusion imaging, from the Council on Cardiovascular Radiology of the American Heart Association. Stroke 34:1084–1104
Lee JM, Grabb MC, Zipfel GJ, Choi DW (2000) Brain tissue responses to ischemia. J Clin Invest 106:723–731
Lin X, Miao P, Wang J, Yuan F, Guan Y, Tang Y, He X, Wang Y, Yang GY (2013) Surgery-related thrombosis critically affects the brain infarct volume in mice following transient middle cerebral artery occlusion. PLoS ONE 8:e75561

Lindahl PA, Moore MJ (2016) Labile low-molecular-mass metal complexes in mitochondria: trials and tribulations of a burgeoning field. Biochemistry 55:4140–4153

Lins BR, Pushie JM, Jones M, Howard DL, Howland JG, Hackett MJ (2016) Mapping alterations to the endogenous elemental distribution within the lateral ventricles and choroid plexus in brain disorders using X-ray fluorescence imaging. PLoS ONE 11:e0158152

Liu F, McCullough LD (2011) Middle cerebral artery occlusion model in rodents: methods and potential pitfalls. J Biomed Biotechnol 2011:464701

Menzie J, Prentice H, Wu JY (2013) Neuroprotective mechanisms of taurine against ischemic stroke. Brain Sci 3:877–907

Mozaffarian D, Benjamin EJ, Go AS, Arnett DK, Blaha MJ, Cushman M, Das SR, de Ferranti S, Després JP, Fullerton HJ, Howard VJ, Huffman MD, Isasi CR, Jiménez MC, Judd SE, Kissela BM, Lichtman JH, Lisabeth LD, Writing Group Members (2016) Heart disease and stroke statistics-2016 update: a report from the American Heart Association. Circulation 133:e38–e360

Nicholson C, Kamali-Zare P, Tao L (2011) Brain extracellular space as a diffusion barrier. Comput Vis Sci 14:309–325

Nowak TS Jr, Jacewicz M (1994) The heat shock/stress response in focal cerebral ischemia. Brain Pathol 4:67–76

Paschen W (1996) Glutamate excitotoxicity in transient global cerebral ischemia. Acta Neurobiol Exp (Wars) 56:313–322

Popescu BF, Frischer JM, Webb SM, Tham M, Adiele RC, Robinson CA, Fitz-Gibbon PD, Weigand SD, Metz I, Nehzati S, George GN, Pickering IJ, Brück W, Hametner S, Lassmann H, Parisi JE, Yong G, Lucchinetti CF (2017) Pathogenic implications of distinct patterns of iron and zinc in chronic MS lesions. Acta Neuropathol 134:45–64

Puljak L, Kilic G (2006) Emerging roles of chloride channels in human diseases. Biochim Biophys Acta 1762:404–413

Pushie MJ, Pickering IJ, Martin GR, Tsutsui S, Jirik FR, George GN (2011) Prion protein expression level alters regional copper, iron and zinc content in the mouse brain. Metallomics 3:206–214

Pushie MJ, Pickering IJ, Korbas M, Hackett MJ, George GN (2014) Elemental and chemically specific X-ray fluorescence imaging of biological systems. Chem Rev 114:8499–8541

Rae TD, Schmidt PJ, Pufahl RA, Culotta VC, O'Halloran TV (1999) Undetectable intracellular free copper: the requirement of a copper chaperone for superoxide dismutase. Science 284:805–808

Robison G, Zakharova T, Fu S, Jiang W, Fulper R, Barrea R, Marcus MA, Zheng W, Pushkar Y (2012) X-ray fluorescence imaging: a new tool for studying manganese neurotoxicity. PLoS ONE 7:e48899

Selim MH, Ratan RR (2004) The role of iron neurotoxicity in ischemic stroke. Ageing Res Rev 3:345–353

Siesjö BK, Katsura K, Kristián T (1996) Acidosis-related damage. Adv Neurol 71:209–233

Smith ML, Auer RN, Siesjö BK (1984) The density and distribution of ischemic brain injury in the rat following 2-10 min of forebrain ischemia. Acta Neuropathol 64:319–332

Song J, Park J, Oh Y, Lee JE (2015) Glutathione suppresses cerebral infarct volume and cell death after ischemic injury: involvement of FOXO3 inactivation and Bcl2 expression. Oxidative Med Cell Longev 2015:426069

Spasojevic I, Mojovic M, Stevic Z, Spasic SD, Jones DR, Morina A, Spasic MB (2010) Bioavailability and catalytic properties of copper and iron for Fenton chemistry in human cerebrospinal fluid. Redox Rep 15:29–35

Sullivan B, Robison G, Osborn J, Kay M, Thompson P, Davis K, Zakharova T, Antipova O, Pushkar Y (2017) On the nature of the Cu-rich aggregates in brain astrocytes. Redox Biol 11:231–239

Thrift AG, Thayabaranathan T, Howard G, Howard VJ, Rothwell PM, Feigin VL, Norrving B, Donnan GA, Cadilhac DA (2017) Global stroke statistics. Int J Stroke 12:13–32

Tombaugh GC, Sapolsky RM (1993) Evolving concepts about the role of acidosis in ischemic neuropathology. J Neurochem 61:793–803

Ward J, Marvin R, O'Halloran T, Jacobsen C, Vogt S (2013) Rapid and accurate analysis of an X-ray fluorescence microscopy data set through gaussian mixture-based soft clustering methods. Microsc Microanal 19:1281–1289

Watson BD, Dietrich WD, Busto R, Wachtel MS, Ginsberg MD (1985) Induction of reproducible brain infarction by photochemically initiated thrombosis. Ann Neurol 17:497–504

West AE, Chen WG, Dalva MB, Dolmetsch RE, Kornhauser JM, Shaywitz AJ, Takasu MA, Tao X, Greenberg ME (2001) Calcium regulation of neuronal gene expression. Proc Natl Acad Sci U S A 98:11024–11031

Winship IR, Murphy TH (2008) In vivo calcium imaging reveals functional rewiring of single somatosensory neurons after stroke. J Neurosci 28:6592–6606

Woodruff TM, Thundyil J, Tang SC, Sobey CG, Taylor SM, Arumugam TV (2011) Pathophysiology, treatment, and animal and cellular models of human ischemic stroke. Mol Neurodegener 6:11

Zhang M, Peng G, Sun D, Xie Y, Xia J, Long H, Hu K, Xiao B (2014) Synchrotron radiation imaging is a powerful tool to image brain microvasculature. Med Phys 41:031907

Zille M, Farr TD, Przesdzing I, Müller J, Sommer C, Dirnagl U, Wunder A (2012) Visualizing cell death in experimental focal cerebral ischemia: promises, problems, and perspectives. J Cereb Blood Flow Metab 32:213–231

Part V
Hypoglycemic Neuronal Injury

Chapter 10
Hypoglycemic Brain Damage

Roland N. Auer

Abstract Hypoglycemic brain damage is a different global brain insult from cardiac arrest encephalopathy. We here follow the path of glucose from blood to the brain interstitial space, into the cell, through glycolysis into the Krebs cycle, including the consequent new homeostasis in amino acid metabolism that gives rise to increased aspartic acid within cells. Leakage of aspartate massively floods the extracellular space to kill neurons, while continued turning of a truncated form of the Krebs cycle keeps most brain cells alive. Endogenous substrates are utilized, chiefly phospholipids and fatty acids. The duration of tolerable insult is much longer for hypoglycemia than ischemia, which also releases more glutamate than aspartate into the brain interstitium. The neuropathology in humans is not always distinguishable, but if there is dentate gyrus destruction, a very late event in global ischemia, the distinction of hypoglycemic from ischemic damage can be made. Hypoglycemic brain damage occurs in hospitals, attempted suicide and homicide.

Keywords Hypoglycemia · Clinical · Experimental · Glucose · EEG · Cortex · Hippocampus · Electron Microscopy

10.1 General Comparison of Hypoglycemic and Ischemic Brain Damage

Hypoglycemia was long considered to produce brain damage similar to ischemia, with respect to both mechanism and result. The former was based on the perceived similarity of oxygen and glucose deprivation on the brain. The latter was based on the similar neuropathology of global ischemia and hypoglycemic brain damage.

Within the hippocampus, hypoglycemic and ischemic nerve cell death, both cause neuronal death in the CA1 pyramidal cells of the hippocampus. The neocortex is also affected by both insults, and thus the structural effects of ischemia and

---

R. N. Auer (✉)
Department of Pathology and Laboratory Medicine, University of Saskatchewan, Saskatoon, SK, Canada

D. G. Fujikawa (ed.), *Acute Neuronal Injury*,
https://doi.org/10.1007/978-3-319-77495-4_10

hypoglycemia were considered identical (Brierley et al. 1971) until the 1980s (Auer and Siesjö 1988).

It is now clear that while the end result of brain damage due to global ischemia and hypoglycemia cannot always be distinguished in humans (Ng et al. 1989; Auer et al. 1989), under controlled experimental conditions in animals it can be easily shown that the distribution of hypoglycemic brain damage (Auer et al. 1984a) is distinct from both ischemic (Smith et al. 1984) and epileptic (Nevander et al. 1985) brain damage. Importantly, despite a common final death pathway and cytologic appearance of neuronal necrosis, the pathophysiology leading to neuronal killing in hypoglycemia is different in many ways from ischemic neuronal death.

10.2 A Brief History of Hypoglycemic Brain Damage

By hypoglycemic brain damage, we mean not the usual hypoglycemia that occurs say, in early metabolic syndrome or when meals are skipped. Hypoglycemic brain damage refers to the permanent brain damage that comes about when nerve cells die. It is an unusual brain insult, difficult to produce because of the requirement for massive amounts of insulin to overcome counter-regulatory hormones cortisol and adrenaline, hormones that increase blood glucose levels.

Prior to the discovery of insulin in 1921, the only endogenous source producing enough insulin to cause hypoglycemic brain damage was a pancreatic tumor, insulinoma, secreting large amounts of insulin into the bloodstream. With the discovery of insulin and its exogenous administration, overdoses came to be seen in the 1920s and 1930s due to over treatment of diabetes and medication error.

But in the late 1930s, the Viennese psychiatrist Dr. Manfred Sakel (1937) published on the use of hypoglycemia to treat schizophrenia. This was the only treatment available for an otherwise untreatable disease that had great life impact. The desired duration of insulin coma in iatrogenic hypoglycemia was 30 min (Sakel 1937). It was well known that a patient in coma for 1 h would never awaken from the coma.

Thus, even in this early historical period, it was already known that something happened between 30 and 60 min of coma which was devastating to the brain. We now know that this time period of 30 min EEG silence is on the cusp of accelerating hypoglycemic brain damage in the form of neuronal necrosis (Auer et al. 1984b, 1985a, b; Kalimo et al. 1985). The sparse neuronal death after only 30 min explains clinical patient survival with relatively intact neurological function after periods of hypoglycemic coma of 30 min or less.

Although easily considered barbaric by today's standards, such treatment for severe psychiatric illness was the only treatment available at the time. And hypoglycemic coma of 30 min duration did indeed have a salubrious effect on the psychiatric condition (Dr. John Menkes, personal communication), perhaps as a result of cortical depolarization and coma itself, rather than the sparse neuronal necrosis that occurs after 30 min of hypoglycemic coma. Nevertheless, criticism of therapeutic hypoglycemia mounted (Mayer-Gross 1951) and the development of phenothiazines in the 1950s sealed the fate of insulin shock therapy for schizophrenia.

10.3 Settings of Hypoglycemic Brain Damage

The three common settings remaining for hypoglycemic brain damage are over treatment of diabetes, suicide via insulin and homicide via insulin administration.

Insulin or sulfonylureas are occasionally given in error to the wrong patient (Auer et al. 1989; Kalimo and Olsson 1980) And in the context of treatment for type 1 or type 2 diabetes, overtreatment with insulin occurs (Anonymous 1993). Many hospitalized patients, especially those in renal failure, have symptomatic hypoglycemia that correlates with other causes of mortality without the hypoglycemia *per se* leading to brain damage or death (Fischer et al. 1986). A new cause of hypoglycemia in the modern era - gastric bypass surgery (Patti et al. 2005), leads to symptomatic hypoglycemia but not to brain damage. Likewise, starvation does not lead to cellular glucose deprivation, due to adaptive brain mechanisms that occur beyond the initial steps of blood-brain glucose transfer and glucose phosphorylation (Crane et al. 1985).

Besides exogenous sources of insulin, endogenous tumors, usually of the islets of Langerhans of the pancreas, account for most of the remaining cases of hypoglycemic brain damage (Auer 1986). Non-adenomatous tumors secreting insulin or congener molecules (Zhou et al. 2005; Lawson et al. 2009), and insulin antibody syndrome (Arzamendi et al. 2014) are rare causes.

With progressive hypoglycemia, the signs and symptoms go through well-defined stages as the blood sugar is progressively lowered. In any setting of hypoglycemia, these progress in an orderly fashion, and the onset of the biochemical changes leading to brain damage, outlined below, do not begin until the EEG goes flat (isoelectric EEG). The clinical counterpart to this flattening of the EEG is coma. We here outline these stages of hypoglycemia in a Table 10.1.

Table 10.1 The stages of hypoglycemia

Blood glucose (mM)[a]	5–8	2–3	≤1
Energy state (as % of normal)[b]	100	100	30
EEG	α and ß—Normal	δ—Slow waves	No waves
	8–20 Hz	1–4 Hz	Isoelectric
Clinical state	Normal	Stupor	Coma
$[\text{Aspartate}]_{\text{ECF}}$[c]	Normal	Normal	1600% of normal
$[\text{Glutamate}]_{\text{ECF}}$[c]	Normal	Normal	400% of normal

[a]To convert mM (millimolar or millimoles per liter) to mg/100 ml, multiply by 18, as the molecular weight of glucose is 180

[b]As defined by Atkinson (1968), energy charge = [ATP] + ½ [ADP]/Σ[ATP + ADP + AMP], normally 0.93

[c]Extracellular fluid

10.4 Delivery of Glucose and Cerebral Blood Flow

The pathophysiology of hypoglycemic brain damage has been detailed in several reviews (Auer and Siesjö 1993; Auer 2004; Suh et al. 2007). Briefly, low blood glucose levels are initially compensated for by an increase in cerebral blood flow (Kety et al. 1948; Abdul-Rahman et al. 1980; Gjedde et al. 1980), which can be conceptualized as the body's attempt to sustain delivery of glucose to the tissue, by increasing the flow of hypoglycemic blood.

Transport of glucose across the blood-brain barrier then occurs by a facilitated carrier, glucose transporter-1 (GLUT-1) (Mueckler et al. 1985; Lloyd et al. 2017), located on the cerebral endothelial cell (Cornford et al. 1993). Eventually, hypoglycemia leads to low interstitial brain levels of glucose, which lag behind blood glucose by roughly ~30 min in humans (Abi-Saab et al. 2002). The 1–2 mM concentration of glucose around neurons is lower than the 5–6 mM serum glucose levels, necessitating a neuronal glucose transporter termed GLUT-3 (Simpson et al. 2008); having roughly 5× the glucose transporting capacity of GLUT-1. The high rate of glucose transport directly into neurons by GLUT-3, perforce, challenges the astrocyte-neuron lactate shuttle hypothesis (Pellerin et al. 1998).

Low brain interstitial fluid levels of glucose then lead to decreased glucose phosphorylation and glycolytic flux, a low cerebral metabolic rate of glucose use (Abdul-Rahman and Siesjö 1980). Glucose is needed not only for synaptic activity, but also for development, evidenced by the microcephaly (Wang et al. 2005) seen in GLUT-1 deficiency, where blood-brain glucose transfer is impaired chronically during brain maturation.

10.5 Biochemistry of Hypoglycemic Brain Damage

Alkalosis, not acidosis, is a feature of hypoglycemia, again contrasting with ischemia. Glycolysis is decreased during hypoglycemia, giving rise to decreased lactate and pyruvate levels within the tissue. Release of any metabolic acid within tissue will cause the pH of the tissue to tend down toward the pK_a of that particular acid. The pK_a of lactic acid is 3.86, making it one of the most powerful metabolic acids reducing tissue pH. Thus, the kind of acidosis seen in *eg* ischemia, Wernicke's encephalopathy (Hakim 1984) and Leigh's disease is absent in hypoglycemic brain damage. This absence of both lactate and ischemia explains the absence of pannecrosis in hypoglycemic brain damage. The absence of a strong metabolic acid, together with the production of the strong base ammonia (Lewis et al. 1974; Agardh et al. 1978), account for the alkalosis of hypoglycemia (Pelligrino et al. 1981).

Energy failure occurs with the onset of EEG silence (Agardh et al. 1978; Feise et al. 1976), the latter being also the harbinger of neuronal necrosis (Auer et al. 1984b). The brain uses endogenous substrates almost immediately, which can be

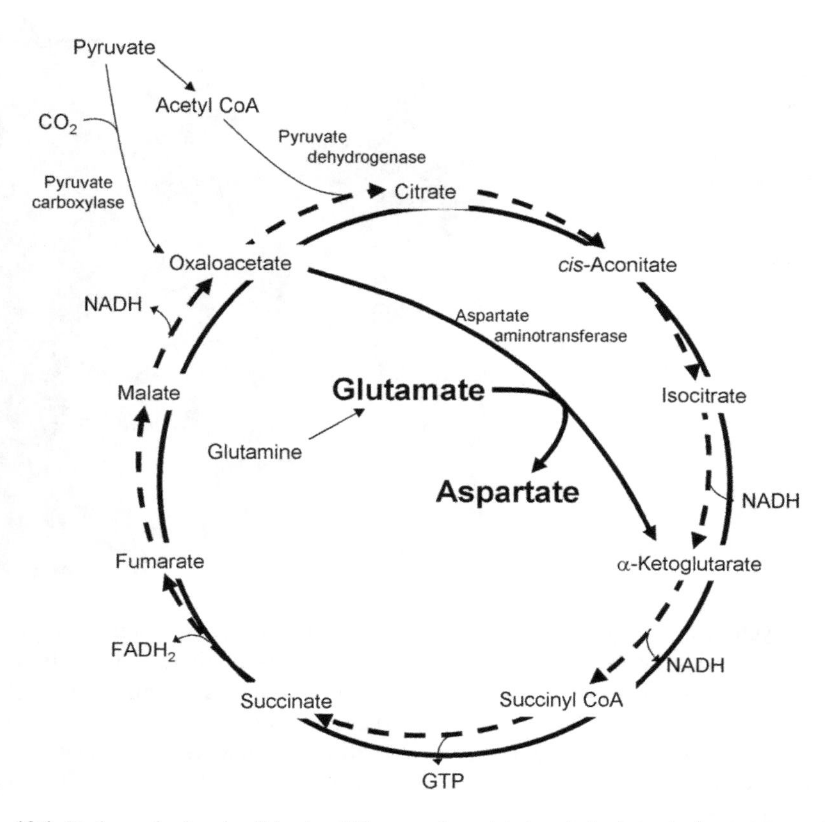

Fig. 10.1 Krebs cycle showing "short cut" from oxaloacetate to α-ketoglutarate due to transamination of glutamate to aspartate during hypoglycemia. The Krebs cycle continues to turn during hypoglycemia with EEG silence, explaining why the majority of brain cells survive

conceptualized as "throwing the furniture into the fire", literally brain fatty acids and phospholipids (Agardh et al. 1981).

Normally, pyruvate is decarboxylated to produce acetate, which enters the Krebs cycle. The release of one molecule of CO_2 from the 3-carbon pyruvate leaves the two carbon acetate, as acetyl coenzyme A to supply carbon atoms for the Krebs cycle which will be released eventually as CO_2. Pyruvate can also be carboxylated to produce the 4-carbon oxaloacetate, an anaplerotic reaction which builds up the molecules of the Krebs cycle itself (Fig. 10.1).

The decrease in glycolysis has consequences for amino acid metabolism. This occurs when reduced glycolytic flux leads to a shortage of acetyl coenzyme A entering the Krebs cycle to condense with oxaloacetate, leading to an increase in the latter. The increase in oxaloacetate drives the aspartate-glutamate transaminase reaction toward aspartate and away from glutamate, due to the law of mass action in chemistry, Le Chetalier's principle, driving the aspartate-glutamate transaminase reaction towards aspartate. When this transamination reaction is written across the

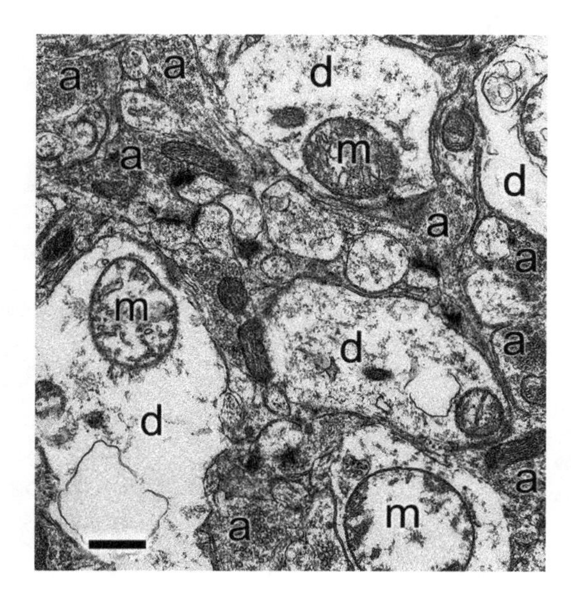

Fig. 10.2 Axon-sparing dendritic swelling characteristic of excitotoxins acting in CNS tissue. Axons (a) are not swollen and contain synaptic vesicles. Dendrites (d) however, are swollen and contain variably swollen mitochondria (m). Bar = 1 μm

Krebs cycle (Fig. 10.1), it is clear that a truncated Krebs cycle obtains in hypoglycemia, bypassing the 6-carbon intermediates citrate, *cis*-aconitate and isocitrate.

As a result, brain tissue aspartic acid levels rise in hypoglycemia (Lewis et al. 1974; Cravioto et al. 1951; Tews et al. 1965) as part of this new metabolic homeostasis. Due to the energy failure (Auer and Siesjö 1993), however, this metabolically derived aspartate leaks across the cell membranes to flood the extracellular space (Sandberg et al. 1986).

The extracellular aspartate then binds to dendritic receptors to produce characteristic axon-sparing lesions (Fig. 10.2). Essentially, the brain maintains a reduced but adequate energy state (Feise et al. 1976) by running a truncated version of the Krebs cycle in a new metabolic homeostasis (Sutherland et al. 2008).

Since the new, truncated, Krebs cycle still turns, fuel can be fed into the Krebs cycle as pyruvate (Suh et al. 2005) or lactate (Nemoto et al. 1974; Won et al. 2012), interconvertible metabolically by the enzyme lactate dehydrogenase (LDH), or as acetate (Urion et al. 1979). However, lactate can only replace one fourth of substrate use during hypoglycemia (Nemoto et al. 1974), and once glucose is given, changes in energy charge (Auer and Siesjö 1993), amino acids and ammonia are rapidly reversed (Agardh et al. 1978).

Thus, once the electroencephalogram goes flat (Auer et al. 1984b), the series of biochemical changes above allows a new homeostasis to occur. Energy failure is incomplete in hypoglycemia, at about 30% of normal (Auer and Siesjö 1993) due to the use of alternative substrates to generate ATP. These include fatty acids and proteins, once carbohydrate stores are exhausted (Agardh et al. 1981). A trickle of glucose supplements these neurochemical measures to maintain the brain's energy state, since glucose in the blood is never absolutely zero. Blood–brain glucose transport via GLUT-1 is maximized at these low blood glucose levels.

10.6 Neuropathology of Hypoglycemic Brain Damage

When over-treatment of diabetes patients with insulin began giving rise to coma and death, autopsies were performed in an attempt to elucidate whether brain damage had occurred (Auer 1986). The question at the time, in the 1920s and 1930s, was whether hypoglycemia had damaged the brain, or whether insulin itself caused brain damage. The therapeutic use of insulin in psychiatry indirectly shed some light on this problem. It was known that 30 min of insulin coma was reversible, whereas 1 h of coma was irreversible and the patient could never again be reawakened. This early clinical observation on the duration of coma has an electrical counterpart in the duration of a flat electroencephalogram, or isoelectric EEG ("flat EEG").

Without a flat EEG, no neuropathology is seen in hypoglycemia, *ie* no cell death, irrespective of the blood glucose levels. Experimentally, it can be shown that the blood glucose levels at which the EEG becomes flat vary by more than one order of magnitude, from 0.12 mM to 1.36 mM (2.16 mg/dL to 24.5 mg/dL) (Auer et al. 1984b).

Once the EEG goes flat, essentially a timer starts. That is to say, the number of dead neurons increases from a zero time point, marked by the onset of a flat EEG. Over time, the number of dead neurons increases in a monotone fashion as the duration of a flat EEG increases. In both animal experiments and in the history of the human therapeutic use of insulin coma in psychiatry, 30 min is a duration of insult that is easily recoverable. By 1 h, however, the number of dead neurons has increased to the point that coma becomes irreversible, explaining the clinical disasters in the historical use of therapeutic insulin-induced hypoglycemia (Auer 1986).

We note that unlike ischemia, the selective neuronal necrosis of hypoglycemia does not augment into infarction. Instead, the number of dead neurons increases, without a transformation of selective neuronal death into an infarct. This is because neither is blood flow occluded nor lactate produced (see above). Pan-necrosis results from either vascular occlusion or lactic acid production at a rate above that which can be cleared from the tissue.

Because hypoglycemia leads to neither vascular occlusion nor lactic acidosis, infarction (pan-necrosis, with death of glia as well as neurons) is not seen in hypoglycemic brain damage. Instead, hypoglycemia leads to selective neuronal necrosis, meaning death of neurons but not glia and other supporting cells of the nervous system. Thus, depending on the degree of selective neuronal necrosis, the brain in hypoglycemic brain damage can be either normal or can show variable degrees of gross atrophy (Fig. 10.3). Cysts are not seen.

The location of the gross atrophy is dictated by the brain areas that are selectively vulnerable in hypoglycemic brain damage. These constitute the cerebral cortex, the striatum, the hippocampus and the thalamus. The metabolic release of aspartate (Sandberg et al. 1986) outlined above, underlies the selective neuronal death in hypoglycemia. Glucose transport from blood to brain is better in the brainstem and cerebellum (LaManna and Harik 1985) leading to a relative sparing of the brainstem

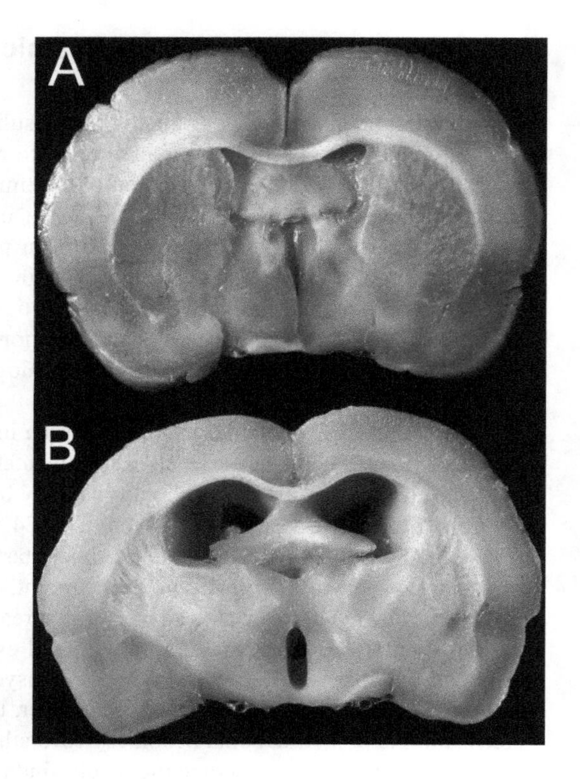

Fig. 10.3 Brain of a normal rat (**a**) compared to a rat with 60 min of flat EEG 1 week previously (**b**), showing hypoglycemia-induced atrophy due to widespread death of neurons and axons. Notably, no cysts are present despite the severe, 1 h insult of a flat EEG

and cerebellum from a metabolic insult during hypoglycemia. Protein synthesis, a high level function utilizing considerable cellular energy, continues unabated in the brainstem even during hypoglycemic coma (Kiessling et al. 1986). The brainstem and cerebellum are thus normal in a pure (without ischemia) primary insult of hypoglycemic brain damage to the nervous system.

Within the hippocampus, the CA1 pyramidal cells are vulnerable (Auer et al. 1985b), in common with ischemia. However, the signature lesion of dentate gyrus necrosis (Fig. 10.4) in the absence of severe necrosis of the CA1 and especially CA3 pyramidal cells, indicates hypoglycemic brain damage as the etiology of the lesion in the hippocampus of the human (Auer et al. 1989). In rats, the lesion of the dentate is seen early, already after 10–20 min of coma, a comparably mild insult. The dentate gyrus lesions of rats and humans are compared (Fig. 10.4), both showing dentate necrosis (Fig. 10.5) as a feature of hypoglycemia. The CA1 necrosis is a feature in common with cerebral ischemia.

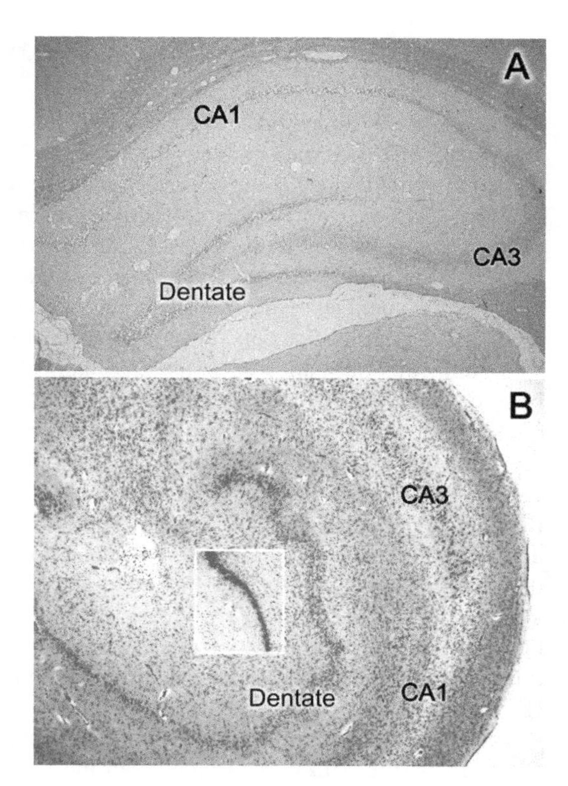

Fig. 10.4 Hippocampus in hypoglycemia. The rat (**a**) shows conspicuous necrosis of the dentate gyrus, as does the human (**b**) hippocampus. CA1 neuronal loss is common to both ischemia and hypoglycemia. The inset shows a stretch of normal dentate gyrus, a dark blue band of cells contrasting with the adjacent, depleted granule cells of the dentate in the surrounding hippocampus damaged by hypoglycemia

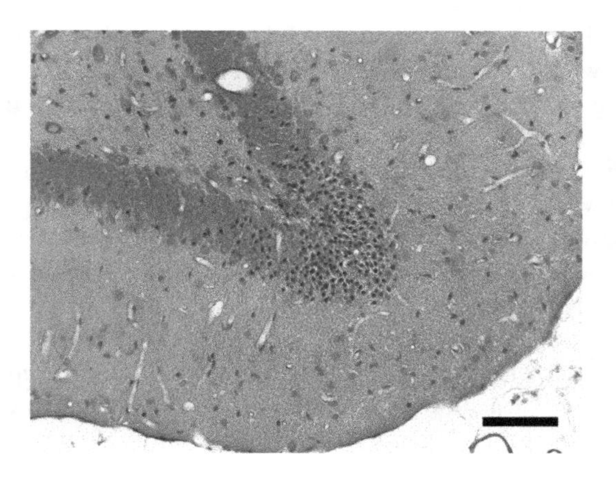

Fig. 10.5 Light microscopic detail of neuronal death in hypoglycemia. The cells of the crest dentate gyrus in the geographic area in the center of the photo are pyknotic, or shrunken. Both nuclei and cytoplasm are small compared to the normal cells continuous to the dead neurons in the band of neurons forming a "V". Neuronal cytoplasm is acidophilic, a sign of cell death. Bar = 100 μm

10.7 Electron Microscopy of Hypoglycemic Brain Damage

The discovery by John Olney of the phenomenon of excitotoxicity in the late 1960s and early 1970s was major (Olney 1969a, b, 1971). Apparently unrelated to hypoglycemic brain damage at the time, the phenomenon of amino acid neurotoxicity to neurons, when the excitatory amino acid is present in the extracellular space in high concentrations, proved to be an essential key to our understanding of how nerve cells die after hypoglycemia. This connection was not initially obvious since hypoglycemia is a deficiency disorder where not enough of a substrate is present in the brain. How a molecular deficiency of glucose could lead to an excess of another metabolite was only apparent after examining the biochemistry of hypoglycemic brain damage outlined above.

Aspartate is released in massive quantities into the extracellular space in hypoglycemic brain damage (Sandberg et al. 1986). But the first clue that hypoglycemic neuronal death may have a connection to excitotoxicity emerged in 1985, when hypoglycemic neuronal death was blocked by an NMDA antagonist of excitatory amino acid receptors (Wieloch 1985) and excitotoxic lesions were seen in the dentate gyrus (Auer et al. 1985c). Historically remarkable is the prescient observation already in 1938 by Arthur Weil, who observed dentate gyrus in a geographic distribution in rabbits, presciently postulating at that time that a molecule in the extracellular space, perhaps a toxin, could kill nerve cells in hypoglycemic brain damage (Weil et al. 1938).

Electron microscopy of hypoglycemic neuronal death shows excitotoxic neuronal necrosis (Fig. 10.2). The lesion under the electron microscope is quite specific: dendritic swelling, sparing intermediate axons, was known to occur in glutamate neurotoxicity. The appearance is due to the selective dendritic location of receptors, sparing axons. Thus, ion fluxes and water fluxes would lead to swelling of dendrites but not axons when examined under the electron microscope. The appearance of axon-sparing dendritic swelling is pathognomonic of the tissue action of excitotoxins.

This dendritic neuronal death can be thought of as a brush fire in the dendritic tree, beginning where synaptic receptors are located and spreading towards the cell body or perikaryon. This is not characteristic of apoptotic cell death. Once the process of membrane leakage and breakage spreads to the perikaryon (cell body) of the neuron, cell death ensues.

The perikaryon of the neuron in hypoglycemic neuronal death shows features of necrosis, with a characteristic coarse chromatin having a tigroid appearance, mitochondrial flocculent densities (Trump et al. 1997) and cell membrane breaks (Fig. 10.6). Following these changes, cell death is fairly rapid, occurring in a few hours in both cortex (Auer et al. 1985a) and hippocampus (Auer et al. 1985b), again contrasting with the delayed neuronal death of ischemia (Kirino 1982).

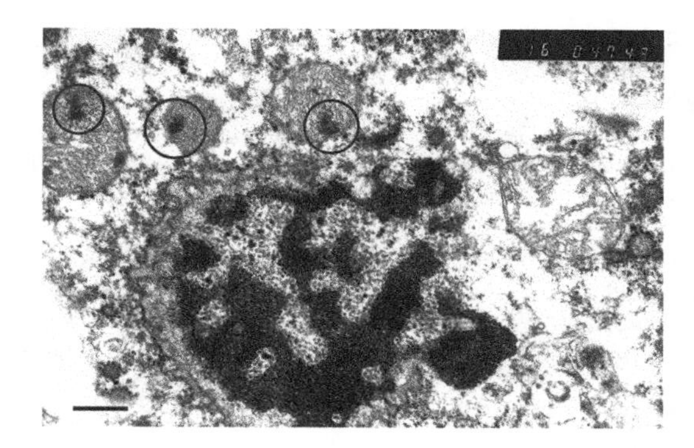

Fig. 10.6 Electron microscopy of neuronal cell death in hypoglycemia. Mitochondrial flocculent densities are seen (circled), representing denatured cytochrome proteins. The mitochondrion on the right is still swollen, but without flocculent densities. The nuclear chromatin has a coarse, tigroid appearance and the nuclear membrane is crenulated. Bar = 1 μm

References

Abdul-Rahman A, Siesjö BK (1980) Local cerebral glucose consumption during insulin-induced hypoglycemia, and in the recovery period following glucose administration. Acta Physiol Scand 110:149–159

Abdul-Rahman A, Agardh C-D, Siesjö BK (1980) Local cerebral blood flow in the rat during severe hypoglycemia and in the recovery period following glucose injection. Acta Physiol Scand 109:307–314

Abi-Saab WM, Maggs DG, Jones T, Jacob R, Srihari V, Thompson J, Kerr D, Leone P, Krystal JH, Spencer DD, During MJ, Sherwin RS (2002) Striking differences in glucose and lactate levels between brain extracellular fluid and plasma in conscious human subjects: effects of hyperglycemia and hypoglycemia. J Cereb Blood Flow Metab 22:271–279

Agardh C-D, Folbergrová J, Siesjö BK (1978) Cerebral metabolic changes in profound insulin-induced hypoglycemia, and in the recovery period following glucose administration. J Neurochem 31:1135–1142

Agardh C-D, Chapman AG, Nilsson B, Siesjö BK (1981) Endogenous substrates utilized by rat brain in severe insulin-induced hypoglycemia. J Neurochem 36:490–500

Anonymous (1993) The effect of intensive treatment of diabetes on the development and progression of long-term complications in insulin-dependent diabetes mellitus. The Diabetes Control and Complications Trial Research Group. N Engl J Med 329:977–986

Arzamendi AE, Rajamani U, Jialal I (2014) Pseudoinsulinoma in a white man with autoimmune hypoglycemia due to anti-insulin antibodies: value of the free C-Peptide assay. Am J Clin Pathol 142:689–693

Atkinson DE (1968) The energy charge of the adenylate pool as a regulatory parameter. Interaction with biofeedback modifiers. Biochemistry 7:4030–4034

Auer RN (1986) Progress review: hypoglycemic brain damage. Stroke 17:699–708

Auer RN (2004) Hypoglycemic brain damage. Forensic Sci Int 146:105–110

Auer RN, Siesjö BK (1988) Biological differences between ischemia, hypoglycemia, and epilepsy. Ann Neurol 24:699–707

Auer RN, Siesjö BK (1993) Hypoglycaemia: brain neurochemistry and neuropathology. Baillières Clin Endocrinol Metab 7:611–625

Auer RN, Wieloch T, Olsson Y, Siesjö BK (1984a) The distribution of hypoglycemic brain damage. Acta Neuropathol 64:177–191

Auer RN, Olsson Y, Siesjö BK (1984b) Hypoglycemic brain injury in the rat. Correlation of density of brain damage with the EEG isoelectric time: a quantitative study. Diabetes 33:1090–1098

Auer RN, Kalimo H, Olsson Y, Siesjö BK (1985a) The temporal evolution of hypoglycemic brain damage. I. Light- and electron-microscopic findings in the rat cerebral cortex. Acta Neuropathol 67:13–24

Auer RN, Kalimo H, Olsson Y, Siesjö BK (1985b) The temporal evolution of hypoglycemic brain damage. II. Light- and electron-microscopic findings in the hippocampal gyrus and subiculum of the rat. Acta Neuropathol 67:25–36

Auer R, Kalimo H, Olsson Y, Wieloch T (1985c) The dentate gyrus in hypoglycemia: pathology implicating excitotoxin-mediated neuronal necrosis. Acta Neuropathol 67:279–288

Auer RN, Hugh J, Cosgrove E, Curry B (1989) Neuropathologic findings in three cases of profound hypoglycemia. Clin Neuropathol 8:63–68

Brierley JB, Brown AW, Meldrum BS (1971) The nature and time course of the neuronal alterations resulting from oligaemia and hypoglycemia in the brain of Macaca mulatta. Brain Res 25:483–499

Cornford EM, Hyman S, Pardridge WM (1993) An electron microscopic immunogold analysis of developmental up-regulation of the blood-brain barrier GLUT1 glucose transporter. J Cereb Blood Flow Metab 13:841–854

Crane PD, Pardridge WM, Braun LD, Oldendorf WH (1985) Two-day starvation does not alter the kinetics of blood–brain barrier transport and phosphorylation of glucose in rat brain. J Cereb Blood Flow Metab 5:40–46

Cravioto RO, Massieu H, Izquierdo JJ (1951) Free amino acids in rat brain during insulin shock. Proc Soc Exp Biol Med 78:856–858

Feise G, Kogure K, Busto R, Scheinberg P, Reinmuth O (1976) Effect of insulin hypoglycemia upon cerebral energy metabolism and EEG activity in the rat. Brain Res 126:263–280

Fischer KF, Lees JA, Newman JH (1986) Hypoglycemia in hospitalized patients. Causes and outcomes. N Engl J Med 315:1245–1250

Gjedde A, Hansen AJ, Siemkowicz E (1980) Rapid simultaneous determination of regional blood flow and blood brain glucose transfer in brain of rat. Acta Physiol Scand 108:321–330

Hakim AM (1984) The induction and reversibility of cerebral acidosis in thiamine deficiency. Ann Neurol 16:673–679

Kalimo H, Olsson Y (1980) Effect of severe hypoglycemia on the human brain. Acta Neurol Scand 62:345–356

Kalimo H, Auer RN, Siesjö BK (1985) The temporal evolution of hypoglycemic brain damage. III. Light and electron microscopic findings in the rat caudoputamen. Acta Neuropathol 67:37–50

Kety SS, Woodford RB, Harmel MH, Freyhan FA, Appel KE, Schmidt CF (1948) Cerebral blood flow and metabolism in schizophrenia. The effect of barbiturate semi-narcosis, insulin coma and electroshock. Am J Psychiat 104:765–770

Kiessling M, Auer RN, Kleihues P, Siesjö BK (1986) Cerebral protein synthesis during long-term recovery from severe hypoglycemia. J Cereb Blood Flow Metab 6:42–51

Kirino T (1982) Delayed neuronal death in the gerbil hippocampus following ischemia. Brain Res 239:57–69

LaManna JC, Harik SI (1985) Regional comparisons of brain glucose influx. Brain Res 326:299–305

Lawson EA, Zhang X, Crocker JT, Wang WL, Klibanski A (2009) Hypoglycemia from IGF2 overexpression associated with activation of fetal promoters and loss of imprinting in a metastatic hemangiopericytoma. J Clin Endocrinol Metab 94:2226–2231

Lewis LD, Ljunggren B, Norberg K, Siesjö BK (1974) Changes in carbohydrate substrates, amino acids and ammonia in the brain during insulin-induced hypoglycemia. J Neurochem 23:659–671

Lloyd KP, Ojelabi OA, De Zutter JK, Carruthers A (2017) Reconciling contradictory findings: Glucose transporter 1 (GLUT1) functions as an oligomer of allosteric, alternating access transporters. J Biol Chem 292:21035–21046
Mayer-Gross W (1951) Insulin coma therapy of schizophrenia: some critical remarks on Dr. Sakel's report. J Ment Sci 97:132–135
Mueckler M, Caruso C, Baldwin SA, Panico M, Blench I, Morris HR, Allard WJ, Lienhard GE, Lodish HF (1985) Sequence and structure of a human glucose transporter. Science 229:941–945
Nemoto EM, Hoff JT, Severinghaus JW (1974) Lactate uptake and metabolism by brain during hyperlactatemia and hypoglycemia. Stroke 5:48–53
Nevander G, Ingvar M, Auer R, Siesjö BK (1985) Status epilepticus in well-oxygenated rats causes neuronal necrosis. Ann Neurol 18:281–290
Ng T, Graham DI, Adams JH, Ford I (1989) Changes in the hippocampus and the cerebellum resulting from hypoxic insults: frequency and distribution. Acta Neuropathol 78:438–443
Olney JW (1969a) Glutamate-induced retinal degeneration in neonatal mice. Electron microscopy of the acutely evolving lesions. J Neuropathol Exp Neurol 28:455–474
Olney JW (1969b) Brain lesions, obesity, and other disturbances in mice treated with monosodium glutamate. Science 164:719–721
Olney JW (1971) Glutamate-induced neuronal necrosis in the infant mouse hypothalamus. J Neuropathol Exp Neurol 30:75–90
Patti ME, McMahon G, Mun EC, Bitton A, Holst JJ, Goldsmith J, Hanto DW, Callery M, Arky R, Nose V, Bonner-Weir S, Goldfine AB (2005) Severe hypoglycaemia post-gastric bypass requiring partial pancreatectomy: evidence for inappropriate insulin secretion and pancreatic islet hyperplasia. Diabetologia 48:2236–2240
Pellerin L, Pellegri G, Bittar PG, Charnay Y, Bouras C, Martin JL, Stella N, Magistretti PJ (1998) Evidence supporting the existence of an activity-dependent astrocyte-neuron lactate shuttle. Dev Neurosci 20:291–299
Pelligrino D, Almquist L-O, Siesjö BK (1981) Effects of insulin-induced hypoglycemia on intracellular pH and impedance in the cerebral cortex of the rat. Brain Res 221:129–147
Sakel M (1937) The methodical use of hypoglycemia in the treatment of psychoses. Am J Psychiatr 94:111–129
Sandberg M, Butcher SP, Hagberg H (1986) Extracellular overflow of neuroactive amino acids during severe insulin-induced hypoglycemia: in vivo dialysis of the rat hippocampus. J Neurochem 47:178–184
Simpson IA, Dwyer D, Malide D, Moley KH, Travis A, Vannucci SJ (2008) The facilitative glucose transporter GLUT3: 20 years of distinction. Am J Physiol Endocrinol Metab 295:E242–E253
Smith ML, Auer RN, Siesjö BK (1984) The density and distribution of ischemic brain injury in the rat following 2-10 min of forebrain ischemia. Acta Neuropathol 64:319–332
Suh SW, Aoyama K, Matsumori Y, Liu J, Swanson RA (2005) Pyruvate administered after severe hypoglycemia reduces neuronal death and cognitive impairment. Diabetes 54:1452–1458
Suh SW, Bergher JP, Anderson CM, Treadway JL, Fosgerau K, Swanson RA (2007) Astrocyte glycogen sustains neuronal activity during hypoglycemia: studies with the glycogen phosphorylase inhibitor CP-316,819 ([R-R*,S*]-5-chloro-N-[2-hydroxy-3-(methoxymethylamino)-3-oxo-1-(phenylmethyl)propyl]-1H-indole-2-carboxamide). J Pharmacol Exp Ther 321:45–50
Sutherland GR, Tyson RL, Auer RN (2008) Truncation of the Krebs cycle during hypoglycemic coma. Med Chem 4:379–385
Tews JK, Carter SH, Stone WE (1965) Chemical changes in the brain during insulin hypoglycemia and recovery. J Neurochem 12:679–683
Trump BF, Berezesky IK, Chang SH, Phelps PC (1997) The pathways of cell death: oncosis, apoptosis, and necrosis. Toxicol Pathol 25:82–88
Urion D, Vreman HJ, Weiner MW (1979) Effect of acetate on hypoglycemic seizures in mice. Diabetes 28:1022–1026
Wang D, Pascual JM, Yang H, Engelstad K, Jhung S, Sun RP, De Vivo DC (2005) Glut-1 deficiency syndrome: clinical, genetic, and therapeutic aspects. Ann Neurol 57:111–118

Weil A, Liebert E, Heilbrunn G (1938) Histopathologic changes in the brain in experimental hyperinsulinism. Arch Neurol Psychiat (Chic) 39:467–481

Wieloch T (1985) Hypoglycemia-induced neuronal damage prevented by an N-methyl-D-aspartate antagonist. Science 230:681–683

Won SJ, Jang BG, Yoo BH, Sohn M, Lee MW, Choi BY, Kim JH, Song HK, Suh SW (2012) Prevention of acute/severe hypoglycemia-induced neuron death by lactate administration. J Cereb Blood Flow Metab 32:1086–1096

Zhou P, Kudo M, Chung H, Minami Y, Ogawa C, Sakaguchi Y, Kitano M, Kawasaki T, Maekawa K (2005) Multiple metastases from a meningeal hemangiopericytoma associated with severe hypoglycemia. J Med Ultrason (2001) 32:187–190

Part VI
Seizure-Induced Neuronal Death

Chapter 11
Activation of Caspase-Independent Programmed Pathways in Seizure-Induced Neuronal Necrosis

Denson G. Fujikawa

Abstract Prolonged epileptic seizures, or status epilepticus (SE), produce morphologically necrotic neurons in many brain regions. In contrast to prior notions of cellular necrosis being a passive process of cell swelling and lysis, SE-induced necrotic neurons show internucleosomal DNA cleavage (DNA laddering), a programmed process requiring endonuclease activation. The underlying mechanisms are triggered by excessive activation of NMDA receptors by glutamate, which allows calcium influx through their receptor-operated cation channels (excitotoxicity). Calcium-dependent enzymes are activated, such as calpain I and neuronal nitric oxide synthase (nNOS), the latter of which, through production of reactive oxygen species (ROS), activates poly(ADP-ribose) polymerase-1 (PARP-1). Calpain I and PARP-1 activation in turn cause translocation of death-promoting mitochondrial proteins and lysosomal enzymes that degrade cytoplasmic proteins and nuclear chromatin, creating irreversible cellular damage. Another programmed necrotic cell death pathway, necroptosis, has been described in cell culture following caspase inhibition, and activation of this pathway has been described following cerebral ischemia and traumatic brain injury *in vivo*. However, whether this pathway interacts with the excitotoxic pathway, while likely, and the specific mechanisms by which this occurs, are at present unknown. Based upon our knowledge of excitotoxic mechanisms, neuroprotective strategies can be devised that could ameliorate neuronal necrosis from refractory SE in humans.

Keywords Calpain I · Excitotoxicity · Glutamate · Necroptosis · Neuronal nitric oxide synthase (nNOS) · Poly(ADP-ribose) polymerase-1 (PARP-1) · Programmed pathways · Status epilepticus

---

D. G. Fujikawa (✉)
Department of Neurology, VA Greater Los Angeles Healthcare System,
North Hills, CA, USA

Department of Neurology and Brain Research Institute, David Geffen School of Medicine,
University of California at Los Angeles, Los Angeles, CA, USA
e-mail: dfujikaw@ucla.edu

D. G. Fujikawa (ed.), *Acute Neuronal Injury*,
https://doi.org/10.1007/978-3-319-77495-4_11

11.1 Excitotoxicity and Seizure-Induced Neuronal Death

Our current concept of how prolonged epileptic seizures (status epilepticus, or SE) kill neurons originated in the 1970s from pioneering work done by John Olney and associates. Olney reported in 1969 that monosodium glutamate killed neurons in the hypothalamic arcuate nucleus, a region that lacks a blood-brain barrier (Olney 1969). Subsequently, Olney and associates found that administration of glutamate (GLU), the most abundant excitatory neurotransmitter in the brain, killed hypothalamic neurons in the infant mouse (Olney 1971), and that systemic administration of a GLU analogue, kainic acid (KA), to the adult rodent resulted in SE and neuronal death (Olney et al. 1974). In 1985, Olney put forth his excitotoxic hypothesis as it applies to SE (Olney 1985). This hypothesis, that excessive presynaptic GLU release results in the death of postsynaptic neurons, has proved to be remarkably robust, and is applicable to a wide variety of acute neuronal insults, as mentioned in the Introduction.

11.1.1 Verification of the Excitotoxic Hypothesis with Respect to Seizures

Microdialysis studies of the hippocampus and piriform cortex, in which microdialysis probes are placed to measure extracellular GLU (GLU_0) concentrations before and during SE, have produced mixed results (Bruhn et al. 1992; Lallement et al. 1991; Lehmann et al. 1985; Millan et al. 1993; Smolders et al. 1997; Tanaka et al. 1996; Wade et al. 1987). At least part of the reason for finding a lack of increased GLU_0 could be that these probes are relatively large, typically 2 mm in diameter, so that only overall GLU_0 is measured, rather than synaptic concentrations. GLU_0 activates the *N*-methyl-D-aspartate (NMDA) subtype of GLU receptor. Despite the uncertain results of the microdialysis studies, experiments in which NMDA-receptor antagonists were given systemically have shown that these agents are remarkably neuroprotective, despite ongoing electrographic seizure discharges (Clifford et al. 1990; Fariello et al. 1989; Fujikawa 1995; Fujikawa et al. 1994).

11.2 The Morphology of Cell Death

A developmental classification of cell death morphology (Clarke 1990) has been adopted by investigators studying acute neuronal injury. In this classification, type I cell death is apoptotic, type II is autophagic, and type III is divided into IIIa (nonlysosomal) and IIIb (cytoplasmic, corresponding to our current understanding of necrotic cell death). However, this classification has been misused by investigators who do not pay detailed attention to morphology. Using surrogate markers is insufficient, because all of the non-ultrastructural surrogate markers studied to date are not specific (for example, TUNEL staining, DNA laddering and annexin V externalization). In addition, biochemical markers have been used to identify the type of cell death. But these markers should not be used to define morphology, because morphology all too often is not sufficiently defined, and mistakes are too often made.

Morphological neuronal apoptosis *in vivo* occurs in fetal or postnatal brain (Ikonomidou et al. 1999, 2000; Ishimaru et al. 1999). By postnatal day 21 (P21), neuronal apoptosis becomes undetectable if naturally occurring and very low if induced by an NMDA-receptor antagonist (Ikonomidou et al. 1999). Almost all *in vitro* cell culture experiments have used fetal neurons, and all have used low magnification fluorescence photomicrographs to label neurons "apoptotic" if they showed nuclear pyknosis (shrinkage). Ironically, this feature makes their neurons conform to necrotic rather than apoptotic morphology (Fujikawa 2000, 2002, 2005; Fujikawa et al. 1999, 2000, 2002, 2007).

Contrary to *in vitro* cell culture experiments, in which cells exposed to an overwhelming stimulus show cell swelling and then lyse, presumably thereafter to disappear, it was shown in the 1970s and 1980s (Evans et al. 1984; Auer et al. 1985a, b; Brown 1977; Griffiths et al. 1983, 1984), and was recently rediscovered that cerebral ischemia and SE *in vivo* produce electron-dense, shrunken neurons with pyknotic (shrunken) nuclei, plasma membrane disruption and cytoplasmic vacuoles, many of which are irreversibly damaged mitochondria, all of which are markers of cellular necrosis (Fujikawa et al. 1999, 2000, 2002; Colbourne et al. 1999). One should be cautious in using the results from *in vitro* experiments of fetal or neonatal cells dispersed in cell culture to explain what happens in the adult brain *in vivo*.

11.2.1 The Morphology of Seizure-Induced Neuronal Death

In the 1980s Meldrum and colleagues showed mitochondrial calcium accumulation in electron-dense, shrunken neurons using the oxalate-pyroantimonate technique in ultrastructural brain sections (Evans et al. 1984; Griffiths et al. 1983, 1984). The shrunken, electron-dense neurons showed nuclear pyknosis and cytoplasmic vacuoles, corresponding to what we now know is neuronal necrosis (Fujikawa et al. 1999, 2000, 2002; Colbourne et al. 1999) (Fig. 11.1). In addition, seizure-induced necrotic neurons show internucleosomal DNA cleavage, or DNA laddering, a programmed process requiring endonuclease activation (Fujikawa et al. 1999, 2000, 2002) (Fig. 11.1). The calcium accumulation in swollen mitochondria predated our understanding of how calcium entered neurons destined to die. Nevertheless, the ultrastructural description of seizure-induced neuronal necrosis was correct.

Our current understanding of the morphology of seizure-induced neuronal death in the adult brain is in remarkable agreement with the ultrastructural morphology of cerebral ischemic and hypoglycemic neuronal death, both of which are excitotoxic in origin, and both of which show unmistakable earmarks of neuronal necrosis

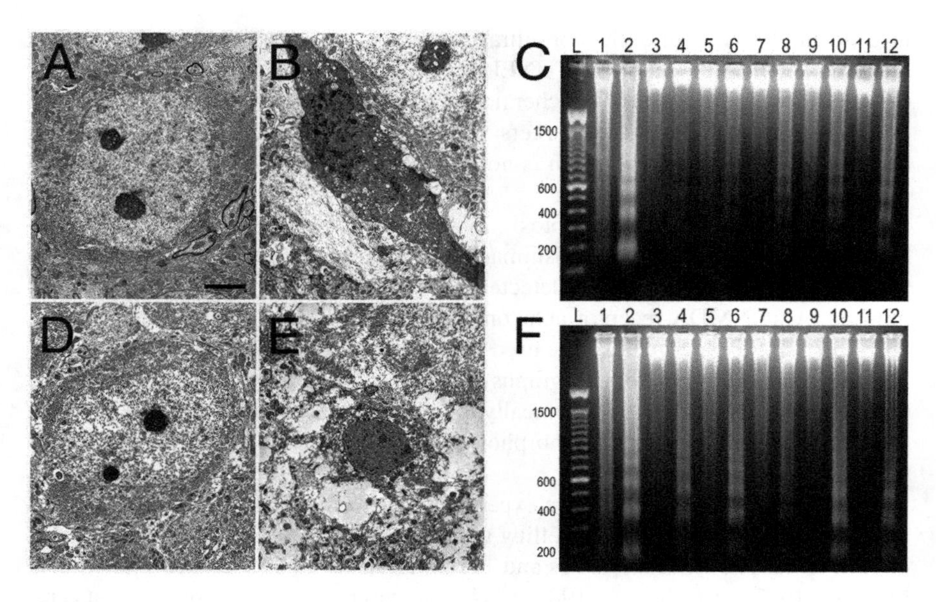

Fig. 11.1 Seizure-induced necrotic neurons show internucleosomal DNA cleavage (DNA laddering). **a** and **d** are electron photomicrographs of ventral hippocampal CA1 neurons in control rats given lithium chloride and normal saline instead of pilocarpine 24 (**a**) and 72 h (**d**) later. **b** and **e** are electron photomicrographs of necrotic neurons 24 (**b**) and 72 h (**d**) after 3-h lithium-pilocarpine-induced status epilepticus (LPCSE). The electron-dense neuronal shrinkage, nuclear pyknosis and cytoplasmic disruption are apparent, as well as the astrocytic end feet swelling surrounding the necrotic neurons. **c** and **f** show DNA agarose gel electrophoresis results 24 and 72 h after normal saline or 3-h LPCSE. Lanes 1 and 2 in **c** and **f** show control and apoptotic thymocytic tissue, controls for DNA laddering. The odd-numbered lanes are control tissue and the even-numbered lanes seizure tissue from dorsal hippocampus (3 and 4), ventral hippocampus (5 and 6), neocortex (7 and 8), amygdala and piriform cortex (9 and 10) and entorhinal cortex (11 and 12). The 180-base-pair DNA laddering can be seen in the seizure lanes at both 24 h and 72 h time points. From Fujikawa (2006), with permission from MIT Press

in vivo (Auer et al. 1985a, b; Colbourne et al. 1999). This suggests that a common mechanism, presynaptic glutamate release from depolarization of neurons, as well as reversal of glutamate uptake by astrocytes, underlies these pathological conditions. After traumatic CNS injury in the adult rodent, ultrastructural photomicrographs (1) have been misinterpreted as showing apoptotic neurons (Colicos and Dash 1996); Fig. 11.2b, c show microglia, Fig. 11.3b shows a microglial cell and Fig. 11.3c shows a necrotic neuron, (2) have suggested both necrotic and apoptotic morphology, predominantly the former, but with DNA laddering (Rink et al. 1995), which we have shown occurs in necrotic neurons, or (3) have suggested neuronal necrosis (Whalen et al. 2008), based on early propidium iodide labeling (Figs. 2, 3, 5, and 6 in ref. Whalen et al. 2008) and ultrastructure (Fig. 9B and 9C in ref. Whalen et al. 2008).

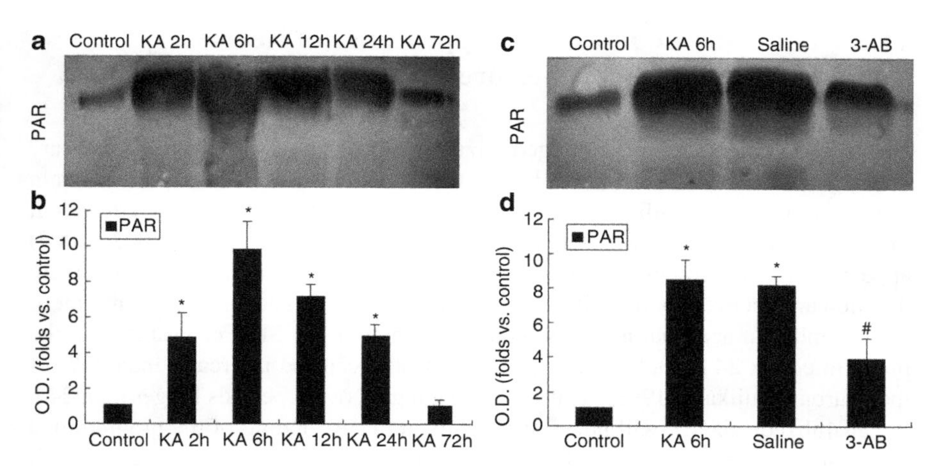

Fig. 11.2 Effects of PARP-1 inhibition on poly(ADP-ribose) (PAR) polymer formation in the rat hippocampus following 60-min SE. (**a**) Shows the time course of PAR formation in rat hippocampus from 2 to 72 h after 60-min SE, and (**c**) shows the effect of 3-aminobenzamide (3-AB), a PARP-1 inhibitor, given i.p. 15 min before intracerebroventricular (i.c.v.) injection of kainic acid, 6 h after SE. Increases in PAR occurred 2, 6, 12 and 24 h after SE, with the maximal increase 6 h after SE (**b**). When compared to SE hippocampus, 6 h after SE, 3-AB inhibited PAR formation by more than 50% (**d**). From Wang et al. (2007), with permission from Lippincott Williams & Wilkins

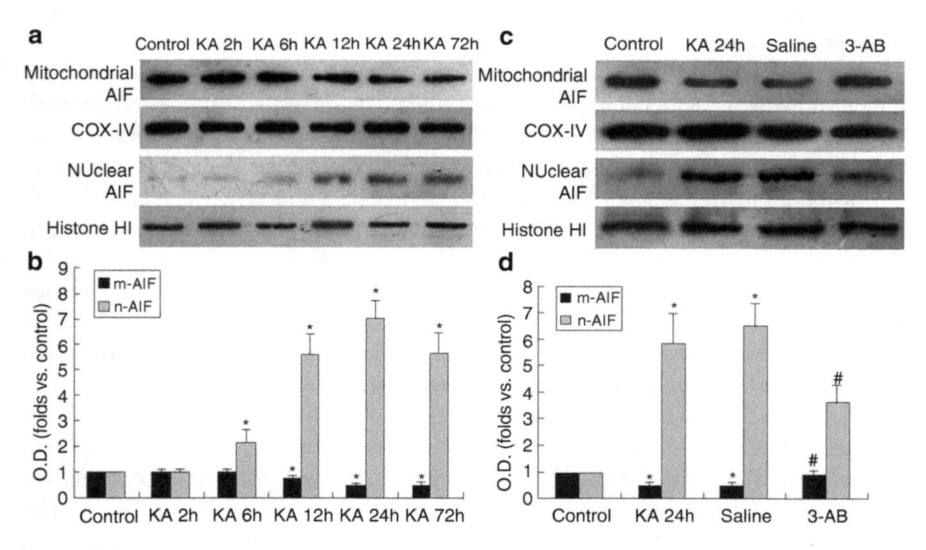

Fig. 11.3 Effects of the PARP-1 inhibitor 3-aminobenzamide (3-AB) on nuclear translocation of AIF following 60-min SE. (**a**) Shows the time course of AIF translocation to hippocampal nuclei from 2 to 2 h after SE, and (**c**) shows the effect of 3-AB, given i.p. 15 min before i.c.v. kainic acid injection, 24 h after SE. Increases in nuclear AIF occurred 6, 12, 24 and 72 h after SE, with the maximal increase 24 h after SE (**b**). 3-AB reduced nuclear translocation of AIF by 50% (**d**). COX-IV is a marker of the mitochondrial fraction and histone H1 is a marker of the nuclear fraction. From Wang et al. (2007), with permission from Lippincott Williams & Wilkins

11.3 The Time Course of the Appearance of Necrotic Neurons Following Seizure Onset

How soon after the onset of prolonged seizures does neuronal death begin to appear? A study of the evolution of neuronal death from seizures has shown that after 40 min of seizures that acidophilic neurons by H&E stain, the light-microscopic equivalent of ultrastructural neuronal necrosis (Fujikawa et al. 1999, 2000, 2002, 2007) appeared in 14 brain regions, with more appearing in more brain regions as seizure duration and recovery periods increased (Fujikawa 1996). The maximum amount of seizure-induced neuronal necrosis occurred 24 h after 3-h SE. For example, in the piriform cortex 24 h after 3-h SE, neuronal death occurred in greater than 75% of the neurons (Fujikawa 1996). Thus, investigating recovery periods longer than 24 h in generalized seizures will not contribute to an understanding of the programmed mechanisms responsible for killing the neurons, because neuronal necrosis will already have occurred in the vast majority of neurons.

11.4 Apoptosis and the Caspase-Dependent Pathways of Cell Death

In general, caspase activation, via either the extrinsic, death-receptor pathway involving activation of caspase-8, or the intrinsic, mitochondrial pathway involving activation of caspase-9, with subsequent downstream activation of caspase-3 by either pathway, is associated with cellular apoptosis. It is also true that in general, caspase activation occurs in immature cells *in vitro* or *in vivo* and in immortalized cell lines. Despite this, there are *in vivo* studies that describe caspase activation in SE in the adult rodent brain (Henshall et al. 2000; Kondratyev and Gale 2000). Kondratyev and colleagues showed neuroprotection with a caspase inhibitor in hippocampus, which did not show caspase-3 activation (Kondratyev and Gale 2000). The reason is that the caspase inhibitor, z-DEVD-fmk, is not specific for caspases (Bizat et al. 2005; Knoblach et al. 2004; Rozman-Pungerčar et al. 2003). With respect to the other study, the reason for the discrepancy is not clear. Although the caspase-3 zymogen and not the active fragment was used for immunofluorescence microscopy, a DEVD cleavage assay for caspase-3-like activity showed an eightfold increase 8 h after 45-min focal seizures (they used 40-min seizures in subsequent studies) (Henshall et al. 2000). The same group has shown upstream caspase activation in both the extrinsic and intrinsic pathways (Henshall et al. 2001a, b). The model that they used, 40 min of focal SE, could possibly explain the difference in results. In this respect, their results were confirmed by another group using their model but with 60 min of SE (Li et al. 2006).

However, there is evidence that caspase activation is age-dependent, and that it occurs in hypoxia-ischemia in the neonatal and not the adult brain (Hu et al. 2000; Liu et al. 2004). Moreover, it has been shown that neither the extrinsic caspase-8 death-receptor pathway nor the intrinsic mitochondrial caspase-9 death receptor pathway is activated in neurons following generalized SE in the adult rat (Fujikawa et al. 2007), and caspase-3, the central downstream effector caspase, is also not activated in neurons after generalized SE in the adult brain (Fujikawa et al. 2002; Araújo et al. 2008; Narkilahti et al. 2003; Takano et al. 2005).

11.5 Caspase-Independent Programmed Mechanisms in Necrotic Cell Death

Chapters by Rieckher and Tavernarakis, Krantic and Susin, Kang et al. and Liu et al. in the first edition of this book (Fujikawa 2010) have pointed out that in pathologically induced cell death that programmed mechanisms other than caspase activation, which occurs most often in morphologically apoptotic cells, are activated. In the nematode *C. elegans* hyperactive mutations of ion channels result in cellular necrosis, which is similar to excitotoxic neuronal death in rodents (Rieckher and Tavernarakis, Chap. 2, ref. Fujikawa 2010). In *C. elegans* lysosomal involvement in cellular necrosis has been shown in mutants defective in lysosomal function (Artal-Sanz et al. 2006). In wild-type nematodes, cathepsins are released, and because of an acidic intracellular pH, brought about by the vacuolar H^+-ATPase in lysosomal membranes, the result is cellular necrosis (Syntichaki et al. 2005). As in transient global ischemia in primates (see Chap. 10, by Tonchev and Yamashima, ref. Fujikawa 2010), calpain I (μ-calpain), the calcium-dependent cytoplasmic cysteine protease, is involved, because downregulation of calpains and cathepsins by RNA interference (RNAi) reduces cell death (Syntichaki et al. 2002). In addition, AIF translocation to nuclei is dependent on calpain I activation in hippocampus in TGI (Cao et al. 2007); (also see Chap. 9, Liu et al. in ref. Fujikawa 2010).

In mouse embryonic fibroblasts in cell culture, Moubarak and colleagues showed, using a large panel of gene knockout cells, that DNA damage by an alkylating agent or inhibition of a nuclear enzyme (topoisomerase II) brought about cellular necrosis by an orderly sequence of events: poly(ADP-ribose) polymerase-1 (PARP-1) activation, calpain I activation, Bax translocation from cytosol to mitochondria, followed by apoptosis-inducing factor (AIF) release from mitochondria and translocation to nuclei, producing necrotic cell death (Moubarak et al. 2007); (also see Chap. 3, by Krantic and Susin in ref. Fujikawa 2010).

Finally, NMDA application to cortical neurons in culture produces an excitotoxic death characterized by PARP-1 activation, which forms poly(ADP-ribose) (PAR) polymers that translocate to mitochondria, releasing AIF, which translocates to nuclei, resulting in neuronal death (Andrabi et al. 2006; Yu et al. 2002, 2006) (also see Chap. 5, by Kang et al. in ref. Fujikawa 2010). The authors call this cell death "parthanatos," from "PAR" and "thanatos," or death. Morphologically, this neuronal death, with shrunken, pyknotic nuclei by fluorescence microscopy, corresponds to neuronal necrosis (Fujikawa et al. 1999, 2000, 2002). Olney's group did not recognize the word "necrosis," saying that it is a word for "death" and is therefore meaningless in terms of cell death classification. However, many investigators world-wide use the word to refer to a particular type of cell death that corresponds to what we have characterized as "neuronal necrosis" and to what the Olney group called "excitotoxic neuronal death." They are one and the same.

11.5.1 Caspase-Independent Programmed Mechanisms in Seizure-Induced Neuronal Necrosis

The controversy regarding the activation and importance of caspases in SE-induced neuronal necrosis has already been addressed. There is no controversy as to the importance of caspase-independent programmed mechanisms in SE-induced neuronal necrosis. The recent study by Moubarak and colleagues (Moubarak et al. 2007) showed the sequential activation of PARP-1 and calpain I, then Bax translocation to mitochondrial membranes, followed by AIF release from mitochondria and translocation to cellular nuclei, a process that they called "programmed necrosis." They showed this in mouse embryonic fibroblasts in cell culture, in which DNA damage was created by exposure to a DNA alkylating agent or by inhibition of the nuclear enzyme topoisomerase II. These results provide a unifying hypothesis that could apply to excitotoxic neuronal death in general, and SE-induced neuronal necrosis in particular.

11.5.1.1 PARP-1, AIF, Calpain I, Cytochrome c and Endonuclease G

In generalized SE both PARP-1 (Wang et al. 2007) and calpain I activation (Araújo et al. 2008; Wang et al. 2008) have been shown to contribute to hippocampal neuronal death. Araújo et al. (2008) showed that following kainic acid-induced seizures, there was Fluoro-Jade evidence of neuronal necrosis in hippocampal CA1 and CA3, calpain I activation and neuroprotection in CA1 with calpain I inhibition (Fig. 11.4). In the studies by Wang and colleagues (Wang et al. 2007, 2008), there was associated translocation of AIF to nuclei, with reduction of AIF translocation with PARP-1 and calpain I inhibitors (Figs. 11.2, 11.3 and 11.5). However, in the studies by Wang and colleagues they did not confirm translocation of AIF to neuronal nuclei, as only western blots were done. In this regard, Narkilahti and colleagues showed that late activation of caspase-3 following SE occurs in almost exclusively in astrocytes, not neurons (Narkilahti et al. 2003). As mentioned previously, Moubarak and colleagues have shown that PARP-1 activation leads to calpain I activation and AIF release from mitochondria and translocation to nuclei, with subsequent cell death (Moubarak et al. 2007).

In addition to the previously cited studies, AIF has been shown to be necessary for SE-induced hippocampal neuronal death in AIF-deficient Harlequin mice

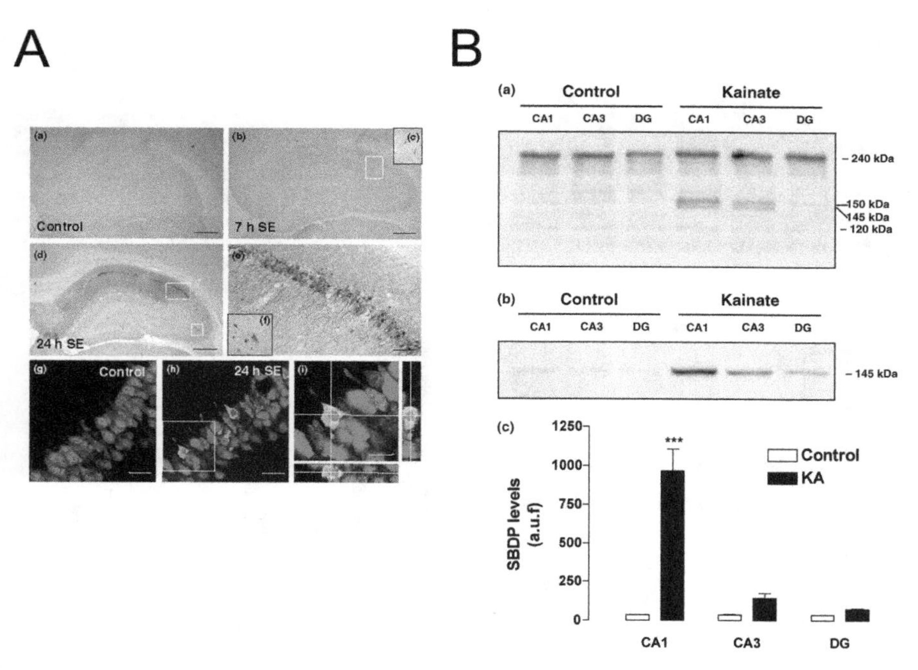

Fig. 11.4 Calpain activation 24 h after the onset of kainic acid-induced SE, assessed by expression of the calpain I-cleaved fragment of αII-spectrin (spectrin breakdown product, or SBDP). In **a**, SBDP expression occurred in dorsal hippocampal CA1 neurons 24 h after seizures began, but not in controls and only minimally 7 h after seizures began (*a*) through (*f*). (*c*) is the inset of the white-boxed area in (*b*), (*e*) is the larger white-boxed area in (*d*), and (*f*) is a higher magnification photomicrograph of pyramidal neurons in the smaller white boxed area in (*d*). Scale bars 400 μm in (*a*), (*b*) and (*d*) and 50 μm in (*e*). (*g–i*) are confocal photomicrographs of hippocampal pyramidal neurons expressing the neuronal marker NeuN (red) and SBDP (green). SBDP appeared 24 h after seizure onset (*h*) and (*i*) but not in a saline-injected control (*g*). Scale bars 20 μm in (*g*) and (*h*) and 8 μm in (*i*). (**b**) Shows western blots of lysates from hippocampal CA1, CA3 and dentate gyrus (DG), demonstrating in (*a*) full-length αII-spectrin (240 kDa) and SBDPs 150, 145 and 120 kDa, with the appearance of the calpain-cleaved 150 and 145 kDa SBDPs but not the caspase-3-cleaved 120 kDa SBDP 24 h after SE. (*b*) Shows results using an antibody that recognizes only the 145 kDa calpain SBDP, and (*c*) shows quantitation of (*b*), with a markedly increased 145 kDa SBDP in CA1. From Araújo et al. (2008), with permission from Blackwell Publishing

(Cheung et al. 2005) and in minocycline-treated mice (Heo et al. 2006). Finally, mitochondrial endonuclease G (endoG) has been shown to translocate to neuronal nuclei 5 d following KASE (Wu et al. 2004). AIF is associated with large-scale (50 kilobase) DNA cleavage (Susin et al. 1999), and endoG produces internucleosomal (180 base-pair) DNA cleavage (Li et al. 2001; Parrish et al. 2001).

There is also a study in which wild-type and calpastatin-deficient mice were subjected to KASE by focal injection of KA into hippocampus (Takano et al. 2005). Calpastatin is an endogenous inhibitor of calpain I. The authors found that AIF and endoG translocation to hippocampal nuclei occurred in the calpastatin-deficient mice, and that caspase-3, the principal effector caspase, was not activated. Other

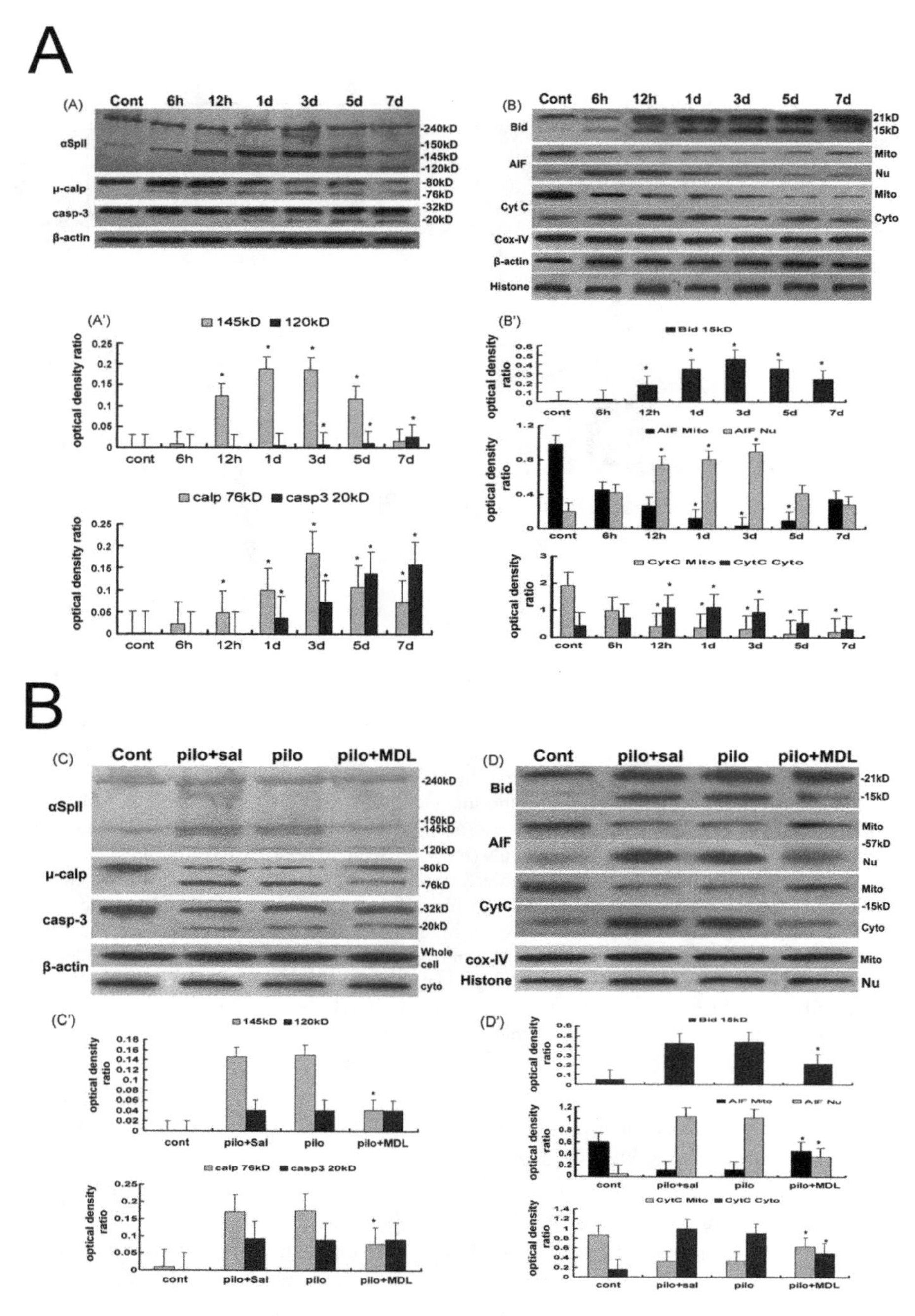

Fig. 11.5 In **a** (*A*) and (*B*) show western blots of αII-spectrin (αSpII), calpain I (μ-calp), caspase-3 (casp-3), Bid, AIF and cytochrome *c* (Cyt C) from 6 h to 7 d after 2-h lithium-pilocarpine-induced SE (LPCSE). (*A′*) Shows quantitation of (*A*) and *B′* shows quantification of (*B*). The calpain-I

studies have also noted a lack of caspase-3 activation following generalized SE (Fujikawa et al. 2002; Araújo et al. 2008; Narkilahti et al. 2003), and neither caspase-9 or caspase-8, initiator caspases in the intrinsic (mitochondrial) and extrinsic pathways, respectively, are activated following lithium-pilocarpine-induced SE (LPCSE) (Fujikawa et al. 2007).

We have found that the mitochondrial proteins cytochrome *c* (cyt *c*) and AIF and the enzyme endoG all translocate to neuronal nuclei in the piriform cortex within the first 60 min of generalized SE, persisting and becoming more widespread as neurons become shrunken and necrotic 6 and 24 h after 3-h lithium-pilocarpine-induced SE (LPCSE) (Zhao and Fujikawa 2010) (Figs. 11.6 and 11.7). We used our 0–3 grading scale, applying it to the degree of nuclear translocation, in which 0 = none, 0.5 = slight (<10%), 1.0 = mild (10–25%), 1.5 = mild-to-moderate (26–45%), 2.0 = moderate (46–54%), 2.5 = moderate-to-severe (55–75%), and 3.0 = severe (>75%) (Zhao and Fujikawa 2010). In the LPCSE seizure model, maximal neuronal necrosis occurs 24 h following 3-h SE (Fujikawa et al. 1999, 2000, 2002, 2007; Fujikawa 1996). Cytochrome *c (in vivo)* has not heretofore been shown to translocate to cellular nuclei, so its presence there quite early, with persistence thereafter, raises interesting questions which we will attempt to answer in future studies.

Translocation of cytochrome *c* to the cytoplasm is known to activate caspase-9, through formation of an "apoptosome" with Apaf-1 and dATP (Li et al. 1997). Cytochrome *c* translocation to cytoplasm has been reported in generalized SE, without evidence of caspase activation (Heo et al. 2006) or with late caspase-3 activation, days after calpain I activation (Wang et al. 2008). Narkilathi et al. (2003) showed that in generalized SE that activated caspase-3 appeared primarily in astrocytes in hippocampus 2 and 7 d after SE. The increasing amount of activated caspase-3 from 1 to 7 d after SE found by Wang et al. (2008) by western blotting could also have occurred in astrocytes, as immunohistochemistry was not done.

A unique *in vitro* study found that DNA damage to HeLa cells or cerebellar granule cells in culture produced nuclear translocation of cyt *c* within 60 min (Nur-E-Kamal et al. 2004). This was associated with translocation of acetylated histone H2A from the nucleus to the cytoplasm and with chromatin condensation. Our *in vivo* results confirm nuclear translocation of cyt *c* within the first 60 min of

Fig. 11.5 (continued) cleaved 145 kDa αSpII fragment increased significantly 12 h and 1, 3 and 5 d after SE, cleaved Bid (15 kDa) was significantly elevated 12 h and 1–7 d after SE, AIF translocated from mitochondria to nuclei 12 h and 1 and 3 d after SE, and Cyt C translocated from mitochondria to cytosol 12 h and 1 and 3 d after SE. Cleaved (activated) calpain I (76 kDa) appeared 12 h and 1–7 d after SE, with maximal expression 3 d after SE. Cleaved (activated) caspase-3 (20 kDa) appeared 1–7 d after SE, with maximal expression 7 d after SE. Narkilahti and colleagues (2003) showed that late activation of caspase-3 occurs in astrocytes, not neurons. In **b** (*C*) and (*D*) and (*C*′) and (*D*′) show that the calpain I inhibitor MDL 28170 reduced significantly the formation of calpain-cleaved αSpII and truncated (15 kDa) Bid, AIF translocation from mitochondria to nuclei and Cyt C translocation from mitochondria to cytosol. B-actin is a loading control, COX-IV is a marker of the mitochondrial fraction and histone is a marker of the nuclear fraction. From Wang et al. (2008), with permission from Elsevier Inc

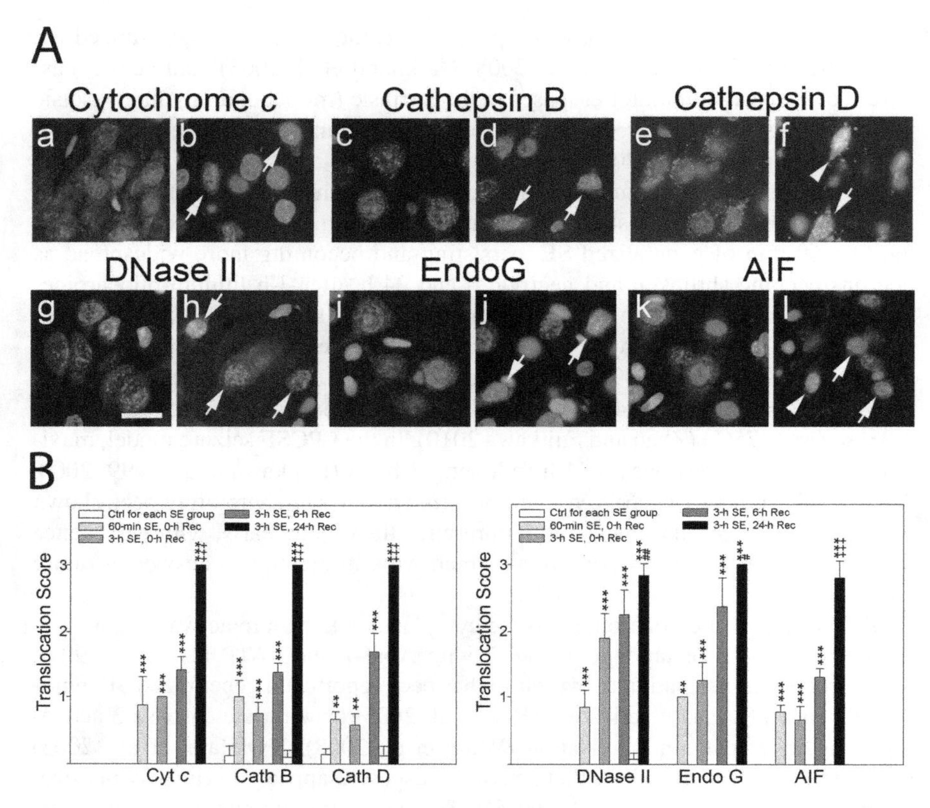

Fig. 11.6 Nuclear translocation of mitochondrial cytochrome *c*, endonuclease G and apoptosis-inducing factor (AIF) 60 min after the onset of LPCSE increases up to 24 h after 3-h SE. (**a**) Shows merged immunofluorescent images of control neurons (*a*, *c*, *e*, *g*, *i* and *k*) and neurons 60 min after the onset of SE in piriform cortex (*b*, *d*, *f*, *h*, *j* and *l*). Nuclei stained with DAPI are blue and cytoplasm staining the six proteins is red in control neurons. Following 60 min of SE, many neurons have purple to purplish-white nuclei, indicating nuclear translocation of the six proteins. Arrows point to nuclei showing translocation and arrowheads point to shrunken nuclei showing translocation. The scale bar is 10 μm. (**b**) Shows nuclear translocation scores for the six proteins in controls, 60-min SE, 3-h SE and 3-h SE with 6 h and 24 h recovery periods. See the text for a detailed description of the scoring method; in general the larger the number the greater the degree of nuclear translocation. ***$p < 0.001$ and **$p < 0.01$ compared to controls, +++$p < 0.001$ in the 24 h recovery group compared to the three other groups, and ##$p < 0.01$ and #$p < 0.05$ in the 24 h recovery group compared to the 60-min and 3-h SE groups. Abbreviations in **a** and **b** are EndoG (endonuclease G), AIF (apoptosis-inducing factor) and in B are cyt *c* (cytochrome *c*), cath B (cathepsin B) and cath D (cathepsin D). (**a**) Shows reformatted images from ref. Zhao and Fujikawa (2010) and in **b** both bar graphs are from ref. Zhao and Fujikawa (2010), which is in the public domain

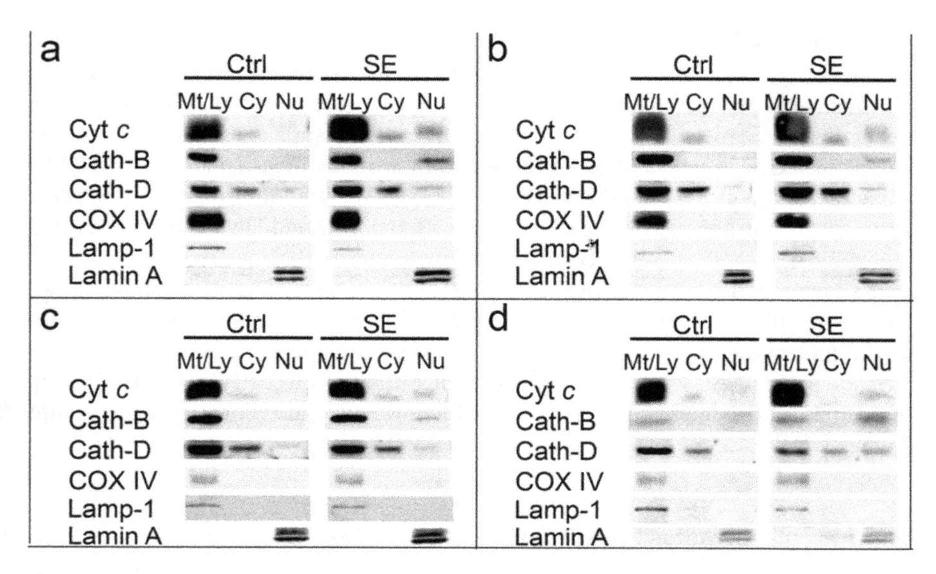

Fig. 11.7 Subcellular fractionation and western blots corroborate the nuclear translocation of cytochrome *c* (Cyt *c*), cathepsin B (Cath-B) and cathepsin D (Cath-D) after 60 min SE (**a**), 3 h SE (**b**), 3-h SE with a 6-h recovery period (**c**) and 3-h SE with a 24-h recovery period (**d**). COX IV is a mitochondrial marker (Mt/Ly), Lamp1 is a lysosomal marker (Mt/Ly) and Lamin A is a nuclear marker. Cath-D is present in the cytosol (Cy) of control and SE lanes at all four time points. Abbreviations are Ctrl (control), Mt/Ly (mitochondrial/lysosomal), Cy (cytosol) and Nu (nuclear). From ref. Zhao and Fujikawa (2010), which is in the public domain

generalized SE (Zhao and Fujikawa 2010). Nuclear translocation of cyt *c* points to a heretofore unexplored pathological mechanism for this mitochondrial respiratory chain protein.

11.5.1.2 Lysosomal Cathepsins, DNase II, Calpain I and Reactive Oxygen Species (ROS)

Calpain I has been shown to bind to lysosomal membranes, with subsequent release of the cysteinyl protease cathepsin B to the cytoplasm and DNase II to the nucleus of hippocampal neurons following transient global ischemia in primates (Tsukada et al. 2001; Yamashima et al. 1996, 1998). It was shown in rat hippocampal slices subjected to oxygen-glucose deprivation that NMDA-receptor activation activates calpain I, which causes lysosomal membrane permeabilization (LMP) and the release of cathepsins B and D to cytosol (Windelborn and Lipton 2008). In this model, superoxide ($\bullet O_2^-$) levels were increased, and reducing reactive oxygen species (ROS) levels blocked LMP. In addition, NMDA-receptor blockade, as well as inhibition of pathways producing arachidonic acid, blocked LMP and ROS production (Windelborn and Lipton 2008).

We have also found that in addition to the mitochondrial proteins AIF and cyt *c* and the enzyme endoG, the lysosomal enzymes cathepsins B and D and DNase II all translocate to neuronal nuclei in the piriform cortex within the first 60 min of generalized SE, persisting and becoming more widespread as neurons become shrunken and necrotic 6 and 24 h after 3-h lithium-pilocarpine-induced SE (LPCSE) (Zhao and Fujikawa 2010) (Figs. 11.6 and 11.7). We used the same grading scale and nuclear translocation scores for cathepsins B and D and DNase II as described above for AIF, cyt *c* and endoG (Zhao and Fujikawa 2010).

As mentioned previously, maximal neuronal necrosis occurs 24 h following 3-h LPCSE (Fujikawa et al. 1999, 2000, 2002, 2007; Fujikawa 1996). Like cyt *c* (*in vivo*), neither cathepsin B or D has been shown to translocate to cellular nuclei *in vitro* or *in vivo*. As with cyt *c*, their presence there quite early, with persistence thereafter, raises interesting questions which we will also investigate in future studies.

11.5.1.3 p53 in Status Epilepticus

p53, a tumor-suppressor gene, is activated by SE, and inhibition of protein synthesis prevents p53 mRNA expression and neuronal death (Sakhi et al. 1994; Schreiber et al. 1993) (also see Chap. 15 by Tan and Schreiber in ref. Fujikawa 2010). p53 has also been shown to regulate the Bax-dependent, mitochondrial caspase cell death pathway (Chipuk et al. 2004). Since we have shown that KASE produces morphologically necrotic neurons, these findings suggest another programmed pathway activated by programmed necrosis.

11.5.1.4 Autophagy in Status Epilepticus

Recently, there has been an increasing amount of attention to type II programmed cell death (Clarke 1990), or macroautophagy, most often referred to simply as "autophagy." Classically, autophagic mechanisms are triggered by starvation, in which internal digestion of cellular components within autophagosomes is an attempt to keep cells alive. Controversy has centered on whether in pathologic insults autophagy is an attempt to protect cells or whether it actively contributes to cell death (Edinger and Thompson 2004; Codogno and Meijer 2005; Shintani and Klionsky 2004; Tsujimoto and Shimizu 2005). A recent report has shown that KASE induces transient autophagic stress in mouse hippocampus, manifested by an increase of LC3-II, a specific marker of autophagic vacuoles, 4–6 h after systemic KA injection (Shacka et al. 2007). This does not, of course, establish whether autophagy contributes to SE-induced neuronal necrosis or whether it is simply a bystander effect. Further studies are warranted.

11.5.1.5 The Role of Necroptosis in Acute Neuronal Injury

Following caspase inhibition with the broad-spectrum caspase inhibitor zVAD.fmk in cell culture, cells still died, but with a necrotic instead of apoptotic morphology (Holler et al. 2000; Vercammen et al. 1998). This form of programmed necrosis was called "necroptosis" (a combination of "necrosis" and "apoptosis") (Degterev et al. 2005). In the same article, following transient focal cerebral ischemia, administration of the receptor interacting protein (RIP) inhibitor 7-Cl-necrostatin-1 (7-Cl-nec-1) reduced the volume of the focal infarct (Degterev et al. 2005). Necrostatin-1 was subsequently found to inhibit RIP-1 (Degterev et al. 2008), and the necroptotic pathway was found to involve phosphorylation of RIP3 by RIP1 (Zhang et al. 2009) and mixed lineage kinase domain-like protein (MLKL) by RIP3 (Sun et al. 2012).

Subsequent studies of transient global cerebral ischemia have shown RIP3 but not RIP1 was activated and translocated to neuronal nuclei (Yin et al. 2015), RIP3 phosphorylation (activation) and interaction with RIP1 (Miao et al. 2015) and nuclear translocation of RIP3 and AIF (Xu et al. 2016), resulting in necroptotic neuronal necrosis. In addition, in non-neuronal cell culture, nuclear translocation of RIP1, RIP3 and MLKL occurred prior to MLKL translocation to plasma membranes, and Nec-1 administration reduced the nuclear translocation and necroptosis (Yoon et al. 2016). This is of interest to us because of our finding of nuclear translocation of mitochondrial cyt-*c*, AIF and endoG and lysosomal cath-B, cath-D and DNase II within the first 60 min of SE (Zhao and Fujikawa 2010).

We have preliminary results showing nuclear translocation of RIP1, RIP3 and MLKL in the piriform cortex within the first 60 min of LPCSE and their persistence in necrotic nuclei 24 h after 3-h SE. We also have preliminary data showing that i.c.v. administration of 7-Cl-O-nec-1 results in a substantial reduction of acidophilic (necrotic) neurons in the dorsal hippocampal hilus 6 h after 3-h LPCSE. This suggests that both necroptotic and excitotoxic pathways are activated by LPCSE and points the way to further studies to determine the interaction between them and how the necroptotic pathway is activated by LPCSE.

11.6 Translating an Understanding of Mechanisms to a Neuroprotective Strategy in Refractory Human Status Epilepticus

The most direct and theoretically most effective neuroprotective approach in excitotoxic neuronal death in general, and SE in particular, would be to administer an NMDA-receptor antagonist. By blocking post-synaptic NMDA receptors, excessive calcium influx would be prevented, thereby preventing activation of downstream biochemical pathways that produce neuronal necrosis. However, there are two caveats to this simplistic approach. The first is that NMDA-receptor antagonists have been shown to produce delayed neuronal necrosis in adult rats (Fix et al. 1993) and neuronal apoptosis in neonatal rats (Ikonomidou et al. 1999). However, in the adult brain, preventing or ameliorating the widespread, severe neuronal necrosis produced by SE should take precedence over any lesser and less widespread adverse effects of NMDA-receptor antagonists.

We advocated the use of NMDA-receptor antagonists following SE more than 20 years ago, based upon their remarkable neuroprotective effects (Fujikawa 1995; Fujikawa et al. 1994). We and others showed years ago that NMDA-receptor antagonists are remarkably neuroprotective in the face of ongoing seizure discharges, which are not eliminated by NMDA-receptor antagonists, including ketamine (Clifford et al. 1990; Fariello et al. 1989; Fujikawa 1995; Fujikawa et al. 1994). The conclusion of these studies is that ketamine and other NMDA-receptor antagonists provide neuroprotection that is independent of an antiepileptic effect. Despite this, ketamine has finally been advocated for use in refractory SE in humans, not because of its neuroprotective properties, but because of a delayed antiepileptic effect (Bleck 2005; Borris et al. 2000). Borris and colleagues (2000) showed that given after 60 min of spontaneous SE from electrical hippocampal stimulation in the rat that ketamine abolished seizure discharges, but that it was ineffective in doing so after 15 min of SE. This is consistent with our data, but as we have shown, after 60 min of SE neuronal necrosis will already have begun to occur in many brain regions (Fujikawa 1996). Immediate use of ketamine should be considered because of its neuroprotective properties, with the added benefit of a potential delayed antiepileptic effect.

The second caveat is that once NMDA-receptor-mediated calcium influx has begun, receptor blockade will have diminishing efficacy the longer after seizure onset it is instituted. The influx of calcium activates enzymes such as calpain I and neuronal nitric oxide synthase (nNOS), the latter of which produces nitric oxide (NO), which in turn forms peroxynitrite ($ONOO^-$), a free radical that can damage DNA, activating PARP-1. Activation of calpain I and PARP-1 results in the release of death-promoting proteins and enzymes from mitochondria and lysosomes that cause the degradation of cytoplasmic proteins and nuclear chromatin. Thus, administration of a cocktail of inhibitors of, for example, NMDA receptors, calpain I, nNOS and PARP-1, to name but a few currently known essential targets, would seem to have the best chance of salvaging still-viable neurons, but with diminishing returns the longer that SE continues.

References

Andrabi SA, Kim S-W, Wang H, Koh DW, Sasaki M, Klaus JA, Otsuka T, Zhang Z, Koehler RC, Hurn PD, Poirier GG, Dawson VL, Dawson TM (2006) Poly(ADP-ribose) (PAR) polymer is a death signal. Proc Natl Acad Sci U S A 103:18308–18313

Araújo IM, Gil JM, Carreira BP, Mohapel P, Petersen A, Pinheiro PS, Soulet D, Bahr BA, Brundin P, Carvalho CM (2008) Calpain activation is involved in early caspase-independent neurodegeneration in the hippocampus following status epilepticus. J Neurochem 105(3):666–676. PubMed PMID: 18088374

Artal-Sanz M, Samara C, Syntichaki P, Tavernarakis N (2006) Lysosomal biogenesis and function is critical for necrotic cell death in Caenorhabditis elegans. J Cell Biol 173(2):231–239. PubMed PMID: 16636145

Auer RN, Kalimo H, Olsson Y, Siesjo BK (1985a) The temporal evolution of hypoglycemic brain damage. II. Light- and electron-microscopic findings in the hippocampal gyrus and subiculum of the rat. Acta Neuropathol (Berl) 67(1–2):25–36. PubMed PMID: 4024869

Auer RN, Kalimo H, Olsson Y, Siesjo BK (1985b) The temporal evolution of hypoglycemic brain damage. I. Light- and electron-microscopic findings in the rat cerebral cortex. Acta Neuropathol (Berl) 67(1–2):13–24. PubMed PMID: 4024866

Bizat N, Galas MC, Jacquard C, Boyer F, Hermel JM, Schiffmann SN, Hantraye P, Blum D, Brouillet E (2005) Neuroprotective effect of zVAD against the neurotoxin 3-nitropropionic acid involves inhibition of calpain. Neuropharmacology 49(5):695–702. PubMed PMID: 15998526

Bleck TP (2005) Refractory status epilepticus. Curr Opin Crit Care 1:117–120

Borris DJ, Bertram EH, Kaipur J (2000) Ketamine controls prolonged status epilepticus. Epilepsy Res 42:117–122

Brown AW (1977) Structural abnormalities in neurones. J Clin Pathol 30(Suppl 11):155–169

Bruhn T, Cobo M, Berg M, Diemer NH (1992) Limbic seizure-induced changes in extracellular amino acid levels in the hippocampal formation: a microdialysis study of freely moving rats. Acta Neurol Scand 86:455–461

Cao G, Xing J, Xiao X, Liou AK, Gao Y, Yin XM, Clark RS, Graham SH, Chen J (2007) Critical role of calpain I in mitochondrial release of apoptosis-inducing factor in ischemic neuronal injury. J Neurosci 27(35):9278–9293. PubMed PMID: 17728442

Cheung EC, Melanson-Drapeau L, Cregan SP, Vanderluit JL, Ferguson KL, McIntosh WC, Park DS, Bennett SA, Slack RS (2005) Apoptosis-inducing factor is a key factor in neuronal cell death propagated by BAX-dependent and BAX-independent mechanisms. J Neurosci 25(6):1324–1334. PubMed PMID: 15703386

Chipuk JE, Kuwana T, Bouchier-Hayes L, Droin NM, Newmeyer DD, Schuler M, Green DR (2004) Direct activation of Bax by p53 mediates mitochondrial membrane permeabilization and apoptosis. Science 303:1010–1014

Clarke PGH (1990) Developmental cell death: morphological diversity and multiple mechanisms. Anat Embryol 181:195–213

Clifford DB, Olney JW, Benz AM, Fuller TA, Zorumski CF (1990) Ketamine, phencyclidine, and MK-801 protect against kainic-acid-induced seizure-related brain damage. Epilepsia 31:382–390

Codogno P, Meijer AJ (2005) Autophagy and signaling: their role in cell survival and cell death. Cell Death Differ 12(Suppl 2):1509–1518. PubMed PMID: 16247498

Colbourne F, Sutherland GR, Auer RN (1999) Electron microscopic evidence against apoptosis as the mechanism of neuronal death in global ischemia. J Neurosci 19:4200–4210

Colicos MA, Dash PK (1996) Apoptotic morphology of dentate granule cells following experimental cortical impact injury in rats: possible role in spatial memory deficits. Brain Res 739:120–131

Degterev A, Huang Z, Boyce M, Li Y, Jagtap P, Mizushima N, Cuny GD, Mitchison TJ, Moskowitz MA, Yuan J (2005) Chemical inhibitor of nonapoptotic cell death with therapeutic potential for ischemic brain injury. Nat Chem Biol 1(2):112–119. https://doi.org/10.1038/nchembio711. PubMed PMID: 16408008

Degterev A, Hitomi J, Germscheid M, Ch'en IL, Korkina O, Teng X, Abbott D, Cuny GD, Yuan C, Wagner G, Hedrick SM, Gerber SA, Lugovskoy A, Yuan J (2008) Identification of RIP1 kinase as a specific cellular target of necrostatins. Nat Chem Biol 4(5):313–321. https://doi.org/10.1038/nchembio.83. PubMed PMID: 18408713

Edinger AL, Thompson CB (2004) Death by design: apoptosis, necrosis and autophagy. Curr Opin Cell Biol 16(6):663–669. PubMed PMID: 15530778

Evans MC, Griffiths T, Meldrum BS (1984) Kainic-acid seizures and the reversibility of calcium loading in vulnerable neurons in the hippocampus. Neuropathol Appl Neurobiol 10:285–302
Fariello RG, Golden GT, Smith GG, Reyes PF (1989) Potentiation of kainic acid epileptogenicity and sparing from neuronal damage by an NMDA receptor antagonist206-213. Epilepsy Res 3:206–213
Fix AS, Horn JW, Wightman KA et al (1993) Neuronal vacuolization and necrosis induced by the noncompetitive N-methyl-D-aspartate (NMDA) antagonist MK(+)801 (dizocilpine maleate): A light and electron microscopic evaluation of the rat retrosplenial cortex. Exp Neurol 123:204–215
Fujikawa DG (1995) The neuroprotective effect of ketamine administered after status epilepticus onset. Epilepsia 36:186–195
Fujikawa DG (1996) The temporal evolution of neuronal damage from pilocarpine-induced status epilepticus. Brain Res 725:11–22
Fujikawa DG (2000) Confusion between neuronal apoptosis and activation of programmed cell death mechanisms in acute necrotic insults. Trends Neurosci 23:410–411
Fujikawa DG (2002) Apoptosis: ignoring morphology and focusing on biochemical mechanisms will not eliminate confusion. Trends Pharmacol Sci 23:309–310
Fujikawa DG (2005) Prolonged seizures and cellular injury: understanding the connection. Epilepsia 7:S3–S11
Fujikawa DG (2006) Neuroprotective strategies in status epilepticus. In: Wasterlain CG, Treiman DM (eds) Status epilepticus: mechanisms and management. MIT Press, Cambridge, MA, pp 463–480
Fujikawa DG (ed) (2010) Acute neuronal injury: the role of excitotoxic programmed cell death mechanisms. Springer, New York. 306 p
Fujikawa DG, Daniels AH, Kim JS (1994) The competitive NMDA-receptor antagonist CGP 40116 protects against status epilepticus-induced neuronal damage. Epilepsy Res 17:207–219
Fujikawa DG, Shinmei SS, Cai B (1999) Lithium-pilocarpine-induced status epilepticus produces necrotic neurons with internucleosomal DNA fragmentation in adult rats. Eur J Neurosci 11:1605–1614
Fujikawa DG, Shinmei SS, Cai B (2000) Kainic acid-induced seizures produce necrotic, not apoptotic, neurons with internucleosomal DNA cleavage: implications for programmed cell death mechanisms. Neuroscience 98:41–53
Fujikawa DG, Ke X, Trinidad RB, Shinmei SS, Wu A (2002) Caspase-3 is not activated in seizure-induced neuronal necrosis with internucleosomal DNA cleavage. J Neurochem 83:229–240
Fujikawa DG, Shinmei SS, Zhao S, Aviles ER Jr (2007) Caspase-dependent programmed cell death pathways are not activated in generalized seizure-induced neuronal death. Brain Res 1135:206–218
Griffiths T, Evans M, Meldrum BS (1983) Intracellular calcium accumulation in rat hippocampus during seizures induced by bicuculline or L-allylglycine. Neuroscience 10:385–395
Griffiths T, Evans MC, Meldrum BS (1984) Status epilepticus: the reversibility of calcium loading and acute neuronal pathological changes in the rat hippocampus. Neuroscience 12:557–567
Henshall DC, Chen J, Simon RP (2000) Involvement of caspase-3-like protease in the mechanism of cell death following focally evoked limbic seizures. J Neurochem 74:1215–1223
Henshall DC, Bonislawski DP, Skradski SL, Araki T, Lan J-Q, Schindler CK, Meller R, Simon RP (2001a) Formation of the Apaf-1/cytochrome c complex precedes activation of caspase-9 during seizure-induced neuronal death. Cell Death Differ 8:1169–1181
Henshall DC, Bonislawski DP, Skradski SL, Lan J-Q, Meller R, Simon RP (2001b) Cleavage of Bid may amplify caspase-8-induced neuronal death following focally evoked limbic seizures. Neurobiol Dis 8:568–580
Heo K, Cho Y-J, Cho K-J, Kim H-W, Kim H-J, Shin HY, Lee BI, Kim GW (2006) Minocycline inhibits caspase-dependent and -independent cell death pathways and is neuroprotective against hippocampal damage after treatment with kainic acid in mice. Neurosci Lett 398:195–200

Holler N, Zaru R, Micheau O, Thome M, Attinger A, Valitutti S, Bodmer JL, Schneider P, Seed B, Tschopp J (2000) Fas triggers an alternative, caspase-8-independent cell death pathway using the kinase RIP as effector molecule. Nat Immunol 1(6):489–495. https://doi.org/10.1038/82732. PubMed PMID: 11101870

Hu BR, Liu CL, Ouyang Y, Blomgren K, Siejö BK (2000) Involvement of caspase-3 in cell death after hypoxia-ischemia declines during brain maturation. J Cereb Blood Flow Metab 20:1294–1300

Ikonomidou C, Bosch F, Miksa M, Bittigau P, Vockler V, Dikranian K, Tenkova TI, Stefovska V, Turksi L, Olney JW (1999) Blockade of NMDA receptors and apoptotic neurodegeneration in the developing brain. Science 283:70–74

Ikonomidou C, Bittigau P, Ishimaru MJ, Wozniak DF, Koch C, Genz K, Price MT, Stefovska V, Horster F, Tenkova T, Dikranian K, Olney JW (2000) Ethanol-induced apoptotic neurodegeneration and fetal alcohol syndrome. Science 287:1056–1060

Ishimaru MJ, Ikonomidou C, Tenkova TI, Der TC, Dikranian K, Sesma MA, Olney JW (1999) Distinguishing excitotoxic from apoptotic neurodegeneration in the developing rat brain. J Comp Neurol 408:461–476

Knoblach SM, Alroy DA, Nikolaeva M, Cernak I, Stoica BA, Faden AI (2004) Caspase inhibitor z-DEVD-fmk attenuates calpain and necrotic cell death in vitro and after traumatic brain injury. J Cereb Blood Flow Metab 24:1119–1132

Kondratyev A, Gale K (2000) Intracerebral injection of caspase-3 inhibitor prevents neuronal apoptosis after kainic acid-evoked status epilepticus. Mol Brain Res 75:216–224

Lallement G, Carpentier P, Collet A, Pernot-Marino I, Baubichon D, Blanchet G (1991) Effects of soman-induced seizures on different extracellular amino acid levels and on glutamate uptake in rat hippocampus. Brain Res 563(1–2):234–240. PubMed PMID: 1786536

Lehmann A, Hagberg H, Jacobson I, Hamberger A (1985) Effects of status epilepticus on extracellular amino acids in the hippocampus. Brain Res 359(1–2):147–151. PubMed PMID: 3000520

Li P, Nijhawan D, Budihardjo I, Srinivasula SM, Ahmad M, Alnemri ES, Wang X (1997) Cytochrome c and dATP-dependent formation of Apaf-1/caspase-9 complex initiates an apoptotic protease cascade. Cell 91(4):479–489. PubMed PMID: 9390557

Li L, Luo X, Wang X (2001) Endonuclease G is an apoptotic DNase when released from mitochondria. Nature 412:95–99

Li T, Lu C, Xia Z, Xiao B, Luo Y (2006) Inhibition of caspase-8 attenuates neuronal death induced by limbic seizures in a cytochrome c-dependent and Smac/DIABLO-independent way. Brain Res 1098(1):204–211. PubMed PMID: 16774749

Liu CL, Siesjö BK, Hu BR (2004) Pathogenesis of hippocampal neuronal death after hypoxia-ischemia changes during brain development. Neuroscience 127:113–123

Miao W, Qu Z, Shi K, Zhang D, Zong Y, Zhang G, Zhang G, Hu S (2015) RIP3 S-nitrosylation contributes to cerebral ischemic neuronal injury. Brain Res 1627:165–176. https://doi.org/10.1016/j.brainres.2015.08.020. PubMed PMID: 26319693

Millan MH, Chapman AG, Meldrum BS (1993) Extracellular amino acid levels in hippocampus during pilocarpine-induced seizures. Epilepsy Res 14(2):139–148. PubMed PMID: 8095893

Moubarak RS, Yuste VJ, Artus C, Bouharrour A, Greer PA, Menissier-de Murcia J, Susin SA (2007) Sequential activation of poly(ADP-ribose) polymerase 1, calpains, and Bax is essential in apoptosis-inducing factor-mediated programmed necrosis. Mol Cell Biol 27(13):4844–4862. PubMed PMID: 17470554

Narkilahti S, Pirtillä TJ, Lukasiuk K, Tuunanen J, Expression PA (2003) activation of caspase 3 following status epilepticus. Eur J Neurosci 18:1486–1496

Nur-E-Kamal A, Gross SR, Pan Z, Balklava Z, Ma J, Liu LF (2004) Nuclear translocation of cytochrome c during apoptosis. J Biol Chem 279:24911–24914

Olney JW (1969) Brain lesions, obesity and other disturbances in mice treated with monosodium glutamate. Science 164:719–721

Olney JW (1971) Glutamate-induced neuronal necrosis in the infant mouse hypothalamus. An electron microscopic study. J Neuropathol Exp Neurol 30(1):75–90. PubMed PMID: 5542543

Olney JW (1985) Excitatory transmitters and epilepsy-related brain damage. In: Smythies JR, Bradley RJ (eds) International review of neurobiology, vol 27. Academic, Orlando, pp 337–362
Olney JW, Rhee V, Ho OL (1974) Kainic acid: a powerful neurotoxic analogue of glutamate. Brain Res 77(3):507–512. PubMed PMID: 4152936
Parrish J, Li L, Klotz K, Ledwich D, Wang X, Xue D (2001) Mitochondrial endonuclease G is important for apoptosis in *C elegans*. Nature 412:90–94
Rink A, Fung KM, Trojanowski JQ, Lee VM-Y, Neugebauer E, McIntosh TK (1995) Evidence of apoptotic cell death after experimental traumatic brain injury in the rat. Am J Pathol 147:1575–1583
Rozman-Pungerčar J, Kopitar-Jerala N, Bogyo M, Turk D, Vasiljeva O, Štefe I, Vandenabeele P, Brőmme D, Pulzdar V, Fonović M, Trstenjak-Prebanda M, Dolenc I, Turk V, Turk B (2003) Inhibition of papain-like cysteine proteases and legumain by caspase-specific inhibitors: when reaction mechanism is more important than specificity. Cell Death Differ 10(8):881
Sakhi S, Bruce A, Sun N, Tocco G, Baudry M, Schreiber SS (1994) p53 induction is associated with neuronal damage in the central nervous system. Proc Natl Acad Sci U S A 91:7525–7529
Schreiber SS, Tocco G, Najm I, Thompson RF, Baudry M (1993) Cycloheximide prevents kainate-induced neuronal death and c-fos expression in adult rat brain. J Mol Neurosci 4:149–159
Shacka JJ, Lu J, Xie ZL, Uchiyama Y, Roth KA, Zhang J (2007) Kainic acid induces early and transient autophagic stress in mouse hippocampus. Neurosci Lett 414(1):57–60. PubMed PMID: 17223264
Shintani T, Klionsky DJ (2004) Autophagy in health and disease: a double-edged sword. Science 306(5698):990–995. PubMed PMID: 15528435
Smolders I, Van Belle K, Ebinger G, Michotte Y (1997) Hippocampal and cerebellar extracellular amino acids during pilocarpine-induced seizures in freely moving rats. Eur J Pharmacol 319(1):21–29. PubMed PMID: 9030893
Sun L, Wang H, Wang Z, He S, Chen S, Liao D, Wang L, Yan J, Liu W, Lei X, Wang X (2012) Mixed lineage kinase domain-like protein mediates necrosis signaling downstream of RIP3 kinase. Cell 148(1–2):213–227. https://doi.org/10.1016/j.cell.2011.11.031. PubMed PMID: 22265413
Susin SA, Lorenzo HK, Zamzami N, Marzo I, Snow BE, Brothers GM, Mangion J, Jacotot E, Costantini P, Loeffler M, Larochette N, Goodlett DR, Aebersold R, Siderovski DP, Penninger JM, Kroemer G (1999) Molecular characterization of mitochondrial apoptosis-inducing factor. Nature 397:441–446
Syntichaki P, Xu K, Driscoll M, Tavernarakis N (2002) Specific aspartyl and calpain proteases are required for neurodegeneration in *C. elegans*. Nature 419:939–944
Syntichaki P, Samara C, Tavernarakis N (2005) The vacuolar H+-ATPase mediates intracellular acidification required for neurodegeneration in C. elegans. Curr Biol 15:1249–1254
Takano J, Tomioka M, Tsubuki S, Higuchi M, Nobuhisa Iwata N, Itohara S, Maki M, Saido TC (2005) Calpain mediates excitotoxic DNA fragmentation via mitochondrial pathways in adult brains: evidence from calpastatin mutant mice. J Biol Chem 280:16175–16184
Tanaka K, Graham SH, Simon RP (1996) The role of excitatory neurotransmitters in seizure-induced neuronal injury in rats. Brain Res 737(1–2):59–63. PubMed PMID: 8930350
Tsujimoto Y, Shimizu S (2005) Another way to die: autophagic programmed cell death. Cell Death Differ 12(Suppl 2):1528–1534. PubMed PMID: 16247500
Tsukada T, Watanabe M, Yamashima T (2001) Implications of CAD and DNase II in ischemic neuronal necrosis specific for the primate hippocampus. J Neurochem 79:1196–1206
Vercammen D, Brouckaert G, Denecker G, Van de Craen M, Declercq W, Fiers W, Vandenabeele P (1998) Dual signaling of the Fas receptor: initiation of both apoptotic and necrotic cell death pathways. J Exp Med 188(5):919–930. PubMed PMID: 9730893; PMCID: PMC2213397
Wade JV, Samson FE, Nelson SR, Pazdernik TL (1987) Changes in extracellular amino acids during soman- and kainic acid-induced seizures. J Neurochem 49(2):645–650. PubMed PMID: 3598590

Wang SJ, Wang SH, Song ZF, Liu XW, Wang R, Chi ZF (2007) Poly(ADP-ribose) polymerase inhibitor is neuroprotective in epileptic rat via apoptosis-inducing factor and Akt signaling. Neuroreport 18(12):1285–1289. PubMed PMID: 17632284

Wang S, Wang S, Shan P, Song Z, Dai T, Wang R, Chi Z (2008) mu-Calpain mediates hippocampal neuron death in rats after lithium-pilocarpine-induced status epilepticus. Brain Res Bull 76:90–96

Whalen MJ, Dalkara T, You Z, Qiu J, Bermpohl D, Mehta N, Suter B, Bhide PG, Lo EH, Ericsson M, Moskowitz MA (2008) Acute plasmalemmal permeability and protracted clearance of injured cells after controlled cortical impact in mice. J Cereb Blood Flow Metab 28:490–505

Windelborn JA, Lipton P (2008) Lysosomal release of cathepsins causes ischemic damage in the rat hippocampal slice and depends on NMDA-mediated calcium influx, arachidonic acid metabolism, and free radical production. J Neurochem 106:56–69. PubMed PMID: 18363826

Wu Y, Dong M, Toepfer NJ, Fan Y, Xu M, Zhang J (2004) Role of endonuclease G in neuronal excitotoxicity in mice. Neurosci Lett 264:203–207

Xu Y, Wang J, Song X, Qu L, Wei R, He F, Wang K, Luo B (2016) RIP3 induces ischemic neuronal DNA degradation and programmed necrosis in rat via AIF. Sci Rep 6:29362. https://doi.org/10.1038/srep29362. PubMed PMID: 27377128; PMCID: PMC4932529

Yamashima T, Saido TC, Takita M, Miyazawa A, Yamano J, Miyakawa A, Nishiyo H, Yamashima J, Kawashima S, Ono T, Yoshioka T (1996) Transient brain ischemia provokes Ca2+, PIP2 and calpain responses prior to delayed neuronal death in monkeys. Eur J Neurosci 8:1932–1944

Yamashima T, Kohda Y, Tsuchiya K, Ueno T, Yamashita J, Yoshioka T, Kominami E (1998) Inhibition of ischaemic hippocampal neuronal death in primates with cathepsin B inhibitor CA-074: a novel strategy for neuroprotection based on 'calpain-cathepsin hypothesis'. Eur J Neurosci 10:1723–1733. PubMed PMID: 9751144

Yin B, Xu Y, Wei RL, He F, Luo BY, Wang JY (2015) Inhibition of receptor-interacting protein 3 upregulation and nuclear translocation involved in Necrostatin-1 protection against hippocampal neuronal programmed necrosis induced by ischemia/reperfusion injury. Brain Res 1609:63–71. https://doi.org/10.1016/j.brainres.2015.03.024. PubMed PMID: 25801119

Yoon S, Bogdanov K, Kovalenko A, Wallach D (2016) Necroptosis is preceded by nuclear translocation of the signaling proteins that induce it. Cell Death Differ 23(2):253–260. https://doi.org/10.1038/cdd.2015.92.. PubMed PMID: 26184911; PMCID: PMC4716306

Yu S-W, Wang H, Poitras MF, Coombs C, Bowers WJ, Federoff HJ, Poirier GG, Dawson TM, Dawson VL (2002) Mediation of poly(ADP-ribose) polymerase-1-dependent cell death by apoptosis-inducing factor. Science 297:259–263

Yu S-W, Andrabi SA, Wang H, Kim NS, Poirier GG, Dawson TM, Dawson VL (2006) Apoptosis-inducing factor mediates poly(ADP-ribose) (PAR) polymer-induced cell death. Proc Natl Acad Sci U S A 103:18314–18319

Zhang DW, Shao J, Lin J, Zhang N, Lu BJ, Lin SC, Dong MQ, Han J (2009) RIP3, an energy metabolism regulator that switches TNF-induced cell death from apoptosis to necrosis. Science 325(5938):332–336. PubMed PMID: 19498109

Zhao S, Aviles ER Jr, Fujikawa DG (2010) Nuclear translocation of mitochondrial cytochrome c, lysosomal cathepsins B and D, and three other death-promoting proteins within the first 60 minutes of generalized seizures. J Neurosci Res 88(8):1727–1737. PubMedPMID: 20077427

Concluding Remarks

Denson G. Fujikawa

The morphology of acutely injured neurons in the adult brain *in vivo* is necrotic, not apoptotic (Fujikawa 2000, 2002; Fujikawa et al. 1999, 2000; Auer et al. 1985a, b; Colbourne et al. 1999). Caspase-independent programmed mechanisms are important in the adult brain *in vivo* (Fujikawa 2010): For example, in status epilepticus (Fujikawa et al. 2002, 2007; Narkilahti et al. 2003; Gao and Geng 2013; Wang et al. 2008) and cerebral ischemia (Vosler et al. 2009; Hu et al. 2000) (see also Chap. 4 in the First Edition of this book and Chap. 8 in this book by Sun and colleagues).

Although this edition of *Acute Neuronal Injury* continues to emphasize the role of excitotoxic mechanisms in producing neuronal necrosis, new topics have been introduced. For example, the topic of synaptic vs. extrasynaptic NMDA receptors and their functions is addressed by Dr. Baudy and colleagues in Chap. 2. Also, in recent years a programmed necrotic cell death pathway other than excitotoxicity, called "necroptosis" (from "necrosis" and "apoptosis"), has been described (Degterev et al. 2005). This was discovered in non-neuronal cells *in vitro* when the broad-spectrum caspase inhibitor z-VAD.fmk to inhibit caspase activation; the cells still died, but with necrotic morphology (Degterev et al. 2005). However, investigators went on to investigate cerebral ischemia *in vivo*, and found that enzymes in the necroptotic pathway, receptor-interaction protein-1 (RIP-1), receptor-interacting protein-3 (RIP3) and mixed lineage kinase domain-like protein (MLKL), were activated, implicating the necroptotic pathway in acute neuronal injury *in vivo* (Yin et al. 2015; Miao et al. 2015; Xu et al. 2016; Liu et al. 2016; Vieira et al. 2014). Carvalho and Vieira address necroptosis in cerebral ischemia in Chap. 9, and Tao includes necroptotic mechanisms in in traumatic brain injury (TBI) in Chap. 5.

At present there is little information about the importance of the necroptotic pathway in acute neuronal injury. The first study of necroptosis in neuronal cells *in vitro* and in transient middle cerebral artery occlusion (MCAO) *in vivo* demonstrated that necrostatin-1 (Nec-1), given intraventricularly, reduced the size of a focal infarct produced by transient middle cerebral artery occlusion (MCAO) (Degterev et al. 2005). The same authors found later that Nec-1 inhibits RIP1 activation (Degterev et al. 2008). In another study, 20-min transient global cerebral ischemia

D. G. Fujikawa (ed.), *Acute Neuronal Injury*,
https://doi.org/10.1007/978-3-319-77495-4

(TGCI) produced RIP3 translocation to hippocampal CA1 pyramidal cell nuclei and Nec-1 protected CA1 neurons and reduced RIP3 nuclear translocation (Yin et al. 2015). They also showed that Nec-1 inhibited cathepsin B (cath-B) release from lysosomes, which is caused by activated calpain 1, which is one of the first studies to show an interaction between the necroptotic and excitotoxic pathways.

In 15-min TGCI, RIP3 S-nitrosylation and RIP3 phosphorylation interacting with RIP1 was associated with neuronal death (Miao et al. 2015). These investigators also found that the non-competitive NMDA-receptor antagonist MK-801 and 7-nitroindazole (7-NI), a neuronal nitric oxide synthase (nNOS) inhibitor, reduced RIP3 S-nitrosylation and neuronal damage, showing for the first time that an NMDA-receptor antagonist and an nNOS (nNOS) inhibitor had an effect on a necroptotic enzyme (Miao et al. 2015). In 20-min TGCI, RIP3 and apoptosis-inducing factor (AIF) translocated from cytosol and mitochondria, respectively, to hippocampal neuronal nuclei, and the RIP1 inhibitor Nec-1 inhibited the nuclear translocation and provided neuroprotection (Xu et al. 2016). This was the first time that RIP3 and AIF were shown to interact and to co-localize together in neuronal nuclei following ischemia and that Nec-1 prevented their interaction and nuclear translocation and was neuroprotective; it is another example of the interaction between the necroptotic and excitotoxic pathways.

Thus, it appears that there is an interaction between the necroptotic and excitotoxic pathways, but the relative contribution of each to the acute neuronal necrosis that is produced is unknown. This is an exciting area of inquiry about which much more is certain to be known in coming years.

References

Auer RN, Kalimo H, Olsson Y, Siesjo BK (1985a) The temporal evolution of hypoglycemic brain damage. II. Light- and electron-microscopic findings in the hippocampal gyrus and subiculum of the rat. Acta Neuropathol 67(1–2):25–36

Auer RN, Kalimo H, Olsson Y, Siesjo BK (1985b) The temporal evolution of hypoglycemic brain damage. I. Light- and electron-microscopic findings in the rat cerebral cortex. Acta Neuropathol 67(1–2):13–24

Colbourne F, Sutherland GR, Auer RN (1999) Electron microscopic evidence against apoptosis as the mechanism of neuronal death in global ischemia. J Neurosci 19:4200–4210

Degterev A, Huang Z, Boyce M, Li Y, Jagtap P, Mizushima N, Cuny GD, Mitchison TJ, Moskowitz MA, Yuan J (2005) Chemical inhibitor of nonapoptotic cell death with therapeutic potential for ischemic brain injury. Nat Chem Biol 1(2):112–119. https://doi.org/10.1038/nchembio711

Degterev A, Hitomi J, Germscheid M, Ch'en IL, Korkina O, Teng X, Abbott D, Cuny GD, Yuan C, Wagner G, Hedrick SM, Gerber SA, Lugovskoy A, Yuan J (2008) Identification of RIP1 kinase as a specific cellular target of necrostatins. Nat Chem Biol 4(5):313–321. https://doi.org/10.1038/nchembio.83

Fujikawa DG (2000) Confusion between neuronal apoptosis and activation of programmed cell death mechanisms in acute necrotic insults. Trends Neurosci 23:410–411

Fujikawa DG (2002) Apoptosis: ignoring morphology and focusing on biochemical mechanisms will not eliminate confusion. Trends Pharmacol Sci 23:309–310

Fujikawa DG (ed) (2010) Acute neuronal injury: the role of excitotoxic programmed cell death mechanisms. Springer, New York. 306 p

Fujikawa DG, Shinmei SS, Cai B (1999) Lithium-pilocarpine-induced status epilepticus produces necrotic neurons with internucleosomal DNA fragmentation in adult rats. Eur J Neurosci 11:1605–1614

Fujikawa DG, Shinmei SS, Cai B (2000) Kainic acid-induced seizures produce necrotic, not apoptotic, neurons with internucleosomal DNA cleavage: implications for programmed cell death mechanisms. Neuroscience 98:41–53

Fujikawa DG, Ke X, Trinidad RB, Shinmei SS, Wu A (2002) Caspase-3 is not activated in seizure-induced neuronal necrosis with internucleosomal DNA cleavage. J Neurochem 83:229–240

Fujikawa DG, Shinmei SS, Zhao S, Aviles ER Jr (2007) Caspase-dependent programmed cell death pathways are not activated in generalized seizure-induced neuronal death. Brain Res 1135:206–218

Gao H, Geng Z (2013) Calpain I activity and its relationship with hippocampal neuronal death in pilocarpine-induced status epilepticus rat model. Cell Biochem Biophys 66(2):371–377. https://doi.org/10.1007/s12013-012-9476-5

Hu BR, Liu CL, Ouyang Y, Blomgren K, Siejö BK (2000) Involvement of caspase-3 in cell death after hypoxia-ischemia declines during brain maturation. J Cereb Blood Flow Metab 20:1294–1300

Liu T, Zhao DX, Cui H, Chen L, Bao YH, Wang Y, Jiang JY (2016) Therapeutic hypothermia attenuates tissue damage and cytokine expression after traumatic brain injury by inhibiting necroptosis in the rat. Sci Rep 6:24547. https://doi.org/10.1038/srep24547

Miao W, Qu Z, Shi K, Zhang D, Zong Y, Zhang G, Zhang G, Hu S (2015) RIP3 S-nitrosylation contributes to cerebral ischemic neuronal injury. Brain Res 1627:165–176. https://doi.org/10.1016/j.brainres.2015.08.020

Narkilahti S, Pirttilä TJ, Lukasiuk K, Tuunanen J, Pitkänen A (2003) Expression and activation of caspase 3 following status epilepticus. Eur J Neurosci 18:1486–1496

Vieira M, Fernandes J, Carreto L, Anuncibay-Soto B, Santos M, Han J, Fernandez-Lopez A, Duarte CB, Carvalho AL, Santos AE (2014) Ischemic insults induce necroptotic cell death in hippocampal neurons through the up-regulation of endogenous RIP3. Neurobiol Dis 68:26–36. https://doi.org/10.1016/j.nbd.2014.04.002

Vosler PS, Sun D, Wang S, Gao Y, Kintner DB, Signore AP, Cao G, Chen J (2009) Calcium dysregulation induces apoptosis-inducing factor release: cross-talk between PARP-1- and calpain-signaling pathways. Exp Neurol 218(2):213–220. https://doi.org/10.1016/j.expneurol.2009.04.032; Epub 2009/05/12; [pii]: S0014-4886(09)00181-2

Wang S, Wang S, Shan P, Song Z, Dai T, Wang R, Chi Z (2008) Mu-calpain mediates hippocampal neuron death in rats after lithium-pilocarpine-induced status epilepticus. Brain Res Bull 76(1–2):90–96. https://doi.org/10.1016/j.brainresbull.2007.12.006

Xu Y, Wang J, Song X, Qu L, Wei R, He F, Wang K, Luo B (2016) RIP3 induces ischemic neuronal DNA degradation and programmed necrosis in rat via AIF. Sci Rep 6:29362. https://doi.org/10.1038/srep29362

Yin B, Xu Y, Wei RL, He F, Luo BY, Wang JY (2015) Inhibition of receptor-interacting protein 3 upregulation and nuclear translocation involved in Necrostatin-1 protection against hippocampal neuronal programmed necrosis induced by ischemia/reperfusion injury. Brain Res 1609:63–71. https://doi.org/10.1016/j.brainres.2015.03.024

CPSIA information can be obtained
at www.ICGtesting.com
Printed in the USA
BVHW01*0147091018
529667BV00002B/4/P

9 783319 774947